BARRON'S

HOW TO PREPARE FOR THE

SAT II

MATH LEVEL IIC

7TH EDITION

Howard P. Dodge
The Wheeler School
Providence, Rhode Island

Richard Ku
North Kingstown High School
North Kingstown, Rhode Island

BARRON'S

All inquiries should be addressed to:
Barron's Educational Series, Inc.
250 Wireless Boulevard
Hauppauge, New York 11788
http://www.barronseduc.com

Answer sheets and self-evaluation charts reprinted from *How to Prepare for the SAT II: Math Level I* © 1993 by Barron's Educational Series, Inc.

Library of Congress Catalog Card No. 2002033233

International Standard Book No. 0-7641-2019-0

Library of Congress Cataloging-in-Publication Data

Dodge, Howard P.
    How to prepare for SAT II : mathematics level IIC / Howard P. Dodge,
Richard Ku.—7th ed.
        p.   cm.
    Rev. ed. of: Barron's how to prepare for SAT II.
    Includes index.
    ISBN 0-7641-2019-0
    1. Mathematics—Examinations, questions, etc.   2. College entrance
achievement tests—United States—Study guides.   I. Title: How to prepare for
SAT 2.   II. Title: How to prepare for SAT two.   III. Dodge, Howard P.
Barron's how to prepare for SAT II.   IV. Title.
QA43.D57   2003
510′.76—dc21

                                                                    2002033233

PRINTED IN THE UNITED STATES OF AMERICA

9 8 7 6

CONTENTS

Acknowledgments v
Introduction vi
  New in the 7th Edition vi
  Organization of This Book vi
  Structure of the Math Level IIC Test vi
  Calculator Issues vii
  How the Test Is Scored vii
  Strategies to Maximize Your Score vii
  Study Plans viii
  Questions Students Ask About SAT II:
    Subject Tests viii

PART 1   DIAGNOSTIC TEST

DIAGNOSTIC TEST 1
  Answer Key / 16
  Answer Explanations / 16

PART 2   REVIEW OF MAJOR TOPICS

CHAPTER 1. Introduction to Functions 25
  1.1 Definitions / 25
  1.2 Function Notation / 26
  1.3 Inverse Functions / 27
  1.4 Odd and Even Functions / 28
  1.5 Multivariable Functions / 30
    Answers and Explanations / 30

CHAPTER 2. Polynomial Functions 32
  2.1 Definitions / 32
  2.2 Linear Functions / 32
  2.3 Quadratic Functions / 34
  2.4 Higher-Degree Polynomial Functions / 37
  2.5 Inequalities / 41
    Answers and Explanations / 42

CHAPTER 3. Trigonometric Functions 45
  3.1 Definitions / 45
  3.2 Arcs and Angles / 47
  3.3 Special Angles / 49
  3.4 Graphs / 50
  3.5 Identities, Equations, and Inequalities / 53
  3.6 Inverse Functions / 59
  3.7 Triangles / 61
    Answers and Explanations / 64

CHAPTER 4. Miscellaneous Relations and
           Functions 68
  4.1 Conic Sections / 68
  4.2 Exponential and Logarithmic Functions / 72
  4.3 Absolute Value / 75
  4.4 Greatest Integer Function / 77
  4.5 Rational Functions and Limits / 77
  4.6 Parametric Equations / 79
  4.7 Polar Coordinates / 80
    Answers and Explanations / 82

CHAPTER 5. Miscellaneous Topics 86
  5.1 Permutations and Combinations / 86
  5.2 Binomial Theorem / 88
  5.3 Probability / 90
  5.4 Sequences and Series / 93
  5.5 Geometry and Vectors / 96
  5.6 Variation / 102
  5.7 Logic / 103
  5.8 Statistics / 105
  5.9 Odds and Ends / 107
    Answers and Explanations /109

PART 3   GRAPHING CALCULATORS

CHAPTER 6. Introduction 117

CHAPTER 7. Basic Operations 119

CHAPTER 8. Specific Applications 122

CHAPTER 9. Programs 132

PART 4   MODEL TESTS

MODEL TEST 1 141
  Answer Key / 151
  Answer Explanations / 151
MODEL TEST 2 159
  Answer Key / 171
  Answer Explanations / 172
MODEL TEST 3 181
  Answer Key / 193
  Answer Explanations / 193

MODEL TEST 4   201
  Answer Key / 211
  Answer Explanations / 212
MODEL TEST 5   221
  Answer Key / 232
  Answer Explanations / 233
MODEL TEST 6   243
  Answer Key / 253
  Answer Explanations / 254

MODEL TEST 7   263
  Answer Key / 273
  Answer Explanations / 273
MODEL TEST 8   281
  Answer Key / 291
  Answer Explanations / 292
APPENDIX—Summary of Formulas   297
Index   301

ACKNOWLEDGMENTS

I would like to express my appreciation to my former colleagues in the mathematics department of Choate Rosemary Hall, Wallingford, Connecticut, for the many helpful suggestions and useful problems used throughout this book. Particular thanks are due to Dean Blanchard and Charles Bodine, formerly of St. George's School, Newport, Rhode Island, for their helpful comments on early drafts of this book, and to my son, Laurence, for his careful checking of the solution to each problem.

Howard P. Dodge

My thanks to Howard Dodge for asking me to do the revisions for the 7th Edition, reviewing the early drafts, and letting me share authorship of the book. Thanks also to my wife, Debra, for her support, encouragement, patience, and understanding while I worked evenings and weekends during the past year. Finally, I would like to thank Barron's editor Pat Hunter for guiding me through the preparation of this new edition.

Richard Ku

INTRODUCTION

The purpose of this book is to prepare you for the SAT II Math Level IIC Subject Test. It can be used as a self-study guide or as a textbook in a test preparation course. It is a self-contained resource for those who want to achieve their maximum possible score.

NEW IN THE 7TH EDITION

Mathematics education has changed in important ways since the introduction of the Math Level IIC Test. The C in the title of the exam recognized the permanent status of calculators in mathematics curricula. Since then, graphing calculators have become commonplace in mathematics classrooms, and in many schools, this technology has changed the way advanced mathematics is taught.

According to the College Board's most recent publication on Subject Tests, a calculator is useful or necessary on approximately 60% of Math Level IIC Test questions. Most students bring graphing calculators to the Math Level IIC Test. The Test is developed based on this assumption and graphing calculators may provide an advantage over scientific calculators on some questions.

This 7th edition includes a new chapter that describes how graphing calculators are used to solve the kinds of problems that you will see on the Math Level IIC Test. In line with this addition, new explanations that reflect graphing calculator solutions have been written for problems in the Diagnostic Test and model tests.

ORGANIZATION OF THIS BOOK

The remainder of this introduction is intended to help you approach the Math Level IIC Test with confidence. It describes the structure and format of the test, presents important calculator issues, offers strategies for maximizing your score, and outlines study plans. More general questions about how colleges use the SAT II Subject Test results, whether to take the Math Level I or II Test, and the logistics of test administration are then addressed.

After reading the rest of this introduction, you should take the Part 1 **Diagnostic Test**, correct it, and go over the explanations that are given at the end of the test. Following each explanation, in square brackets are cross-references to topics that are reviewed in Part 2. The explanations begin with a code indicating whether calculators are inactive (i), neutral (n), or active (a) (see Calculator Issues below). You may need to refer to Part 3 **Graphing Calculators** for clarification on language that

is used in explanations where the primary solution involves graphing calculators.

The topic reviews in Part 2 describe algebraic methods and formulas for solving Math Level IIC problems. As such, it contains no reference to the use of graphing calculators. The use of graphing calculators on the Math Level IIC Test is covered in Part 3. This part provides reference material for test-takers who took courses in which graphing calculators were routinely used as problem-solving tools. It provides enough detail, however, to be used in an instructional mode for those who may not have used graphing calculators in courses, but may wish to avail themselves of this technology on the Math Level IIC Test.

Finally, Part 4 contains eight model tests, each consisting of 50 multiple-choice items. Answer sheets, directions, and formulas are included. As noted above, calculator codes and cross-references to topic reviews in Part 2 are also given in all of the answer explanations. These model tests will enable you to practice, practice, and practice some more!

STRUCTURE OF THE MATH LEVEL IIC TEST

The Math Level IIC Test is one hour in length and consists of 50 multiple-choice questions, each with five answer choices, of which only one is correct. The test is aimed at students who have had more than three to four years of high school mathematics courses. It is developed each year for talented students who have had two years of algebra, one year of geometry, and trigonometry/elementary functions. In short, most of the topics will have been covered by the time a student completes a "precalculus" course.

According to the most recent College Board information, the questions on the Math Level IIC Test are divided into the following six categories and percentages:

- Algebra—18%
- Geometry
  - Solid geometry—8%
  - Coordinate geometry (2 and 3 dimensions)—12%
- Trigonometry—20%
  - Emphasis on properties and graphs of the trig functions, inverse trig functions, trig equations and identities, and laws and sines and cosines
- Functions—24%
  - Algebraic, exponential, and logarithmic functions

- Statistics/Probability—6%

  Probability, mean, median, and mode, counting, data interpretation, permutations, combinations, and standard deviation

- Miscellaneous—12%

  Logic, elementary number theory, sequences, and limits

To give you a realistic experience with the breadth of problems you will encounter on an actual test, these topics have been distributed in approximately the same way in the eight model tests.

CALCULATOR ISSUES

Three calculator codes are noted in the explanations of answers to problems on the eight model tests. *Inactive* (i) questions are those for which a calculator is of no use. These questions typically involve variables only, or the computations are very simple. *Active* (a) questions are those that require a calculator. The code g refers to graphing calculators. When this code is used, the primary explanation involves graphing calculators, with secondary explanations based on the availability of scientific calculators only.

When a calculator is used for calculator-active questions, intermediate results should be left in the calculator's memory. If they are written down, and rounded values are rekeyed, the final result may not be among the answer choices. This should not be a problem for almost all test-takers. With features such as parentheses and the ability to handle composite functions, all graphing calculators, and most scientific calculators, can evaluate even the most complex expressions without the user having to write down and rekey intermediate results.

Certain types of technology are explicitly prohibited from testing sites. These include calculators with QWERTY (typewriter-like) keyboards, calculators with paper tape, calculators that "talk" or otherwise make noise, calculators that require an electrical outlet, pocket organizers, handheld and laptop computers, and electronic writing pads, and pen-input devices.

You must be sure that your calculator is in good working order. You may bring spare batteries or a backup calculator. However, the test center will not have substitute calculators or batteries for you to borrow. Students may not share calculators. If your calculator malfunctions and you don't have a spare, you may cancel your score (on this test) by informing the test supervisor of the malfunction.

HOW THE TEST IS SCORED

There are 50 questions on the Math Level IIC Test. Your raw score is the number of correct answers *minus* one-fourth of the number of incorrect answers. For example, if you answered all 50 questions and got 12 wrong, your raw score would be 35 (38 right minus one-fourth of 12). If you simply didn't answer the 12 questions you got wrong, your raw score would have been 38.

Under this system of scoring, guessing blindly is not advisable. If you can eliminate only one choice, you are not likely to raise your score. You should guess only when you feel confident eliminating *more than one* answer choice.

Raw scores are transformed into reported scores between 200 and 800, and the formula for this transformation changes from year to year. In some years, a raw score of 42 might transform to a reported score of 780. However, it is possible to get a reported score of 800 with a raw score as low as 42. On a recent Math Level IIC test, raw scores of 20, 30, and 40 transformed into reported scores of 610, 700, and 780, respectively.

STRATEGIES TO MAXIMIZE YOUR SCORE

- **Budget your time.** Although most testing centers have wall clocks, you would be wise to have a watch on your desk. Since there are 50 items on a one-hour test, you have a little over a minute per item. Typically, test items are easier near the beginning of a test, and they get progressively more difficult. Don't linger over difficult questions. Work the problems you are confident of first, and then return later to the ones that are difficult for you.

- **Guess intelligently.** As noted above, you are likely to get a higher score if you can confidently eliminate two or more answer choices, and a lower score if you can't eliminate any.

- **Read the questions carefully.** Answer the question asked, not the one you may have expected. For example, you may have to solve an equation to answer the question, but the solution itself may not be the answer.

- **Mark answers clearly and accurately.** Since you may skip questions that are difficult, be sure to mark the correct number on your answer sheet. If you change an answer, erase cleanly and leave no stray marks. Mark only one answer; an item will be graded as incorrect if more than one answer choice is marked.

- **Change an answer only if you have a good reason for doing so.** It is usually not a good idea to change an answer on the basis of a hunch or whim.

- **As you read a problem, think about possible computational shortcuts to obtain the correct answer choice.** Even though calculators simplify the computational process, you may save time by identifying a pattern that leads to a shortcut.

- **Substitute numbers to determine the nature of a relationship.** If a problem contains only variable quantities, it is sometimes helpful to substitute numbers to understand the relationships implied in the problem.

- **Think carefully about whether to use a calculator.** As noted above, the College Board's guideline is that a calculator is useful or necessary in about 60% of the problems on the Math Level IIC Test. An appropriate percentage for you may differ significantly from this, depending on your experience with calculators, especially graphing calculators. Even if you learned the material in a highly calculator-active environment, you may discover that a problem can be done more efficiently without a calculator than with one.
- **Check the answer choices.** If the answer choices are in decimal form, the problem is likely to require the use of a calculator.

STUDY PLANS

Your first step is to take the Diagnostic Test. This should be taken under test conditions: timed, quiet, without interruption. Correct the test and identify areas of weakness using the cross-references to the Part 2 review. Use the review to strengthen your understanding of the concepts involved. Check the Part 3 graphing calculator applications to see if this approach is optimal for you on particular problems.

Ideally, you would start preparing for the Math Level IIC Test two to three months in advance. Each week, you would be able to take one sample test, following the same procedure as for the Diagnostic Test. Depending on how well you do, it might take you anywhere between 15 minutes and an hour to complete the work after you take the test. Obviously, if you have less time to prepare, you would have to intensify your efforts to complete the eight sample tests, or do fewer of them.

The best way to use Parts 2 and 3 of this book is as reference material. You should look through this material quickly before you take the sample tests, just to get an idea of the range of topics covered and the level of detail. However, these parts of the book are more effectively used *after* you've taken and corrected a sample test.

QUESTIONS STUDENTS ASK ABOUT SAT II: SUBJECT TESTS*

What is an SAT II Subject Test?

An SAT II: Subject Test tests your knowledge of a subject and your ability to apply that knowledge. Unlike the SAT I test, an SAT II test is curriculum-based and is intended to measure how much you have learned rather than how much learning ability you have.

*Adapted from *How to Prepare for the SAT II: Mathematics Level IC*, 6th Ed., by James J. Rizzuto, Barron's Educational Series, Inc., Hauppauge, New York, 1994. Reprinted with permission.

How is it used?

Colleges use SAT II: Subject Tests to predict an admission candidate's future success on the level of work typical at that college. Applicants to colleges use the tests to demonstrate their acquired knowledge in subject areas important to their future goals.

Some colleges, especially the most selective schools, have found subject test results to be better predictors of success in related college courses than SAT I scores and, frequently, the high school record. Because grading systems, standards, and course offerings differ among secondary schools, a student's SAT II: Subject Test score may provide the college with more and better information about what the student knows than is available from his or her high school transcript.

Colleges also use SAT II: Subject Test results to place students in appropriate courses after they have been admitted.

High schools use SAT II: Subject Tests to examine the success of their curriculum and to indicate strengths and weaknesses.

Should I take a math SAT II: Subject Test?

Depending on the secondary school you attend and the college to which you apply, you may not have much choice. Most private colleges and many state universities require applicants to take three SAT II: Subject Tests, though often not specifying which ones. According to a counselor at a selective independent secondary school, "Ninety-nine percent of the colleges our students apply to require three achievement [SAT II: Subject] tests, so we ask all of our students to take them regardless of where they apply."

Many secondary school counselors believe that an application is improved by SAT II: Subject Test scores even when a college does not require them, because the tests allow the candidate to demonstrate proficiency in the areas of his or her greatest strength.

Which tests you take will depend on your most successful courses in secondary school and your intended major in college. You should not take an SAT II: Subject Test unless you have been successful in that subject or have made a special effort to prepare for the test (it is best if you satisfy both conditions).

Examine the description of the contents of the Level IIC test on pages vi–vii to see whether you have studied the topics tested.

Many secondary school counselors believe that the subject tests most helpful to your application are an English test, a math test, and a third test in your best third subject. If you intend to major in a science or in any other math-related field (engineering, economics, etc.), you definitely should take a math SAT II: Subject Test.

How important are SAT II: Subject Tests in college admissions?

College admissions committees generally regard achievement test scores as "confirming" evidence to be compared with high school grades, the level of courses taken, and SAT I scores. Grades (especially because of grade inflation) and courses offered vary greatly from high school to high school (and even from teacher to teacher within the same high school). A grade in "intermediate math," for example, at a selective independent school may represent something entirely different from the same grade in a course of the same name at a small rural public high school.

No other factor on an application is so uniform an assessment of all applicants as a standardized achievement test.

The scores are used not only as a measure of knowledge but also as a means of interpreting the high school record. Low achievement test scores accompanied by high grades, for example, may suggest an inflated grading system. High scores with average grades may mean the applicant is attending a more demanding high school that offers a challenging program to competitive students chosen for special abilities.

In general, test scores are more important if they disagree with the high school record and less important if they agree. They are also more important if they lie outside the range of scores of students normally selected by the college and less important if they fall within this range. And, finally, they are more important if they conflict with the scores on the SAT I than if they confirm these scores.

In other words, if the SAT II: Subject Test scores don't say the same thing as the rest of the record, it's the new message that is important. If the new message is positive, it will help. But in the majority of cases, the achievement test scores confirm the rest of the data on the application.

When and where are SAT II: Subject Tests given?

These tests are usually given in the morning of the first Saturday in November, December, May, and June and the last Saturday in January. There are hundreds of test centers around the country. Dates vary from year to year. For other information on dates and test sites, write to the College Board Admission Testing Program for its current schedule.

To whom do I write for information and registration?

Write to:

> College Board SAT Program
> P.O. Box 6200
> Princeton, New Jersey 08541-6200

or call:

> (609) 771-7600.

The College Entrance Examination Board, an association of colleges and secondary schools, is headquartered in New York City. To arrange for testing, use *only* the Princeton address.

What should I bring with me to the test?

You should bring your ticket of admission, some positive form of identification, two or more no. 2 pencils (with erasers, or bring an eraser), a watch (though test centers *should* have clearly visible clocks), and a simple twist-type pencil sharpener (so that you need not waste time walking back and forth to and from a pencil sharpener if your point breaks).

No books, calculators (except for the Math Level IIC Subject Test), rulers, scratch paper, slide rules, compasses, protractors, or other devices are allowed in the examination room.

DIAGNOSTIC TEST

PART

1

ANSWER SHEET FOR DIAGNOSTIC TEST

Determine the correct answer for each question. Then, using a no. 2 pencil, blacken completely the oval containing the letter of your choice.

1. Ⓐ Ⓑ Ⓒ Ⓓ Ⓔ
2. Ⓐ Ⓑ Ⓒ Ⓓ Ⓔ
3. Ⓐ Ⓑ Ⓒ Ⓓ Ⓔ
4. Ⓐ Ⓑ Ⓒ Ⓓ Ⓔ
5. Ⓐ Ⓑ Ⓒ Ⓓ Ⓔ
6. Ⓐ Ⓑ Ⓒ Ⓓ Ⓔ
7. Ⓐ Ⓑ Ⓒ Ⓓ Ⓔ
8. Ⓐ Ⓑ Ⓒ Ⓓ Ⓔ
9. Ⓐ Ⓑ Ⓒ Ⓓ Ⓔ
10. Ⓐ Ⓑ Ⓒ Ⓓ Ⓔ
11. Ⓐ Ⓑ Ⓒ Ⓓ Ⓔ
12. Ⓐ Ⓑ Ⓒ Ⓓ Ⓔ
13. Ⓐ Ⓑ Ⓒ Ⓓ Ⓔ
14. Ⓐ Ⓑ Ⓒ Ⓓ Ⓔ
15. Ⓐ Ⓑ Ⓒ Ⓓ Ⓔ
16. Ⓐ Ⓑ Ⓒ Ⓓ Ⓔ
17. Ⓐ Ⓑ Ⓒ Ⓓ Ⓔ

18. Ⓐ Ⓑ Ⓒ Ⓓ Ⓔ
19. Ⓐ Ⓑ Ⓒ Ⓓ Ⓔ
20. Ⓐ Ⓑ Ⓒ Ⓓ Ⓔ
21. Ⓐ Ⓑ Ⓒ Ⓓ Ⓔ
22. Ⓐ Ⓑ Ⓒ Ⓓ Ⓔ
23. Ⓐ Ⓑ Ⓒ Ⓓ Ⓔ
24. Ⓐ Ⓑ Ⓒ Ⓓ Ⓔ
25. Ⓐ Ⓑ Ⓒ Ⓓ Ⓔ
26. Ⓐ Ⓑ Ⓒ Ⓓ Ⓔ
27. Ⓐ Ⓑ Ⓒ Ⓓ Ⓔ
28. Ⓐ Ⓑ Ⓒ Ⓓ Ⓔ
29. Ⓐ Ⓑ Ⓒ Ⓓ Ⓔ
30. Ⓐ Ⓑ Ⓒ Ⓓ Ⓔ
31. Ⓐ Ⓑ Ⓒ Ⓓ Ⓔ
32. Ⓐ Ⓑ Ⓒ Ⓓ Ⓔ
33. Ⓐ Ⓑ Ⓒ Ⓓ Ⓔ
34. Ⓐ Ⓑ Ⓒ Ⓓ Ⓔ

35. Ⓐ Ⓑ Ⓒ Ⓓ Ⓔ
36. Ⓐ Ⓑ Ⓒ Ⓓ Ⓔ
37. Ⓐ Ⓑ Ⓒ Ⓓ Ⓔ
38. Ⓐ Ⓑ Ⓒ Ⓓ Ⓔ
39. Ⓐ Ⓑ Ⓒ Ⓓ Ⓔ
40. Ⓐ Ⓑ Ⓒ Ⓓ Ⓔ
41. Ⓐ Ⓑ Ⓒ Ⓓ Ⓔ
42. Ⓐ Ⓑ Ⓒ Ⓓ Ⓔ
43. Ⓐ Ⓑ Ⓒ Ⓓ Ⓔ
44. Ⓐ Ⓑ Ⓒ Ⓓ Ⓔ
45. Ⓐ Ⓑ Ⓒ Ⓓ Ⓔ
46. Ⓐ Ⓑ Ⓒ Ⓓ Ⓔ
47. Ⓐ Ⓑ Ⓒ Ⓓ Ⓔ
48. Ⓐ Ⓑ Ⓒ Ⓓ Ⓔ
49. Ⓐ Ⓑ Ⓒ Ⓓ Ⓔ
50. Ⓐ Ⓑ Ⓒ Ⓓ Ⓔ

The diagnostic test is designed to help you pinpoint the weak spots in your background. The answer explanations that follow the test are keyed to sections of the book.

To make the best use of this diagnostic test, set aside between 1 and 2 hours so you will be able to do the whole test at one sitting. Tear out the preceding answer sheet and indicate your answers in the appropriate spaces. Do the problems as if this were a regular testing session. Review the suggestions on pages vii–viii.

When finished, check your answers with those at the end of the test. For those that you got wrong, note the sections containing the material that you must review. If you do not fully understand how you arrived at some of the correct answers, you should review those sections also.

Finally, fill out the self-evaluation sheet on page 21 in order to pinpoint the topics that gave you the most difficulty.

50 questions   1 hour

TEST DIRECTIONS

<u>Directions</u>: Decide which answer choice is best. If the exact numerical value is not one of the answer choices, select the closest approximation. Fill in the oval on the answer sheet that corresponds to your choice.

Notes:
(1) You will need to use a scientific or graphing calculator to answer some of the questions.
(2) You will have to decide whether to put your calculator in degree or radian mode for some problems.
(3) All figures that accompany problems are plane figures unless otherwise stated. Figures are drawn as accurately as possible to provide useful information for solving the problem, except when it is stated in a particular problem that the figure is not drawn to scale.
(4) Unless otherwise indicated, the domain of a function is the set of all real numbers for which the functional value is also a real number.

<u>Reference Information</u>. The following formulas are provided for your information.

Volume of a right circular cone with radius r and height h: $V = \frac{1}{3}\pi r^2 h$

Lateral area of a right circular cone if the base has circumference c and slant height is l: $S = \frac{1}{2}cl$

Volume of a sphere of radius r: $V = \frac{4}{3}\pi r^3$

Surface area of a sphere of radius r: $S = 4\pi r^2$

Volume of a pyramid of base area B and height h: $V = \frac{1}{3}Bh$

---

1. A linear function, f, has a slope of -2. $f(1) = 2$ and $f(2) = q$. Find q.

   USE THIS SPACE FOR SCRATCH WORK

   (A) 0

   (B) 4

   (C) $\frac{3}{2}$

   (D) $\frac{5}{2}$

   (E) 3

GO ON TO THE NEXT PAGE

2. A function is said to be even if $f(x) = f(-x)$. Which of the following is *not* an even function?

   (A) $y = |x|$
   (B) $y = \sec x$
   (C) $y = \log x^2$
   (D) $y = x^2 + \sin x$
   (E) $y = 3x^4 - 2x^2 + 17$

3. What is the radius of a sphere, with center at the origin, that passes through point (2,3,4)?

   (A) 3.32
   (B) 5.39
   (C) 3
   (D) 3.31
   (E) 5.38

4. If a point (x,y) is in the second quadrant, which of the following must be true?

   I. $x < y$
   II. $x + y > 0$
   III. $\dfrac{x}{y} < 0$

   (A) only I
   (B) only II
   (C) only III
   (D) only I and II
   (E) only I and III

5. If $f(x) = x^2 - ax$, then $f(a) =$

   (A) a
   (B) $a^2 - a$
   (C) 0
   (D) 1
   (E) $a - 1$

6. The average of your first three test grades is 78. What grade must you get on your fourth and final test to make your average 80?

   (A) 80
   (B) 82
   (C) 84
   (D) 86
   (E) 88

7. $\log_7 9 =$

   (A) 0.89
   (B) 0.95
   (C) 1.13
   (D) 1.21
   (E) 7.61

USE THIS SPACE FOR SCRATCH WORK

GO ON TO THE NEXT PAGE

8. If $\log_2 m = x$ and $\log_2 n = y$, then $mn =$
   (A) 2^{x+y}
   (B) 2^{xy}
   (C) 4^{xy}
   (D) 4^{x+y}
   (E) cannot be determined

9. How many integers are there in the solution set of $|x - 2| \le 5$?
   (A) 11
   (B) 0
   (C) an infinite number
   (D) 9
   (E) 7

10. If $f(x) = \sqrt{x^2}$, then $f(x)$ can also be expressed as
    (A) x
    (B) $-x$
    (C) $\pm x$
    (D) $|x|$
    (E) $f(x)$ cannot be determined because x is unknown.

11. The graph of $(x^2 - 1)y = x^2 - 4$ has
    (A) one horizontal and one vertical asymptote
    (B) two vertical but no horizontal asymptotes
    (C) one horizontal and two vertical asymptotes
    (D) two horizontal and two vertical asymptotes
    (E) neither a horizontal nor a vertical asymptote

12. $\lim\limits_{x \to \infty} \left(\dfrac{3x^2 + 4x - 5}{6x^2 + 3x + 1} \right) =$
    (A) $\dfrac{1}{2}$
    (B) 1
    (C) -5
    (D) $\dfrac{1}{5}$
    (E) This expression is undefined.

13. A linear function has an x-intercept of $\sqrt{3}$ and a y-intercept of $\sqrt{5}$. The graph of the function has a slope of
    (A) 0.77
    (B) -1.29
    (C) 2.24
    (D) 1.29
    (E) -0.77

14. If $f(x) = \sin x$, then $f^{-1}\left(\dfrac{\pi}{4}\right) =$
    (A) 52
    (B) 41
    (C) 0.90
    (D) 0.71
    (E) none of the above

GO ON TO THE NEXT PAGE

15. The plane $2x + 3y - 4z = 5$ intersects the x-axis at $(a,0,0)$, the y-axis at $(0,b,0)$, and the z-axis at $(0,0,c)$. The value of $a + b + c$ is

(A) 5

(B) $\dfrac{35}{12}$

(C) $\dfrac{65}{12}$

(D) 1

(E) 9

16. Given the set of data 1, 1, 2, 2, 2, 3, 3, 4, which one of the following statements is true?

(A) mean ≤ median ≤ mode

(B) median ≤ mean ≤ mode

(C) median ≤ mode ≤ mean

(D) mode ≤ mean ≤ median

(E) The relationship cannot be determined because the median cannot be calculated.

17. If $\dfrac{x - 3y}{x} = 7$, what is the value of $\dfrac{x}{y}$?

(A) $-\dfrac{8}{3}$

(B) -2

(C) $-\dfrac{1}{2}$

(D) $\dfrac{3}{8}$

(E) 2

18. $\dfrac{\sin 120° \cdot \cos \dfrac{2\pi}{3}}{\tan 315°} =$

(A) $\dfrac{\sqrt{3}}{2}$

(B) $-\dfrac{\sqrt{3}}{4}$

(C) $\dfrac{\sqrt{6}}{4}$

(D) $-\dfrac{\sqrt{3}}{2}$

(E) $\dfrac{\sqrt{3}}{4}$

19. If $f(x) = \dfrac{1}{2}x^2 - 8$ is defined when $-4 \le x \le 4$, the maximum value of the graph of $|f(x)|$ is

(A) –8

(B) 0

(C) 8

(D) 4

(E) 2

GO ON TO THE NEXT PAGE

20. If $\tan \theta = \dfrac{2}{3}$, then $\sin \theta =$

    (A) $\dfrac{2\sqrt{13}}{13}$

    (B) $\pm\dfrac{2\sqrt{13}}{13}$

    (C) $\dfrac{3\sqrt{13}}{13}$

    (D) $\pm\dfrac{2}{5}$

    (E) $\dfrac{2\sqrt{5}}{5}$

21. If a circle has a central angle of 75° that intercepts an arc of length 75 feet, the number of feet in the radius is

    (A) 63.7
    (B) 57.3
    (C) 44.1
    (D) 75.0
    (E) 28.6

22. The area of a triangle with sides 3, 5, and 7 is

    (A) 7.5
    (B) 6.5
    (C) 3.75
    (D) 13.0
    (E) 2.4

23. If $f(x) = i$, where i is an integer such that $i \leq x < i + 1$, and $g(x) = f(x) - |f(x)|$, what is the maximum value of $g(x)$?

    (A) 0
    (B) 1
    (C) –1
    (D) 2
    (E) i

24. If $f(x) = \dfrac{1}{\sec x}$, then

    (A) $f(x) = f(-x)$

    (B) $f(\dfrac{1}{x}) = -f(x)$

    (C) $f(-x) = -f(x)$

    (D) $f(x) = f(\dfrac{1}{x})$

    (E) $f(x) = \dfrac{1}{f(x)}$

GO ON TO THE NEXT PAGE

25. The polar coordinates of a point P are $(2,240°)$. The Cartesian (rectangular) coordinates of P are

   (A) $\left(-1,-\sqrt{3}\right)$

   (B) $\left(-1,\sqrt{3}\right)$

   (C) $\left(-\sqrt{3},-1\right)$

   (D) $\left(-\sqrt{3},1\right)$

   (E) none of the above

26. The height of a cone is equal to the radius of its base. The radius of a sphere is equal to the radius of the base of the cone. The ratio of the volume of the *cone* to the volume of the *sphere* is

   (A) $\dfrac{1}{3}$

   (B) $\dfrac{1}{4}$

   (C) $\dfrac{1}{12}$

   (D) $\dfrac{1}{1}$

   (E) $\dfrac{4}{3}$

27. In how many different ways can the seven letters in the word MINIMUM be arranged, if all the letters are used each time?

   (A) 7

   (B) 42

   (C) 420

   (D) 840

   (E) 5040

28. From a deck of 52 different cards, how many different hands, each consisting of three cards, can be drawn?

   (A) 132,600

   (B) 1.3×10^{67}

   (C) 22,100

   (D) 1.4×10^{10}

   (E) 2652

29. What is the probability of getting at least three heads when flipping four coins?

   (A) $\dfrac{3}{4}$

   (B) $\dfrac{1}{4}$

   (C) $\dfrac{7}{16}$

   (D) $\dfrac{5}{16}$

   (E) $\dfrac{3}{16}$

GO ON TO THE NEXT PAGE

30. The positive zero of $y = 3x^2 - 4x - 5$ is, to the nearest tenth, equal to

   (A) 0.8
   (B) $0.7 + 1.1i$
   (C) 0.7
   (D) 2.1
   (E) 2.2

31. In the figure at the right, S is the set of all points in the shaded region. Which of the following represents the set consisting of all points $(2x, y)$, where (x, y) is a point in S?

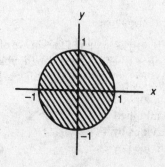

(A)

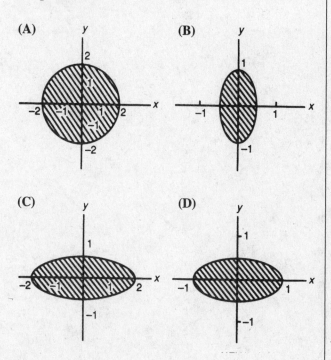

(B)

(C)

(D)

(E)

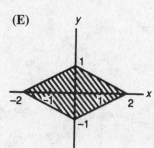

32. If a square prism is inscribed in a right circular cylinder of radius 3 and height 6, the volume inside the cylinder but outside the prism is

   (A) 169.6
   (B) 3.14
   (C) 115.6
   (D) 2.14
   (E) 61.6

33. If y varies jointly as x, w, and the square of z, what is the effect on w when x, y, and z are doubled?

   (A) w is doubled
   (B) w is multiplied by 4
   (C) w is multiplied by 8
   (D) w is divided by 4
   (E) w is divided by 2

34. Given the statement "All girls play tennis," which of the following negates this statement?

   (A) All boys play tennis.
   (B) Some girls play tennis.
   (C) All boys do not play tennis.
   (D) At least one girl doesn't play tennis.
   (E) All girls do not play tennis.

35. $\displaystyle\sum_{j=1}^{5} 2\left(\frac{3}{2}\right)^{j-1} =$

   (A) $26\dfrac{3}{8}$

   (B) $26\dfrac{5}{8}$

   (C) $26\dfrac{1}{2}$

   (D) $26\dfrac{1}{8}$

   (E) $26\dfrac{7}{8}$

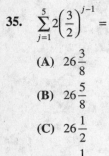

36. If $f(x) = \dfrac{k}{x}$ for all nonzero real numbers, for what value of k does $f(f(x)) = x$?

   (A) only 1
   (B) only 0
   (C) all real numbers
   (D) all real numbers except 0
   (E) no real numbers

37. $F(x) = \begin{cases} \dfrac{3x^2 - 3}{x - 1}, & \text{when } x \neq 1 \\[2ex] k, & \text{when } x = 1 \end{cases}$

For what value(s) of k is F a continuous function?

(A) 1
(B) 2
(C) 3
(D) 6
(E) no value of k

38. If $f(x,y) = 2x^2 - y^2$ and $g(x) = 2^x$, the value of $g(f(1,2)) =$

(A) 1
(B) 4
(C) $\dfrac{1}{4}$
(D) -4
(E) 0

39. What is the amplitude of the graph of the function $y = \cos^4 x - \sin^4 x$?

(A) $\dfrac{1}{2}$

(B) $\dfrac{\sqrt{2}}{2}$

(C) 1

(D) $1 + \dfrac{\sqrt{2}}{2}$

(E) 2

40. Which of the following could be the equation of one cycle of the graph in the figure on the right?

  I. $y = \sin 4x$

  II. $y = \cos\left(4x - \dfrac{\pi}{2}\right)$

  III. $y = -\sin(4x + \pi)$

(A) only I
(B) only I and II
(C) only II and III
(D) only II
(E) I, II, and III

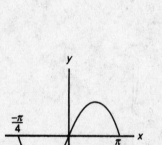

41. If $2 \cdot \sin^2 x - 3 = 3 \cdot \cos x$ and $90° < x < 270°$, the number of values that satisfy the equation is

(A) 0
(B) 1
(C) 2
(D) 3
(E) 4

GO ON TO THE NEXT PAGE

42. If $A = \text{Arctan}\left(-\dfrac{3}{4}\right)$ and $A + B = 315°$, then $B =$

    (A) 278.13°
    (B) 351.87°
    (C) –8.13°
    (D) 171.87°
    (E) 233.13°

43. The units digit of 1567^{93} is

    (A) 1
    (B) 3
    (C) 7
    (D) 9
    (E) none of the above

44. The vertex angle of an isosceles triangle is 35°. The length of the base is 10 centimeters. How many centimeters are in the perimeter?

    (A) 17.4
    (B) 44.9
    (C) 20.2
    (D) 16.6
    (E) 43.3

45. If the graph below represents the function $f(x)$, which of the following could represent the equation of the inverse of f?

    (A) $x = y^2 - 8y - 1$
    (B) $x = y^2 + 11$
    (C) $x = (y - 4)^2 - 3$
    (D) $x = (y + 4)^2 - 3$
    (E) $x = (y + 4)^2 + 3$

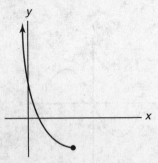

46. The figure on the right most closely resembles the graph whose equation is

    (A) $r = 2 \cdot \cos 2\theta + 2$
    (B) $r = 4 \cdot \cos \theta$
    (C) $r^2 = 4 \cdot \cos^2 2\theta$
    (D) $r^2 = 4 \cdot \cos 2\theta$
    (E) $r = \sin \theta + 4$

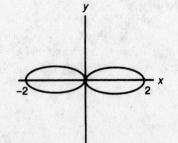

47. If $f(x) = \log_b x$ and $f(2) = 0.231$, the value of b is

    (A) 1.3
    (B) 20.1
    (C) 0.3
    (D) 13.2
    (E) 32.5

USE THIS SPACE FOR SCRATCH WORK

48. If $f_{n+1} = f_{n-1} + 2 \cdot f_n$ for $n = 2, 3, 4, \ldots$, and $f_1 = 1$ and $f_2 = 1$, then $f_5 =$

    (A) 7
    (B) 41
    (C) 11
    (D) 21
    (E) 17

49. In a plane, the *homogeneous coordinates* of a point P, whose rectangular coordinates are (x,y), are any three numbers a, b, and c for which $\dfrac{a}{c} = x$ and $\dfrac{b}{c} = y$. If the coordinates of P are $(3,4)$ and a, b, and c are integers, then the sum of a, b, and c could be

    (A) 2
    (B) 5
    (C) 8
    (D) 11
    (E) 14

50. If $[x]$ is defined to represent the greatest integer less than or equal to x, and $f(x) = \left| x - [x] - \dfrac{1}{2} \right|$, what is the period of $f(x)$?

    (A) 1
    (B) $\dfrac{1}{2}$
    (C) 2
    (D) 4
    (E) f is not a periodic function.

ANSWER KEY

1. A	6. D	11. C	16. C	21. B	26. B	31. C	36. D	41. D	46. D
2. D	7. C	12. A	17. C	22. B	27. C	32. E	37. D	42. B	47. B
3. B	8. A	13. B	18. E	23. A	28. C	33. D	38. C	43. C	48. E
4. E	9. A	14. C	19. C	24. A	29. D	34. D	39. C	44. E	49. C
5. C	10. D	15. B	20. B	25. A	30. D	35. A	40. E	45. C	50. A

ANSWER EXPLANATIONS

The following explanations are keyed to the review portions of this book. For example, material covered in Question 1 is reviewed in Sections 1.2 and 2.2. Several explanations also include graphing calculator applications that are described in Part 3. If you had trouble understanding these explanations, you should peruse these sections of the book.

In these solutions the following notation is used:

a: active—Calculator use is necessary or, at a minimum, extremely helpful.

n: neutral—Answers may be found without a calculator, but a calculator may help.

i: inactive—Calculator use is not helpful and may even be a hindrance.

1. i **A** $f(1) = 2$ means that the line goes through point $(1,2)$. $f(2) = q$ means that the line goes through point $(2,q)$. $-2 = $ slope $= \dfrac{\Delta y}{\Delta x} = \dfrac{q-2}{2-1}$ implies $-2 = \dfrac{q-2}{1}$ and $q = 0$. [2.2, 1.2].

2. a **D** Even functions are symmetric about the y-axis. Graph each answer choice to see that Choice D is not symmetric about the y-axis.

   An alternative solution is to use the fact that $\sin x \neq \sin(-x)$, from which you deduce the correct answer choice. [1.4].

TIP: Properties of even and odd functions:
Even + even is always an even function.
Odd + odd is always an odd function.
Odd × even is always an odd function.

---

3. a **B** Since the radius of a sphere is the distance between the center, $(0,0,0)$, and a point on the surface, $(2,3,4)$, use the distance formula in three dimensions to get

$$\sqrt{(2-0)^2 + (3-0)^2 + (4-0)^2} = \sqrt{29}$$

Use your calculator to find $\sqrt{29} \approx 5.39$. [2.2, 5.5].

4. i **E** A point in the second quadrant has a negative x-coordinate and a positive y-coordinate. Therefore, $x < y$, and $\dfrac{x}{y} < 0$ must be true, but $x + y$ can be less than or equal to zero. The correct answer is E. [1.3].

5. i **C** $f(a)$ means to replace x in the formula with an a. Therefore, $f(a) = a^2 - a \cdot a = 0$. [1.2].

6. n **D** Since the average of your first three test grades is 78, each test grade could have been a 78. If x represents your final test grade, the average of the four test grades is $\dfrac{78 + 78 + 78 + x}{4}$, which is to be equal to 80. Therefore, $\dfrac{234 + x}{4} = 80$. $234 + x = 320$. So $x = 86$. [5.8].

7. a **C** Use the change-of-base theorem and your calculator to get:

$$\log_7 9 = \frac{\log_{10} 9}{\log_{10} 7} \approx \frac{0.9542}{0.8451} \approx 1.13. \quad [4.2].$$

8. i **A** Add the two equations: $\log_2 m + \log_2 n = x + y$, which becomes $\log_2 mn = x + y$ (basic property of logs). $2^{x+y} = mn$. [4.2].

9. a **A** Plot the graph of $y = abs(x - 2) - 5$ in the standard window that includes both x-intercepts. You can count 11 integers between -3 and 7 if you include both endpoints.

   The inequality $|x - 2| \leq 5$ means that x is less than or equal to 5 units away from 2. Therefore, $-3 \leq x \leq 8$, and there are 11 integers in this interval. [2.5, 4.3].

10. i **D** $\sqrt{x^2}$ indicates the need for the *positive* square root of x^2. Therefore, $\sqrt{x^2} = x$ if $x \geq 0$ and $\sqrt{x^2} = -x$ if $x < 0$. This is just the definition of absolute value, and so $\sqrt{x^2} = |x|$ is the only answer for all values of x. [4.3].

11. a **C** Solve for y, and plot the graph of $y = \dfrac{x^2 - 4}{x^2 - 1}$ in the standard window to observe two vertical and one horizontal asymptote.

An alternative solution is to use the facts that $y = \dfrac{x^2 - 4}{x^2 - 1}$ has vertical asymptotes when the denominator is zero, i.e., when $x = \pm 1$, and a horizontal asymptote of $y = 1$ as $x \to \infty$. [4.5].

12. a **A** Enter the given expression into Y_t and key in TBLSET with TblStart = 0 and ΔTbl = 10. Observe Y_t approach 0.5 as x gets larger.

Divide the numerator and denominator of the expression by x^2 and observe that the expression approaches $\dfrac{1}{2}$ as $x \to \infty$. [4.5].

13. a **B** $y = mx + b$. Use the x-intercept to get $0 = \sqrt{3}m + b$, and the y-intercept to get $\sqrt{5} = 0 \cdot m + b$. Therefore, $0 = \sqrt{3}m + \sqrt{5}$ and $m = -\dfrac{\sqrt{5}}{\sqrt{3}}$.

Use your calculator to get $m \approx -1.29$. [2.2].

14. a **C** Use your calculator in radian mode to find the value of the inverse sin of $\dfrac{\pi}{4}$, which is approximately 0.90. [3.6].

15. i **B** Substituting the points into the equation gives $a = \dfrac{5}{2}, b = \dfrac{5}{3}$, and $c = -\dfrac{5}{4}$. [5.5].

16. i **C** Mode = 2, median $= \dfrac{2+2}{2} = 2$, mean $= \dfrac{2+6+6+4}{8} = \dfrac{18}{8} = 2.25$.

Thus, median $\leq$ mode $\leq$ mean. [5.8].

17. n **C** Multiply $\dfrac{x - 3y}{x} = 7$ through by x to get $x - 3y = 7x$. Combine like terms to get $-3y = 6x$. Divide through by $6y$ so that $\dfrac{x}{y}$ will be on one side of the equals sign. This gives $\dfrac{x}{y} = -\dfrac{3}{6} = -\dfrac{1}{2}$. [2.2].

18. i **E** $\sin 120° = \dfrac{\sqrt{3}}{2}$, $\cos \dfrac{2\pi}{3} = -\dfrac{1}{2}$, $\tan 315° = -1$. [3.3]

TIP: Any angle greater than 90° has an acute reference angle associated with it, formed by the terminal side of the large angle and the **x-axis**. The reference angle has the same trig function value as the original angle, with the possible exception of the sign. Thus, you need to learn the values of the trig functions for only a handful of special acute angles.

Example: 120° has a reference angle of 60°. Therefore, $|\sin 120°| = |\sin 60°| = \dfrac{\sqrt{3}}{2}$. To remember which functions have which signs in which quadrant, the following statement is helpful:

"**A**ll **S**tudents **T**ake **C**alculus."

All functions are positive in quadrant I.
Sin (and its reciprocal, csc) is positive in quadrant II.
Tan (and its reciprocal, cot) is positive in quadrant III.
Cos (and its reciprocal, sec) is positive in quadrant IV.

Therefore, $\sin 120° = +\dfrac{\sqrt{3}}{2}$.

---

19. a **C** Plot the graph of $y = abs((1/2)x^2 - 8)$ in an $x \varepsilon [-4,4]$ and $y \varepsilon [-10,10]$ window and observe that the maximum value is -8 (at $x = 0$).

An alternative solution is to recognize that the graph of f is a parabola that is symmetric about the y – axis and opens up. The minimum value $y = -8$ occurs when $x = 0$, so 8 is the maximum value of $|f(x)|$. [2.3, 4.3].

20. i **B** In a right triangle, $\tan \theta = \dfrac{\text{opposite side}}{\text{adjacent side}}$. Draw the right triangle below and use the Pythagorean theorem to compute the hypotenuse. Since in a right triangle $\sin \theta = \dfrac{\text{opposite side}}{\text{hypotenuse}}$, $\sin \theta = \dfrac{2}{\sqrt{13}}$. Since $\tan \theta$ is positive, θ could lie in either quadrant I or quadrant III.

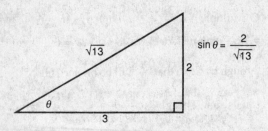

Therefore, $\sin\theta = \pm\dfrac{2\sqrt{13}}{13}$ since θ could be in quadrant I or III, where $\tan\theta$ is positive. [3.1].

Don't use a calculator on this problem because the answer choices are not in decimal form.

21. a B $s = r\cdot\theta^R$, $75° = \dfrac{75\pi}{180}$. Therefore, 75 feet $=$

$r\cdot\dfrac{75\pi}{180}$ and $r = \dfrac{180}{\pi}$ feet $\approx \dfrac{180}{3.1416} \approx 57.3$ feet. [3.2].

22. a B According to the Law of Cosines, $7^2 = 5^2 + 3^2 - 2(5)(3)\cos\theta$, where θ is the angle opposite the side of length 7. Therefore, the area of the triangle is

$\dfrac{1}{2}(5)(3)\sin\theta = \dfrac{15}{2}\sin(\cos^{-1}\theta) \approx 6.5$.

23. a A The function f is the greatest integer function. Plot the graph of $y = \text{int}(x) - \text{abs}(\text{int}(x))$ in the standard window and observe that the maximum value of y is 0.

An alternative solution is to observe that $f(x) = |f(x)|$ if $i \geq 0$, and $f(x) = -|f(x)|$ if $i < 0$. In the latter case, $g(x) = 2i < 0$, so the maximum is 0. [4.4].

24. i A Since $\sec x = \dfrac{1}{\cos x}$, $\cos x = \dfrac{1}{\sec x}$, so $f(x) = \cos x$, which is an even function. Therefore, $f(x) = f(-x)$. [1.4].

25. i A From the figure $\left(-1, -\sqrt{3}\right)$. [4.7].

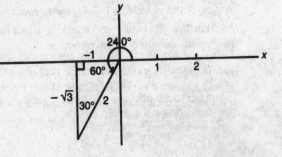

26. i B Since $r = h$ in the cone,

$\dfrac{\text{Volume of cone}}{\text{Volume of sphere}} = \dfrac{\frac{1}{3}\pi r^2 h}{\frac{4}{3}\pi r^3} = \dfrac{\frac{1}{3}\pi r^3}{\frac{4}{3}\pi r^3} = \dfrac{1}{4}$. [5.5].

27. a C Permutation with repetitions is $\dfrac{7!}{3!2!} = 420$. [5.1].

28. a C Since the order in which the cards are drawn does not make a difference, this is a combination.

$_{52}C_3 = \dfrac{52!}{49!\cdot3!}$.

Use your calculator to find that the answer is 22,100 [5.1].

29. i D There are 16 outcomes in the sample space. $\dbinom{4}{3} = 4$ ways to get 3 heads. $\dbinom{4}{4} = 1$ way to get 4 heads. [5.3].

Alternative Solution: There are $2^4 = 16$ outcomes when flipping 4 coins. List those that contain 3 or 4 heads: HHHT, HHTH, HTHH, THHH, HHHH. Therefore, $P(3 \text{ or } 4 \text{ heads}) = \dfrac{5}{16}$.

30. a D Use the quadratic formula program on your graphing calculator to get both zeros and choose the positive one. [2.3].

31. i C Since the y values remain the same but the x values are doubled, the circle is stretched along the x-axis. [5.5].

32. a E Volume of cylinder $= \pi r^2 h = \pi\cdot9\cdot6 = 54\pi$. Volume of square prism $= Bh$, where B is the area of the square base, which is $3\sqrt{2}$ on a side. Thus, $Bh = (3\sqrt{2})^2\cdot6 = 112$. Therefore, the desired volume is $54\pi - 108$, which (using your calculator) is approximately 61.6. [5.5].

33. i D $\dfrac{y}{xwz^2} = K$. The doubling of y and x cancel each other out. The doubling of z is squared, and so w has to be divided by 4. [5.6].

34. i D The statement, "All G are T." is negated by the statement, "At least one G is not T." [5.7].

35. a A Use the LIST/OPS/seq command to generate the sequence $2\left(\dfrac{3}{2}\right)^{j-2}$ as follows: seq(2(3/2)^(x – 1), x,1,5). Then use LIST/MATH/sum (Ans) to get the correct answer choice. This result can be achieved with the single command LIST/MATH/sum(LIST/OPS/seq(2(3/2)∧(x – 1), x,1,5)). [5.4].

An alternative solution is to evaluate each sum and add them on your calculator.

36. i D $f(f(x)) = \dfrac{k}{f(x)} = k \div \dfrac{k}{x} = k \cdot \dfrac{x}{k} = x$. Since the k's divide out, k can equal any real number except zero (since you cannot divide by zero). [1.2].

> **TIP:** Always keep an eye out for places *throughout* your solution where division by zero might occur.

37. a D Enter $(3x^2 – 3)/(x – 1)$ into Y_1 and key TBLSET. Set Indpnt to Ask and key TABLE. Enter values of x progressively closer to 1 (e.a. .9, .99, .999, etc.) and observe that Y_1 gets progressively closer to 6, so choose $k = 6$.

An alternative solution is to factor the numerator to $3(x + 1)(x – 1)$, divide out $x – 1$ from the numerator and denominator. As $x \to 1$, $3(x + 1) \to 6$. [4.5, 2.4].

38. i C $f(1,2) = 2(1^2) – 2^2 = –2$.

$g(f(1,2)) = g(–2) = 2^{-2} = \dfrac{1}{4}$. [1.2, 1.5, 4.2].

39. a C Plot the graph of $y = \cos(x)^\wedge 4 – \sin(x)^\wedge 4$ using ZTrig and observe that the amplitude is 1.

An alternative solution is to factor y as $(\cos^2 x + \sin^2 x)(\cos^2 x – \sin^2 x) = 1 \cos 2x$, which has an amplitude of 1. [3.4].

40. a E Plot the graphs of all three functions in an $x\varepsilon\left[-\dfrac{\pi}{4}, \dfrac{\pi}{4}\right]$ and $y\varepsilon[–2,2]$ window and observe that they coincide.

An alternative solution is to deduce facts about the graphs from the equations. All three equations indicate graphs that have period $\dfrac{\pi}{2}$. The graph of equation I is a normal sine curve. The graph of equation II is a cosine curve with a phase shift right of $\dfrac{\pi}{8}$, one-fourth of the period. Therefore, it fits a normal sine curve. The graph of equation III is a sine curve that has a phase shift left of $\dfrac{\pi}{4}$, one-half the period, and reflected through the x-axis. This also fits a normal sine curve. [3.4].

41. a D With your calculator in degree mode, plot the graphs of $y = 2\sin(x)^2 – 3$ and $y = 3\cos x$ in an $x\varepsilon[90,270]$ and $y\varepsilon[–3,0]$ window and observe that the graphs intersect in places.

An alternative solution is to distribute and transform the equation to read: $2\cos^2 x + 3\cos x + 1 = 0$. This factors to $(2\cos x + 1)(\cos x + 1) = 0$, so $\cos x = -\dfrac{1}{2}$ or $\cos x = –1$. For $90° < x < 270°$, there are three solutions: $x = 120°, 240°, 180°$. [3.1].

42. a B With your calculator in degree mode, evaluate $315° – \tan^{-1}(–3/4)$ to get the correct answer choice. [3.6].

43. i C When 1567 is multiplied by itself successively, the units digit takes on values of 7, 9, 3, 1, 7, 9, 3, 1, ... Every fourth multiplication ends in 1. Therefore, 1567^{92} ends in 1, and so 1567^{93} ends in a 7. [5.9].

44. a E Drop the altitude from the vertex to the base. The altitude bisects both the vertex angle and the base, cutting the triangle into two congruent right triangles. Since $\sin 17.5° = \dfrac{5}{\text{leg}}$, the leg $= \dfrac{5}{\sin 17.5°} \approx \dfrac{5}{0.3007} \approx 16.628$ cm and the perimeter = 43.3 cm. [3.7, 6.3].

45. i C The given graph looks like the left half of a parabola with vertex $(4,–3)$ (using the values given in the answer choices as guides) that opens up. The equation of such a parabola is $y = (x – 4)^2 – 3$. The vertex of the inverse is $(–3,4)$, so its equation is $x = (y – 4)^2 – 3$. [1.3].

46. a D Sketch the graph of each answer choice to see that the correct answer choice is D. By letting $\theta = 0$, answer choices A, B, and E can be ruled out because their graphs do not cross the x – axis at $r = \pm2$. [4.7].

47. a **B** $f(2) = \log_b 2 = 0.231$. Therefore, $b^{0.231} = 2$, and so $b = 2^{1/0.231}$, which (using your calculator) is approximately 20.1. [4.2].

48. i **E** Let $n = 2$, $f_3 = 3$; then let $n = 3$, $f_4 = 7$; and finally let $n = 4$, $f_5 = 17$. [5.4].

49. i **C** Since $\dfrac{a}{c} = 3$ and $\dfrac{b}{c} = 4$, therefore $a = 3c$ and $b = 4c$. If $c = 1$, then $a = 3$ and $b = 4$. [5.9].

50. a **A** Plot the graph of $y = abs(x - int(x) - 1/2)$ in an $x\varepsilon[-3,3]$ and $y\varepsilon[-1,1]$ window. Inspection of the graph leads to the conclusion that the period is 1.

An alternative solution is to plot a few points using integral values of x and observe that the period is 1. [4.4, 4.3].

SELF-EVALUATION CHART FOR DIAGNOSTIC TEST

SUBJECT AREA	QUESTIONS	NUMBER OF RIGHT WRONG OMITTED

Mark correct answers with C, wrong answers with X, and omitted answers with O.

Algebra
(9 questions)

7	8	9	10	11	17	30	33	50

Review section

4.2	4.2	2.5	4.3	4.5	2.2	2.3	5.6	4.4

___ ___ ___

Solid geometry
(4 questions)

3	15	26	32

Review section

2.2	5.5	5.5	5.5

___ ___ ___

Coordinate geometry
(6 questions)

13	24	25	31	46	49

Review section

2.3	3.1	4.7	5.5	4.7	5.9

___ ___ ___

Trigonometry
(10 questions)

14	18	20	21	22	39	40	41	42	44

Review section

3.6	3.3	3.1	3.2	3.7	3.4	3.4	3.1	3.6	3.7

___ ___ ___

Functions
(12 questions)

1	2	4	5	19	23	36	37	38	45	47	48

Review section

2.2	1.4	1.3	1.2	2.3	4.4	1.2	4.5	1.2	1.3	4.2	5.4

___ ___ ___

Miscellaneous
(9 questions)

6	12	16	27	28	29	34	35	43

Review section

5.8	4.5	5.8	5.1	5.1	5.3	5.7	5.4	5.9

TOTALS   ___ ___ ___

Raw score = (number right) – $\frac{1}{4}$ (number wrong) = _____

Round your raw score to the nearest whole number = _____

Evaluate Your Performance
Diagnostic Test

Rating	Number Right
Excellent	41–50
Very good	33–40
Above average	25–32
Average	15–24
Below average	Below 15

REVIEW OF MAJOR TOPICS

PART

2

This part reviews the mathematical concepts and techniques for the topics covered in the Math Level IIC Examination. A sound understanding of these concepts certainly will improve your score on the exam, whether you use a scientific calculator or a graphing calculator. The techniques discussed will enable you to save time solving some of the problems without a calculator at all. For problems requiring computational power, techniques are described that will help you use your calculator in the most efficient manner. Problems for which a graphing calculator should be the primary route to a solution are discussed in Part 3.

Your classroom experience will guide your decisions about whether to use a graphing calculator and how best to use one. If you have been through a secondary mathematics program that attached equal importance to graphical, tabular, and algebraic presentations, then you probably will rely on your graphing calculator as your primary tool to help you find solutions. However, if you went through a more traditional mathematics program, where algebra and algebraic techniques were stressed, it may be more natural for you to use a graphing calculator only after considering other approaches.

INTRODUCTION TO FUNCTIONS

CHAPTER

1

1.1 DEFINITIONS

A *relation* is a set of ordered pairs. A *function* is a relation such that for each first element there is one and only one second element. The set of numbers that make up all first elements of the ordered pairs is called the *domain* of the function, and the resulting set of second elements is called the *range* of the function.

EXAMPLE 1: {(1,2),(3,4),(5,6),(6,1),(2,2)}
This is a function because every ordered pair has a different first element. Domain = {1,2,3,5,6}. Range = {1,2,4,6}.

EXAMPLE 2: $f(x) = 3x + 2$
This is a function because for each value substituted for x there is one and only one value for $f(x)$. Domain = {all real numbers}. Range = {all real numbers}.

EXAMPLE 3: {(1,2),(3,2),(1,4)}
This is a relation but *not* a function because when the first element is 1, the second element can be either 2 or 4. Domain = {1,3}. Range = {2,4}.

EXAMPLE 4: $\{(x,y):y^2 = x\}$
[This should be read, "The set of all ordered pairs (x,y) such that $y^2 = x$."] This is a relation but *not* a function because, for each nonnegative number that is substituted for x, there are two values for y. For example, $x = 4$, $y = +2$, or $y = -2$. Domain = {all nonnegative real numbers}. Range = {all real numbers}.

EXERCISES

1. If {(3,2),(4,2),(3,1),(7,1),(2,3)} is to be a function, which one of the following must be removed from the set?

   (A) (3,2)
   (B) (4,2)
   (C) (2,3)
   (D) (7,1)
   (E) none of the above

2. For $f(x) = 3x^2 + 4$, $g(x) = 2$, and h {(1,1), (2,1), (3,2)},

   (A) f is the only function
   (B) h is the only function
   (C) f and g are the only functions
   (D) g and h are the only functions
   (E) f, g, and h are all functions

3. What value(s) must be excluded from the domain of
   $$f = \left\{(x, y): y = \frac{x+2}{x-2}\right\}?$$

   (A) 2
   (B) −2
   (C) 0
   (D) 2 and −2
   (E) no value

1.2 FUNCTION NOTATION

The expressions $f = \{(x,y):y = x^2\}$ and $f(x) = x^2$ both name the same function; f is the rule that pairs any number with its square. Thus, $f(x) = x^2$, $f(a) = a^2$, $f(z) = z^2$ all name the same function. The symbol $f(2)$ is the value of the function f when $x = 2$. Thus, $f(2) = 4$.

If f and g name two functions, the following rules apply:

$$(f + g)(x) = f(x) + g(x)$$
$$(f \cdot g)(x) = f(x) \cdot g(x)$$
$$\frac{f}{g}(x) = \frac{f(x)}{g(x)}$$

if and only if $g(x) \neq 0$

$$(f \circ g)(x) = f(x) \circ g(x) = f(g(x))$$

(this is called the *composition* of functions).

EXAMPLE: If $f(x) = 3x - 2$ and $g(x) = x^2 - 4$, then indicate the function that represents (A) $(f + g)(x)$ (B) $(f - g)(x)$ (C) $(f \cdot g)(x)$ (D) $\frac{f}{g}(x)$ (E) $(f \circ g)(x)$ (F) $(g \circ f)(x)$

(A) $(f + g)(x) = f(x) + g(x)$
$$= (3x - 2) + (x^2 - 4) = x^2 + 3x - 6$$
(B) $(f - g)(x) = f(x) - g(x)$
$$= (3x - 2) - (x^2 - 4) = -x^2 + 3x + 2$$
(C) $(f \cdot g)(x) = f(x) \cdot g(x)$
$$= (3x - 2)(x^2 - 4)$$
$$= 3x^3 - 2x^2 - 12x + 8$$
(D) $\frac{f}{g}(x) = \frac{f(x)}{g(x)} = \frac{3x - 2}{x^2 - 4}$ and $x \neq \pm 2$
(E) $(f \circ g)(x) = f(x) \circ g(x)$
$$= f(g(x)) = 3(g(x)) - 2$$
$$= 3(x^2 - 4) - 2 = 3x^2 - 14$$
(F) $(g \circ f)(x) = g(x) \circ f(x)$
$$= g(f(x)) = (f(x))^2 - 4$$
$$= (3x - 2)^2 - 4 = 9x^2 - 12x$$

(Notice that the composition of functions is not commutative.)

EXERCISES

1. If $f(x) = 3x^2 - 2x + 4$, $f(-2) =$
    (A) -2
    (B) 20
    (C) -4
    (D) 12
    (E) -12

2. If $f(x) = 4x - 5$ and $g(x) = 3^x$, then $f(g(2)) =$
    (A) 27
    (B) 9
    (C) 3
    (D) 31
    (E) none of the above

3. If $f(g(x)) = 4x^2 - 8x$ and $f(x) = x^2 - 4$, then $g(x) =$
    (A) $2x - 2$
    (B) x
    (C) $4x$
    (D) $4 - x$
    (E) x^2

4. What values must be excluded from the domain of $\frac{f}{g}(x)$ if $f(x) = 3x^2 - 4x + 1$ and $g(x) = 3x^2 - 3$?
    (A) no values
    (B) 0
    (C) 3
    (D) 1
    (E) both ± 1

5. If $g(x) = 3x + 2$ and $g(f(x)) = x$, then $f(2) =$
    (A) 2
    (B) 6
    (C) 0
    (D) 8
    (E) 1

6. If $p(x) = 4x - 6$ and $p(a) = 0$, then $a =$
    (A) -6
    (B) 2
    (C) $\frac{3}{2}$
    (D) $\frac{2}{3}$
    (E) $-\frac{3}{2}$

7. If $f(x) = e^x$ and $g(x) = \sin x$, then the value of $f \circ g\left(\sqrt{2}\right)$ is
    (A) -0.8
    (B) -0.01
    (C) 0.34
    (D) 2.7
    (E) 1.8

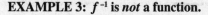

1.3 INVERSE FUNCTIONS

The *inverse* of a function f, denoted by f^{-1}, is a relation that has the property that $f(x) \circ f^{-1}(x) = f^{-1}(x) \circ f(x) = x$, where f^{-1} is not necessarily a function.

EXAMPLE 1: $f(x) = 3x + 2$. **Is** $y = \dfrac{x-2}{3}$ **the inverse of** f?

To answer this question assume that $f^{-1}(x) = \dfrac{x-2}{3}$ and verify that $f(x) \circ f^{-1}(x) = x$.

To verify this, proceed as follows:

$$f(x) \circ f^{-1}(x) = f(f^{-1}(x))$$

$$= f\left(\frac{x-2}{3}\right) = 3\left(\frac{x-2}{3}\right) + 2 = x$$

and

$$f^{-1}(x) \circ f(x) = f^{-1}(f(x)) = f^{-1}(3x+2)$$

$$= \frac{(3x+2)-2}{3} = x.$$

Since $f(x) \circ f^{-1}(x) \circ f^{-1}(x) \circ f(x) = x$, $\dfrac{x-2}{3}$ is the inverse of f.

EXAMPLE 2: $f = \{(1,2),(2,3),(3,2)\}$ **Find the inverse.**

$$f^{-1} = \{(2,1),(3,2),(2,3)\}$$

To verify this, check $f \circ f^{-1}$ and $f^{-1} \circ f$ term by term.

$$f \circ f^{-1} = f(f^{-1}(x)); \quad \text{when } x = 2, f(f^{-1}(2)) = f(1) = 2$$

$$\text{when } x = 3, f(f^{-1}(3)) = f(2) = 3$$

$$\text{when } x = 2, f(f^{-1}(2)) = f(3) = 2$$

Thus, for each x, $f(f^{-1}(x)) = x$.

$$f^{-1} \circ f = f^{-1} \circ f(x); \quad \text{when } x = 1, f^{-1}(f(1)) = f^{-1}(2) = 1$$

$$\text{when } x = 2, f^{-1}(f(2)) = f^{-1}(3) = 2$$

$$\text{when } x = 3, f^{-1}(f(3)) = f^{-1}(2) = 3$$

Thus, for each x, $f^{-1}(f(x)) = x$. In this case f^{-1} is *not* a function.

If the point with coordinates (a,b) belongs to a function f, then the point with coordinates (b,a) belongs to the inverse of f. Because this is true of a function and its inverse, the graph of the inverse is the reflection of the graph of f about the line $y = x$.

EXAMPLE 3: f^{-1} **is** *not* **a function.**

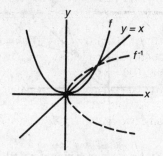

EXAMPLE 4: f^{-1} **is a function.**

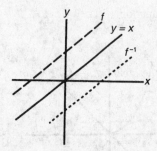

As can be seen from the above examples, if the graph of f is given, the graph of f^{-1} is the image obtained by folding the graph of f about the line $y = x$. Algebraically the equation of the inverse of f can be found by interchanging the variables.

EXAMPLE 5: $f = \{(x,y): y = 3x + 2\}$. **Find** f^{-1}.
In order to find f^{-1}, interchange x and y and solve for y: $x = 3y + 2$, which becomes $y = \dfrac{x-2}{3}$. Thus,

$$f^{-1} = \left\{(x,y): y = \frac{x-2}{3}\right\}.$$

EXAMPLE 6: $f = \{(x,y): y = x^2\}$. **Find** f^{-1}.
Interchange x and y: $x = y^2$.

Solve for $y = y = \pm\sqrt{x}$.

Thus, $f^{-1} = \left\{(x,y): y = \pm\sqrt{x}\right\}$, which is *not* a function.

The inverse of any function f can always be made a function by limiting the domain of f. In Example 6 the domain of f could be limited to all nonnegative numbers or all nonpositive numbers. In this way f^{-1} would become either $y = +\sqrt{x}$ or $y = -\sqrt{x}$, both of which are functions.

EXAMPLE 7: $f = \{(x,y): y = x^2 \text{ and } x \geq 0\}$. **Find** f^{-1}.
$f^{-1} = \{(x,y): x = y^2 \text{ and } y \geq 0\}$, which can also be written as

$$f^{-1} = \left\{(x,y): y = +\sqrt{x}\right\}.$$

Here f^{-1} is a function.

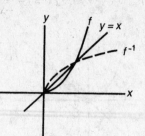

EXAMPLE 8: $f = \{(x,y):y = x^2 \text{ and } x \leq 0\}$. find f^{-1}.
$f^{-1} = \{(x,y):x = y^2 \text{ and } y \leq 0\}$, which can also be written as

$$f^{-1} = \left\{(x,y) : y = -\sqrt{x}\right\}.$$

Here f^{-1} is a function.

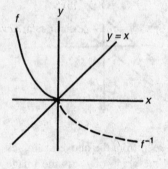

EXERCISES

1. If $f(x) = 2x - 3$, the inverse of f, f^{-1}, could be represented by

   (A) $f^{-1}(x) = 3x - 2$

   (B) $f^{-1}(x) = \dfrac{1}{2x - 3}$

   (C) $f^{-1}(x) = \dfrac{x - 2}{3}$

   (D) $f^{-1}(x) = \dfrac{x + 2}{3}$

   (E) $f^{-1}(x) = \dfrac{x + 3}{2}$

2. If $f(x) = x$, the inverse of f, f^{-1}, could be represented by

   (A) $f^{-1}(x) = x$

   (B) $f^{-1}(x) = 1$

   (C) $f^{-1}(x) = \dfrac{1}{x}$

   (D) $f^{-1}(x) = y$

   (E) f^{-1} does not exist

3. The inverse of $f = \{(1,2),(2,3),(3,4),(4,1),(5,2)\}$ would be a function if the domain of f is limited to

   (A) $\{1,3,5\}$
   (B) $\{1,2,3,4\}$
   (C) $\{1,5\}$
   (D) $\{1,2,4,5\}$
   (E) $\{1,2,3,4,5\}$

4. Which of the following could represent the equation of the inverse of the graph in the figure?

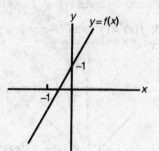

   (A) $y = -2x + 1$

   (B) $y = 2x + 1$

   (C) $y = \dfrac{1}{2}x + 1$

   (D) $y = \dfrac{1}{2}x - 1$

   (E) $y = \dfrac{1}{2}x - \dfrac{1}{2}$

1.4 ODD AND EVEN FUNCTIONS

A relation is said to be *even* if $(-x,y)$ is in the relation whenever (x,y) is. If the relation is defined by an equation, it is even if $(-x,y)$ satisfies the equation whenever (x,y) does. If the relation is a function f, it is even if $f(-x) = f(x)$ for all x in the domain of f. The graph of an even relation or function is symmetric with respect to the y axis.

EXAMPLE 1: $\{(1,0),(-1,0),(3,0),(-3,0),(5,4),(-5,4)\}$ is an even relation because $(-x,y)$ is in the relation whenever (x,y) is.

EXAMPLE 2: $x^4 + y^2 = 10$ is an even relation because $(-x)^4 + y^2 = x^4 + y^2 = 10$.

EXAMPLE 3: $f(x) = x^2$ and $f(-x) = (-x)^2 = x^2$.

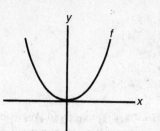

EXAMPLE 4: $f(x) = |x|$ and $f(-x) = |-x| = |-1 \cdot x| = |-1| \cdot |x| = |x|$.

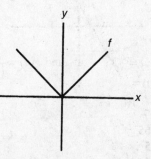

A relation is said to be *odd* if $(-x,-y)$ is in the relation whenever (x,y) is. If the relation is defined by an equation, it is odd if $(-x,-y)$ satisfies the equation whenever (x,y) does. If the relation is a function f, it is odd if $f(-x) = -f(x)$ for all x in the domain of x. The graph of an odd relation or function is symmetric with respect to the origin.

EXAMPLE 5: $\{(5,3),(-5,-3),(2,1),(-2,-1),(-10,8),$ $(10,-8)\}$ is an off relation because $(-x,-y)$ is in the relation whenever (x,y) is.

EXAMPLE 6: $x^4 + y^2 = 10$ is an odd relation because $(-x)^4 + (-y)^2 = x^4 + y^2 = 10$. Note that $x^4 + y^2 = 10$ is both even and odd.

EXAMPLE 7: $f(x) = x^3$ and $f(-x) = (-x)^3 = -x^3$. Therefore, $-f(-x) = x^3 = f(x)$.

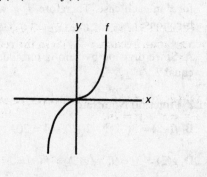

EXAMPLE 8: $f(x) = \dfrac{1}{x}$ and $f(-x) = \dfrac{1}{-x}$.

Therefore, $-f(-x) = \dfrac{1}{x} = f(x)$.

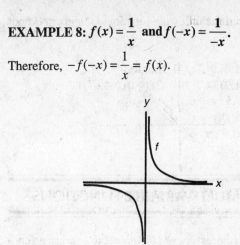

EXERCISES

1. Which of the following relations are *even*?

    I. $y = 2$
    II. $f(x) = x$
    III. $x^2 + y^2 = 1$

    (A) only I
    (B) only I and II
    (C) only II and III
    (D) only I and III
    (E) I, II, and III

2. Which of the following relations are *odd*?

    I. $y = 2$
    II. $y = x$
    III. $x^2 + y^2 = 1$

    (A) only II
    (B) only I and II
    (C) only I and III
    (D) only II and III
    (E) I, II, and III

3. Which of the following relations are both *odd* and *even*?

    I. $x^2 + y^2 = 1$
    II. $x^2 - y^2 = 0$
    III. $x + y = 0$

    (A) only III
    (B) only I and II
    (C) only I and III
    (D) only II and III
    (E) I, II, and III

4. Which of the following functions is neither *odd* nor *even*?

(A) $\{(1,2),(4,7),(-1,2),(0,4),(-4,7)\}$
(B) $\{(1,2),(4,7),(-1,-2),(0,0),(-4,-7)\}$
(C) $\{(x,y):y = x^3 - 1\}$
(D) $\{(x,y):y = x^2 - 1\}$
(E) $f(x) = -x$

1.5 MULTIVARIABLE FUNCTIONS

At times it is necessary or convenient to express a function or relation in terms of more than one variable. For instance, the area of a triangle is expressed in terms of the length of its base, b, and its altitude, h. Thus, in function notation, $A(b,h) = \frac{1}{2}bh$, where b and h are the independent variables.

The formula for the surface area of a rectangular solid expresses the surface area in terms of three independent variables, l, w, and h: $S(l,w,h) = 2lw + 2lh + 2wh$.

EXAMPLE 1: If $f(x,y,z) = (x + y)(y + z)$, what does $f(1,-2,3) = ?$
Substituting 1 for x, -2 for y, and 3 for z gives

$$f(1,-2,3) = (1 - 2)(-2 + 3) = (-1)(1) = -1.$$

EXAMPLE 2: If $f(x,y) = 2x + 3y - 4$, what does $f(2x,-3y + 1) = ?$
Substituting $2x$ for x and $-3y + 1$ for y gives

$$f(2x,-3y + 1) = 2(2x) + 3(-3y + 1) - 4$$
$$= 4x - 9y - 1.$$

EXAMPLE 3: If $f(x,y) = x^2 + y^2 + y$ and $g(z) = z - 2$, what does $f(2, g(1)) = ?$
Substituting 2 for x and $g(1) = 1 - 2 = -1$ for y gives

$$f(2,g(1)) = 2^2 + (g(1))^2 + g(1)$$
$$= 4 + (-1)^2 + (-1) = 4.$$

EXERCISES

1. If $f(x,y) = 3x + 2y - 8$ and $g(z) = z^2$ for all real numbers x, y, and z, then $f(3,g(4)) =$

(A) 9
(B) 33
(C) 81
(D) 7
(E) 5

2. If $f(x,y,z) = x^2 + y^2 - 2z$ for all real numbers x, y, and z, then $f(-1,1,-1) =$

(A) 0
(B) 2
(C) 4
(D) -2
(E) 1

3. If $f(x,y) = 3x + 2y$ and $g(x,y) = x^2 - y^2$ for all real numbers x and y, then $f(g(1,2)3) =$

(A) 40
(B) -3
(C) 9
(D) 6
(E) 3

4. If $f(x,y) = \sqrt{2x^2 - y^2}$ and $g(x) = 2^x$, the value of $g(f(2,1)) =$

(A) 2.65
(B) 14.65
(C) 6.25
(D) 5.65
(E) 7.05

ANSWERS AND EXPLANATIONS

In these solutions the following notation is used:

a: active—Calculator use is necessary or, at a minimum, extremely helpful.

n: neutral—Answers may be found without a calculator, but a calculator may help.

i: inactive—Calculator use is not helpful and may even be a hindrance.

Part 1.1 Definitions

1. i   **A** Either (3,2) or (3,1), which is not an answer choice, must be removed so that 3 will be paired with only one number.

2. i   **E** For each value of x there is only one value for y in each case. Therefore, f, g, and h are all functions.

3. i   **A** Since division by zero is forbidden, x cannot equal 2.

Part 1.2 Function Notation

1. i   **B** $f(-2) = 3(-2)^2 - 2(-2) + 4 = 20$.

2. i   **D** $g(2) = 3^2 = 9$. $f(g(2)) = f(9) = 31$.

3. i A To get from $f(x)$ to $f(g(x))$, x^2 must become $4x^2$. Therefore, the answer must contain $2x$ since $(2x)^2 = 4x^2$.

4. i E $g(x)$ cannot equal 0. Therefore, $x \neq \pm 1$.

5. i C Since $f(2)$ implies that $x = 2$, $g(f(2)) = 2$. Therefore, $g(f(2)) = 3(f(2)) + 2 = 2$. Therefore, $f(2) = 0$.

6. i C $p(a) = 0$ implies $4a - 6 = 0$.

7. a D $f \circ g(\sqrt{2}) = f(g(\sqrt{2})) = f(\sin\sqrt{2}) = e^{\sin\sqrt{2}} \approx$ 2.7.

Part 1.3 Inverse Functions

1. i E If $y = 2x - 3$, the inverse is $x = 2y - 3$, which equals $y = \dfrac{x+3}{2}$.

2. i A By definition.

3. i B The inverse is $\{(2,1),(3,2),(4,3),(1,4),(2,5)\}$, which is not a function because of $(2,1)$ and $(2,5)$. Therefore, the domain of the original function must lose either 1 or 5.

4. i E If this line were reflected about the line $y = x$ to get its inverse, the slope would be less than 1 and the y-intercept would be less than zero. The only possibilities are Choices D and E. Choice D can be excluded because since the x-intercept of $f(x)$ is greater than -1, the y-intercept of its inverse must be greater than -1.

Part 1.4 Odd and Even Functions

Exercises

1. i D Use the appropriate test for determining whether a relation is even.

   I. The graph of $y = 2$ is a horizontal line, which is symmetric about the y axis.

   II. Since $f(-x) = -x \neq x = f(x)$ unless $x = 0$, this function is not even.

   III. Since $(-x)^2 + y^2 = 1$ whenever $x^2 + y^2 = 1$ does, this relation is even.

2. i D Use the appropriate test for determining whether a relation is odd.

   I. The graph of $y = 2$ is a horizontal line, which is not symmetric about the origin.

   II. Since $f(-x) = -x = -f(x)$, this function is odd.

   III. Since $(-x)^2 + (-y)^2 = 1$ whenever $x^2 + y^2 = 1$ does, this relation is odd.

3. i B The analysis of relation III in the above examples indicate that I and II are both even and odd. Since $-x + y \neq 0$ when $x + y = 0$ unless $x = 0$, III is not even, and is therefore not both even and odd.

4. i C A is even, B is odd, D is even, and E is odd. C is not even because $(-x)^3 - 1 = -x^3 - 1$, which is neither $x^3 - 1$ nor $-x^3 + 1$.

Part 1.5 Multivariable Functions

1. i B $g(4) = 16$. $f(3,16) = 9 + 32 - 8 = 33$.

2. i C $f(-1,1,-1) = (-1)^2 + (1)^2 + 2(1)^2 = 4$.

3. i B $g(1,2) = 1^2 - 2^2 = -3$. $f(-3,3) = 3(-3) + 2(3) = -3$.

4. a C $f(2,1) = \sqrt{2(2^2) - 1^2} = \sqrt{7}$, and so $g(f(2,1)) = g(\sqrt{7}) = 2^{\sqrt{7}} \approx 6.25$.

POLYNOMIAL FUNCTIONS

CHAPTER

2

2.1 DEFINITIONS

A *polynomial* is an algebraic expression of the form

$$a_n x^n + a_{n-1} x^{n-1} + \cdots + a_1 x + a_0$$

where x is a variable, n is a nonnegative integer, and the coefficients $a_n, a_{n-1}, \ldots, a_1, a_0$ are complex numbers. If the coefficients are all real numbers, the expression is called a real polynomial. If the coefficients are all rational numbers, the expression is called a rational polynomial.

2.2 LINEAR FUNCTIONS

Linear functions are polynomials in which the largest exponent is 1. The graph is always a straight line. Although the general form of the equation is $Ax + By + C = 0$, where A, B, and C are constants, the most useful form occurs when the equation is solved for y. This is known as the *slope-intercept* form and is written $y = mx + b$. The slope of the line is represented by m and is defined to be the ratio of $\dfrac{y_1 - y_2}{x_1 - x_2}$, where (x_1, y_1) and (x_2, y_2) are any two points on the line. The y-intercept is b (the point where the graph crosses the y-axis).

Parallel lines have the same slope. The slopes of two perpendicular lines are negative reciprocals of one another.

EXAMPLE 1: The equation of line l_1 is $y = 2x + 3$, and the equation of line l_2 is $y = 2x - 5$.
These lines are parallel because the slope of each line is 2, and the y-intercepts are different.

EXAMPLE 2: The equation of line l_1, is $y = \dfrac{5}{2} x - 4$, and the equation of line l_2 is $y = -\dfrac{2}{5} x + 9$.

These lines are perpendicular because the slope of l_2, $-\dfrac{2}{5}$, is the negative reciprocal of the slope of l_1, $\dfrac{5}{2}$.

The distance between two points P and Q whose coordinates are (x_1, y_1) and (x_2, y_2) is given by the formula

$$\text{Distance} = \sqrt{(x_1 - x_2)^2 + (y_1 - y_2)^2}$$

and the midpoint, M, of the segment $\overline{PQ}$ has coordinates $\left(\dfrac{x_1 + x_2}{2}, \dfrac{y_1 + y_2}{2} \right)$. TI-83 programs to find the distances between two points and the midpoint of a segment can be found in Part 3, Chapter 9.

EXAMPLE 3: Given point (2,–3) and point (–5,4). Find the length of $\overline{PQ}$ and the coordinates of the midpoint, M.

$$PQ = \sqrt{(2-(-5))^2 + (-3-4)^2} = \sqrt{(7)^2 + (-7)^2}$$

$$= \sqrt{98} = 7\sqrt{2}$$

$$M = \left(\frac{2+(-5)}{2}, \frac{-3+4}{2}\right) = \left(\frac{-3}{2}, \frac{1}{2}\right)$$

The perpendicular distance between a line $Ax + By + C = 0$ and a point $P(x_1, y_1)$ not on the line is given by the formula

$$\text{Distance} = \frac{|Ax_1 + By_1 + C|}{\sqrt{A^2 + B^2}}$$

The angle θ between two lines, l_1 and l_2, can be found by using the formula

$$\text{Tan } \theta = \frac{m_1 - m_2}{1 + m_1 m_2}$$

where m_1 is the slope of l_1, and m_2 is the slope of l_2. If $\tan \theta > 0$, θ is the acute angle formed by the two lines. If $\tan \theta < 0$, θ is the obtuse angle formed by the two lines.

EXAMPLE 4: Find the distance between the line $3x + 4y = 5$ and the origin.

$$d = \frac{|3 \cdot 0 + 4 \cdot 0 - 5|}{\sqrt{9+16}} = \frac{5}{\sqrt{25}} = 1$$

EXAMPLE 5: Find the acute angle formed by lines $2x + 3y + 5 = 0$ and $3x – 5y + 8 = 0$.

When these two equations are written in the slope-intercept form, the slope of $2x + 3y + 5 = 0$ is found to be $-\frac{2}{3}$ and the slope of $3x - 5y + 8 = 0$ is found to be $\frac{3}{5}$.

$$\text{Tan } \theta = \frac{-\dfrac{2}{3} - \dfrac{3}{5}}{1 + \left(-\dfrac{2}{3}\right)\left(\dfrac{3}{5}\right)} = \frac{\left(-\dfrac{2}{3} - \dfrac{3}{5}\right) \cdot 15}{\left(1 - \dfrac{2}{5}\right) \cdot 15}$$

$$= \frac{-2 \cdot 5 - 3 \cdot 3}{15 - 2 \cdot 3} = \frac{-10 - 9}{15 - 6} = \frac{-19}{9}$$

Therefore, if θ is acute, $\tan \theta = \frac{19}{9}$, and if θ is obtuse,

$\tan \theta = \frac{-19}{9}$. Therefore, $\theta = \text{Tan}^{-1}\left(\frac{19}{9}\right) \approx 64.65°$.

> **TIP:** When using a calculator to find the angle represented by an inverse trig function, always evaluate the positive number to obtain the acute reference angle. You can then easily determine the appropriate angle in the correct quadrant.

EXERCISES

1. The slope of the line through points $A(3,-2)$ and $B(-2,-3)$ is

   (A) –5

   (B) $-\dfrac{1}{5}$

   (C) $\dfrac{1}{5}$

   (D) 1

   (E) 5

2. The slope of line $8x + 12y + 5 = 0$ is

   (A) 2

   (B) $\dfrac{2}{3}$

   (C) 3

   (D) $-\dfrac{3}{2}$

   (E) $-\dfrac{2}{3}$

3. The slope of the line perpendicular to line $3x - 5y + 8 = 0$ is

   (A) $\dfrac{3}{5}$

   (B) $\dfrac{5}{3}$

   (C) $-\dfrac{3}{5}$

   (D) $-\dfrac{5}{3}$

   (E) 3

4. The y-intercept of the line through the two points whose coordinates are $(5,-2)$ and $(1,3)$ is

   (A) $\dfrac{5}{4}$

   (B) $-\dfrac{5}{4}$

   (C) 17

   (D) $\dfrac{17}{4}$

   (E) 7

5. The equation of the perpendicular bisector of the segment joining the points whose coordinates are (1,4) and (–2,3) is

(A) $3x - 2y + 5 = 0$
(B) $x - 3y + 2 = 0$
(C) $3x + y - 2 = 0$
(D) $x - 3y + 11 = 0$
(E) $x + 3y - 10 = 0$

6. The length of the segment joining the points with coordinates (–2,4) and (3,–5) is

(A) 2.8
(B) 10.3
(C) 3,7
(D) 10.0
(E) none of these

7. The slope of the line parallel to the line whose equation is $2x + 3y = 8$ is

(A) $\dfrac{2}{3}$

(B) $-\dfrac{2}{3}$

(C) -2

(D) $-\dfrac{3}{2}$

(E) $\dfrac{3}{2}$

8. If point $P(m,2m)$ is 5 units from the line $12x + 5y = 1$, m could equal

(A) $\dfrac{43}{11}$

(B) -3

(C) $-\dfrac{65}{22}$

(D) 5

(E) 3

9. If θ is the angle between lines $2x - 3y + 4 = 0$ and $2x - y - 3 = 0$, θ could equal

(A) 131°
(B) 97°
(C) 76°
(D) 30°
(E) 146°

10. The distance between point (2,4) and the line $3x - 7y = 8$ is

(A) 1.8
(B) 3.6
(C) 3.9
(D) 5.5
(E) 5.9

11. If the graph of $\pi x + \sqrt{2}y + \sqrt{3} = 0$ is perpendicular to the graph of $ax + 3y + 2 = 0$, then $a =$

(A) –4.5
(B) 1.35
(C) 0.45
(D) –1.35
(E) –2.22

12. The lines $3x - 4y + 8 = 0$ and $9x + 6y - 4 = 0$ intersect at point P. One angle formed at point P contains

(A) 36.9°
(B) 89.5°
(C) 56.3°
(D) 86.8°
(E) 123.7°

13. If $x + 3y = 6$ and $2x - y = 3$, then $\dfrac{x}{y} =$

(A) 1.67
(B) 0.43
(C) 0.60
(D) 2.14
(E) 1.29

14. The distance between line $11x + 7y = 5$ and the origin is

(A) 0.14
(B) 0.45
(C) 1.0
(D) 1.76
(E) 0.38

2.3 QUADRATIC FUNCTIONS

Quadratic functions are polynomials in which the largest exponent is 2. The graph is always a parabola. The general form of the equation is $y = ax^2 + bx + c$. If $a > 0$, the parabola opens up and has a minimum value. If $a < 0$, the parabola opens down and has a maximum value. The x-coordinate of the vertex of the parabola is equal to $-\dfrac{b}{2a}$, and the axis of symmetry is the vertical line whose equation is $x = -\dfrac{b}{2a}$.

To find the minimum (or maximum) value of the function, substitute $-\dfrac{b}{2a}$ for x to determine y. Thus, in the general case the coordinates of the vertex are $\left(-\dfrac{b}{2a}, c - \dfrac{b^2}{4a} \right)$ and the minimum (or maximum) value of the function is $c - \dfrac{b^2}{4a}$.

Unless specifically limited, the domain of a quadratic function is all real numbers, and the range is all values of y greater than or equal to the minimum value (or all values of y less than or equal to the maximum value) of the function.

EXAMPLE 1: Determine the coordinates of the vertex and the equation of the axis of symmetry of $y = 3x^2 + 2x - 5$. Does the quadratic function have a minimum or maximum value? If so, what is it?
The equation of the axis of symmetry is

$$x = -\frac{b}{2a} = -\frac{2}{2 \cdot 3} = -\frac{1}{3}$$

and the y-coordinate of the vertex is

$$y = 3\left(-\frac{1}{3}\right)^2 + 2\left(-\frac{1}{3}\right) - 5 = \frac{1}{3} - \frac{2}{3} - 5 = -5\frac{1}{3}$$

The vertex is, therefore, at $\left(-\dfrac{1}{3}, -5\dfrac{1}{3} \right)$.

The function has a minimum value because $a = 3 > 0$.

The minimum value is $-5\dfrac{1}{3}$. The graph of $y = 3x^2 + 2x - 5$ is shown below.

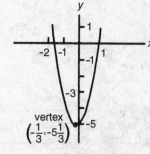

vertex
$\left(-\dfrac{1}{3}, -5\dfrac{1}{3}\right)$

The points where the graph crosses the x-axis are called the *zeros* of the function and occur when $y = 0$. To find the zeros of $y = 3x^2 + 2x - 5$, solve the quadratic equation

$3x^2 + 2x - 5 = 0$. The roots of this equation are the zeros of the polynomial. Often the quickest way to do this is to factor the polynomial. However, if it is not immediately obvious how to do the factoring, substitute the coefficients into the general quadratic formula.

Every quadratic equation can be changed into the form $ax^2 + bx + c = 0$ (if it is not already in that form), which can be solved by completing the square. The solutions are $x = \dfrac{-b \pm \sqrt{b^2 - 4ac}}{2a}$, the *general quadratic formula*. In the case of $3x^2 + 2x - 5 = 0$, factoring is the best method to use; $3x^2 + 2x - 5 = (3x + 5)(x - 1) = 0$. Thus, $3x + 5 = 0$ or $x - 1 = 0$, which leads to $x = -\dfrac{5}{3}$ or 1. The zeros of the polynomials are $-\dfrac{5}{3}$ and 1.

EXAMPLE 2: Find the zeros of $y = 2x^2 + 3x - 4$.
Solve the equation $2x^2 + 3x - 4 = 0$. This does not factor easily. Using the general quadratic formula, where $a = 2$, $b = 3$, and $c = -4$, gives

$$x = \frac{-3 \pm \sqrt{9 + 32}}{4} = \frac{-3 \pm \sqrt{41}}{4}.$$

Thus, the zeros are $\dfrac{-3 + \sqrt{41}}{4}$ and $\dfrac{-3 - \sqrt{41}}{4}$.

It is interesting to note that the sum of the two zeros, $\dfrac{-b + \sqrt{b^2 - 4ac}}{2a}$ and $\dfrac{-b - \sqrt{b^2 - 4ac}}{2a}$, equals $-\dfrac{b}{a}$, and their product equals $\dfrac{c}{a}$. This information can be used to check whether the correct zeros have been found. In Example 2, the sum and product of the zeros can be determined by inspection from the equations. Sum $= -\dfrac{3}{2}$ and Product $= \dfrac{-4}{2} = -2$. Adding the zeros $\dfrac{-3 + \sqrt{41}}{4}$ and $\dfrac{-3 - \sqrt{41}}{4}$ gives $\dfrac{-6}{4} = -\dfrac{3}{2}$. Multiplying the zeros $\dfrac{-3 + \sqrt{41}}{4}$ and $\dfrac{-3 - \sqrt{41}}{4}$ gives $\dfrac{9 - 41}{16} = \dfrac{-32}{16} = -2$. Thus, the zeros are correct.

At times it is necessary to determine only the *nature* of the roots of a quadratic equation, not the roots themselves. Because $b^2 - 4ac$ of the general quadratic formula is under the radical, it determines the nature of the roots and is called the *discriminant* of a quadratic equation.

(i) If $b^2 - 4ac = 0$, the two roots become $\dfrac{-b+0}{2a}$ and $\dfrac{-b-0}{2a}$, which are the same, and the graph of the function is tangent to the x-axis.

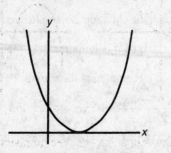

(ii) If $b^2 - 4ac < 0$, there is a negative number under the radical, which gives two complex numbers (of the form $p + qi$ and $p - qi$, where $i = \sqrt{-1}$) as roots, and the graph of the function does not intersect the x-axis.

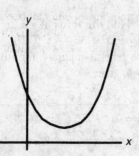

(iii) If $b^2 - 4ac > 0$, there is a positive number under the radical, which gives two different real roots, and the graph of the function intersects the x-axis at two points.

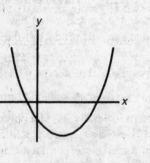

EXERCISES

1. The coordinates of the vertex of the parabola whose equation is $y = 2x^2 + 4x - 5$ are

   (A) $(2, 11)$
   (B) $(-1, -7)$
   (C) $(1, 1)$
   (D) $(-2, -5)$
   (E) $(-4, 11)$

2. The range of the function
   $f = \{(x,y) : y = 5 - 4x - x^2\}$ is

   (A) $\{y : y \le 0\}$
   (B) $\{y : y \ge -9\}$
   (C) $\{y : y \le 9\}$
   (D) $\{y : y \ge 0\}$
   (E) $\{y : y \le 1\}$

3. The equation of the axis of symmetry of the function $y = 2x^2 + 3x - 6$ is

   (A) $x = \dfrac{1}{3}$

   (B) $x = -\dfrac{3}{4}$

   (C) $x = -\dfrac{1}{3}$

   (D) $x = \dfrac{3}{4}$

   (E) $x = -\dfrac{3}{2}$

4. Find the zeros of $y = 2x^2 + x - 6$.

   (A) 3 and 2

   (B) -3 and 2

   (C) $\dfrac{1}{2}$ and $\dfrac{3}{2}$

   (D) $-\dfrac{3}{2}$ and 1

   (E) $\dfrac{3}{2}$ and -2

5. The sum of the zeros of $y = 3x^2 - 6x - 4$ is

   (A) $\dfrac{4}{3}$

   (B) 2

   (C) 6

   (D) -2

   (E) $-\dfrac{4}{3}$

6. $x^2 + 2x + 3 = 0$ has

   (A) two real rational roots
   (B) two real irrational roots
   (C) two equal real roots
   (D) two equal rational roots
   (E) two complex conjugate roots

7. If $f(x) = ax^2 + bx + c$, $f(1.54) = -7.3$, and $f(-1.54) = 7.3$, what is the ratio of a to c?

(A) $-2.37:1$
(B) $1.54:1$
(C) $-0.42:1$
(D) $-0.21:1$
(E) $0.73:1$

8. A parabola with a vertical axis has its vertex at the origin and passes through point (7,7). The parabola intersects line $y = 6$ at two points. The length of the segment joining these points is

(A) 14
(B) 12
(C) 13
(D) 8.6
(E) 6.5

2.4 HIGHER-DEGREE POLYNOMIAL FUNCTIONS

Polynomial functions of degree greater than two (largest exponent greater than 2) are usually treated together since there are no simple formulas, such as the general quadratic formula, that aid in finding zeros.

Five facts about the graphs of polynomial functions:

(1) They are always continuous curves. (The graph can be drawn without removing the pencil from the paper.)
(2) If the largest exponent is an even number, both ends of the graph leave the coordinate system either at the top or at the bottom:

EXAMPLES:

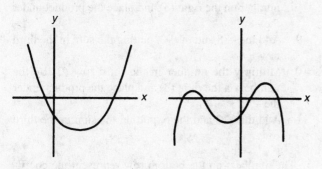

(3) If the largest exponent is an odd number, the ends of the graph leave the coordinate system at opposite ends.

EXAMPLES:

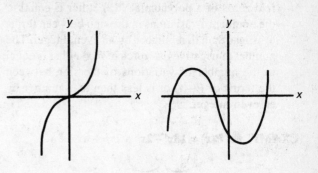

(4) If all the exponents are even numbers, the polynomial is an *even function* and is symmetric about the y-axis.

EXAMPLE: $y = 3x^4 + 2x^2 - 8$

(5) If all the exponents are odd numbers and there is no constant term, the polynomial is an *odd function* and is symmetric about the origin of the coordinate system.

EXAMPLE: $y = 4x^5 + 2x^3 - 3x$

To find the zeros (if possible) of a higher-degree polynomial function, $a_n x^n + a_{n-1} x^{n-1} + \cdots + a_1 x + a_0$, set $y = 0$ and attempt to solve the resulting polynomial equation.

Ten facts useful in solving polynomial equations:

(1) Remainder theorem—If a polynomial $P(x)$ is divided by $x - r$ (where r is any constant), then the remainder is $P(r)$.
(2) Factor theorem—r is a zero of the polynomial $P(x)$ if and only if $x - r$ is a divisor of $P(x)$.
(3) Every polynomial of degree n has exactly n zeros.
(4) Rational zero (root) theorem—If $\dfrac{p}{q}$ is a rational zero (reduced to lowest terms) of a polynomial $P(x)$ with integral coefficients, then p is a factor of a_0 (the constant term) and q is a factor of a_n (the leading coefficient).
(5) If $P(x)$ is a polynomial with rational coefficients, then irrational zeros occur as conjugate pairs. (For example, if $p + \sqrt{q}$ is a zero, then $p - \sqrt{q}$ is also a zero.)
(6) If $P(x)$ is a polynomial with real coefficients, then complex zeros occur as conjugate pairs. (For example, if $p + qi$ is a zero, then $p - qi$ is also a zero.)

(7) Descartes' rule of signs—The number of positive real zeros of a polynomial $P(x)$ either is equal to the number of variations in the sign between terms or is less than that number by an even integer. The number of negative real zeros of $P(x)$ either is equal to the number of variations of the sign between the terms of $P(-x)$ or is less than that number by an even integer.

EXAMPLE: $P(x) = 18x^4 - 2x^3 + 7x^2 + 8x - 5$

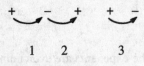

Three sign changes indicate there will be either one or three positive zeros of $P(x)$.

$$P(-x) = 18x^4 + 2x^3 + 7x^2 - 8x - 5$$

One sign change indicates there will be exactly one negative zero of $P(x)$.

(8) Relation between zeros and coefficients—In a polynomial $P(x)$ with zeros $z_1, z_2, z_3, \ldots, z_n$,

$$-\frac{a_{n-1}}{a_n} = \text{sum of the zeros} = z_1 + z_2 + z_3 + \cdots + z_n$$

$$\frac{a_{n-2}}{a_n} = \text{sum of the products of the zeros taken two}$$
at a time $= z_1z_2 + z_1z_3 + \cdots + z_1z_n + z_2z_3 + z_2z_4 + \cdots + z_2z_n + \cdots z_{n-1}z_n$

$$-\frac{a_{n-3}}{a_n} = \text{sum of the products of the zeros taken three}$$
at a time $= z_1z_2z_3 + z_1z_2z_4 + \cdots + z_1z_2z_n + z_2z_3z_4 + \cdots + z_2z_3z_n + \cdots + z_{n-2}z_{n-1}z_n$

$$\vdots$$

$(-1)^k, \dfrac{a_0}{a_n} = \text{product of all the zeros} = z_1z_2z_3 \cdots z_n$, where k is the number of zeros in each product

EXAMPLE: $P(x) = 18x^4 - 2x^3 + 7x^2 + 8x - 5$ **has four zeros.**

$$z_1 + z_2 + z_3 + z_4 = -\frac{-2}{18} = \frac{1}{9}$$

$$z_1z_2 + z_1z_3 + z_1z_4 + z_2z_3 + z_2z_4 + z_3z_4 = \frac{7}{18}$$

$$z_1z_2z_3 + z_1z_2z_4 + z_1z_3z_4 + z_2z_3z_4 = -\frac{8}{18} = -\frac{4}{9}$$

$$z_1z_2z_3z_4 = \frac{-5}{18}$$

(9) Synthetic division—Synthetic division greatly decreases the amount of work necessary when dividing a polynomial by a divisor of the form $x - r$. The easiest way to learn how to use synthetic division is through examples.

EXAMPLE: Divide $P(x) = 3x^5 - 4x^4 - 15x^2 - 88x - 12$ **by** $x - 3$.
Write just the coefficients of $P(x)$, inserting a zero for any missing term. Put the value of r to the right of the row of coefficients.

3	−4	0	−15	−88	−12	⌐3
↓	9	15	45	90	6	
3	5	15	30	2	−6 = remainder = $P(3)$	

Procedure to follow:
1. Bring the first coefficient (3) down to the third row.
2. Multiply the number in the third row (3) by the number on the right (3) and place the product under the −4.
3. Add the −4 and the 9, putting the sum in the third row.
4. Multiply the number in the third row (5) by the number on the right (3) and place the product under the 0.
5. Add the 0 and the 15, putting the sum in the third row.
6. Multiply the number in the third row (15) by the number on the right (3) and place the product under the −15.
7. Add the −15 and the 45, putting the sum in the third row.
8. Multiply the number in the third row (30) by the number on the right (3) and place the product under the −88.
9. Add the −88 and the 90, putting the sum in the third row.
10. Multiply the number in the third row (2) by the number on the right (3) and place the product under the −12.
11. Add the −12 and the 6, putting the sum in the third row.

The numbers on the bottom row represent the coefficients of the quotient, and the −6 represents the remainder. Since a fifth-degree polynomial was divided by a first-degree polynomial, the quotient must have degree 4. Therefore, $\dfrac{3x^5 - 4x^4 - 15x^2 - 88x - 12}{x - 3} = 3x^4 + 5x^3 + 15x^2 + 30x + 2$ with a remainder of −6. From the remainder theorem it is determined that $P(3) = -6$.

EXAMPLE: Divide $P(x) = x^4 + 8$ by $x + 2$.
Since the divisor should be in the form $x - r$, $x + 2$ is changed to $x - (-2)$. In the synthetic division format zeros must be inserted as coefficients for the x^3, x^2, and x terms that are missing.

$$
\begin{array}{rrrrr|r}
1 & 0 & 0 & 0 & 8 & \underline{-2} \\
\downarrow & -2 & 4 & -8 & 16 & \\
\hline
1 & -2 & 4 & -8 & 24 & = P(-2)
\end{array}
$$

Procedure to follow:
1. Bring the first coefficient (1) down to the third row.
2. Multiply the number in the third row by the number on the right (–2) and add the product to the next number in the top row.
3. Repeat step 2 until the third row is filled.

The resulting quotient will be $x^3 - 2x^2 + 4x - 8$ with a remainder of 24. Thus, from the remainder theorem $P(-2) = 24$.

(10) **Upper and lower bounds on zeros**—If $P(x)$ is divided by $x - r$, where r is a positive number, and if the numbers in the third row of the synthetic division are all positive or zero, there is no zero of the polynomial greater than r. Thus, r is an upper bound of all the zeros.
If $P(x)$ is divided by $x - r$, where r is a negative number, and if all the numbers in the third row of the synthetic division are alternately positive and negative (or zero), there is no zero of the polynomial less than r. Thus, r is a lower bound of all the zeros.

Each of the following examples illustrates one or more of these 10 facts.

EXAMPLE 1: What is the remainder when $3x^3 + 2x^2 - 5x - 8$ is divided by $x + 2$?

Method 1: If $P(x) = 3x^3 + 2x^2 - 5x - 8$, the remainder is given to be $P(-2)$ and $P(-2) = 3(-2)^3 + 2(-2)^2 - 5(-2) - 8 = -24 + 8 + 10 - 8 = -14$. Therefore, the remainder = –14.

Method 2: Using synthetic division:

$$
\begin{array}{rrrr|r}
3 & 2 & -5 & -8 & \underline{-2} \\
\downarrow & -6 & 8 & -6 & \\
\hline
3 & -4 & 3 & -14 & = P(-2)
\end{array}
$$

Therefore, the remainder = –14.

EXAMPLE 2: Is $x - 99$ a factor of $P(x) = x^4 - 100x^3 + 97x^2 + 200x - 197$?
If $x - 99$ is a factor, the remainder upon division by $x - 99$ will be zero. It would be tedious to determine whether $P(99) = 0$ directly (as in method 1 above) even if you use your calculator, but synthetic division is quite easy.

$$
\begin{array}{rrrrr|r}
1 & -100 & 97 & 200 & -197 & \underline{99} \\
\downarrow & 99 & -99 & -198 & 198 & \\
\hline
1 & -1 & -2 & 2 & 1 & = P(99)
\end{array}
$$

Therefore, $x - 99$ is not a factor of $p(x)$.

EXAMPLE 3: If $3 + 2i$, 2, and $2 - 3i$ are all zeros of $P(x) = 3x^5 - 36x^4 + 2x^3 - 8x^2 + 9x - 338$, what are the other zeros?
Since $P(x)$ must have five zeros because it is a fifth-degree polynomial, and since complex zeros come in conjugate pairs, the two remaining zeros must be $3 - 2i$ and $2 + 3i$.

EXAMPLE 4: What are all the possible rational zeros of $P(x) = 3x^3 + 2x^2 + 4x - 6$?
Note that this question does not ask what the zeros are, only which rational numbers might be zeros. From the rational root theorem, the rational roots, $\dfrac{p}{q}$, are such that p is a factor of 6 and q is a factor of 3. Thus, $p \in \{\pm 1, \pm 2, \pm 3, \pm 6,\}$ and $q \in \{\pm 1, \pm 3\}$. Forming all possible fractions gives $\dfrac{p}{q} \in \left\{ \pm 1, \pm 2, \pm 3, \pm 6, \pm \dfrac{1}{3}, \pm \dfrac{2}{3} \right\}$.
Thus, these 12 numbers are the only possible rational numbers that could be zeros of $P(x)$. It could turn out that none of them actually is a zero, meaning that the three zeros are irrational and/or complex numbers.

EXAMPLE 5: How many positive real zeros and how many negative real zeros could you expect to find for the polynomial $P(x) = 3x^3 + 2x^2 + 4x - 6$?
Since there is only one sign change in $P(x)$, there is exactly one positive real zero.

$$P(-x) = \underbrace{-3x^3 + 2x^2}_{1} \underbrace{- 4x - 6}_{2}$$

Since there are two sign changes in $P(-x)$, there will be either two negative real zeros or no negative real zero. Since $P(x)$ is a third-degree polynomial and has three zeros, there will be either one positive real zero and two negative real zeros, or one positive real zero and two complex conjugate zeros.

EXAMPLE 6: What is the smallest positive integer that is an upper bound of the zeros of $P(x) = 3x^3 + 2x^2 + 4x - 6$?
Use synthetic division, dividing by successive integers:

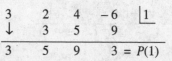

$$
\begin{array}{rrrr|r}
3 & 2 & 4 & -6 & \underline{1} \\
\downarrow & 3 & 5 & 9 & \\
\hline
3 & 5 & 9 & 3 & = P(1)
\end{array}
$$

Since all the numbers in the third row are positive, there are no zeros greater than 1. Therefore, 1 is the smallest positive integer that is an upper bound for the zeros of $P(x)$.

EXAMPLE 7: Find the zeros of $P(x) = x^6 - x^5 - 4x^4 - x^3 + 5x^2 + 8x + 4$.

The only possible rational zeros are ± 1, ± 2, ± 4. By Descartes' rule of signs there are two or no positive real zero and four, two, or no negative real zero. Using synthetic division, divide by $x - 1$:

$$
\begin{array}{rrrrrrr|l}
1 & -1 & -4 & -1 & 5 & 8 & 4 & \underline{1} \\
\downarrow & 1 & 0 & -4 & -5 & 0 & 8 & \\
\hline
1 & 0 & -4 & -5 & 0 & 8 & 12 = P(1) &
\end{array}
$$

Thus, $x - 1$ is not a factor of $P(x)$ and 1 is not a zero. Divide by $x - 2$.

$$
\begin{array}{rrrrrrr|l}
1 & -1 & -4 & -1 & 5 & 8 & 4 & \underline{-2} \\
\downarrow & 2 & 2 & -4 & -10 & -10 & -4 & \\
\hline
1 & 1 & -2 & -5 & -5 & -2 & 0 = P(2) &
\end{array}
$$

Thus, $x - 2$ is a factor, and 2 is a zero. $P(x)$ can be factored into $(x - 2)Q_1(x)$, where $Q_1(x) = x^5 + x^4 - 2x^3 - 5x^2 - 5x - 2$, using the numbers in the third row of the synthetic division. Since all future factors of $P(x)$ must also be factors of $Q_1(x)$, it is sufficient to search for factors (and zeros) of $Q_1(x)$. Since $Q_1(x) = x^5 + x^4 - 2x^3 + 5x^2 - 5x - 2$ was obtained from a polynomial of higher degree, it is called a *depressed equation*. Continuing the synthetic division by dividing $Q_1(x)$ by $x - 2$:

$$
\begin{array}{rrrrrr|l}
1 & 1 & -2 & -5 & -5 & -2 & \underline{2} \\
\downarrow & 2 & 6 & 8 & 6 & 2 & \\
\hline
1 & 3 & 4 & 3 & 1 & 0 = Q_1(2) = P(2) &
\end{array}
$$

we find that both positive zeros of $P(x)$ are 2, and 2 is said to be a *double zero* or a *zero of multiplicity* 2. In factored form $P(x) = (x - 2)^2(x^4 + 3x^3 + 4x^2 + 3x + 1)$. Descartes' rule of signs indicates that there are no more positive zeros.

Attempting to find negative zeros, divide $Q_2(x) = x^4 + 3x^3 + 4x^2 + 3x + 1$ by $x + 1$.

$$
\begin{array}{rrrrr|l}
1 & 3 & 4 & 3 & 1 & \underline{-1} \\
\downarrow & -1 & -2 & -2 & -1 & \\
\hline
1 & 2 & 2 & 1 & 0 = Q_2(-1) = P(-1) &
\end{array}
$$

Thus, $x + 1$ is a factor, and -1 is a zero. Continuing the synthetic division by dividing $Q_3(x) = x^3 + 2x^2 + 2x + 1$ by $x + 1$:

$$
\begin{array}{rrrr|l}
1 & 2 & 2 & 1 & \underline{-1} \\
\downarrow & -1 & -1 & -1 & \\
\hline
1 & 1 & 1 & 0 = Q_3(-1) = P(-1) &
\end{array}
$$

we find that -1 is also a zero of multiplicity 2. In factored form, $P(x) = (x - 2)^2(x + 1)^2(x^2 + x + 1)$. Since the final quotient is a second-degree polynomial, its zeros can be found by using the general quadratic formula; $x^2 + x + 1 = 0$, where $a = 1$, $b = 1$, and $c = 1$. Therefore,

$$x = \frac{-1 \pm \sqrt{1 - 4}}{2} = \frac{-1 \pm i\sqrt{3}}{2}$$

and the entire list of zeros of $P(x) = x^6 - x^5 - 4x^4 - x^3 + 5x^2 + 8x + 4$ is $2, 2, -1, -1, \dfrac{-1 + i\sqrt{3}}{2}$, and $\dfrac{-1 - i\sqrt{3}}{2}$.

EXAMPLE 8: Is $2x + 1$ a factor of $P(x) = 2x^3 + 3x^2 - 3x - 2$?

Since $2x + 1$ is not in the form $x - r$, it appears that synthetic division cannot be used. However, if $2x + 1$ is written as $2\left(x + \dfrac{1}{2}\right)$, synthetic division can be used with the factor $\left(x + \dfrac{1}{2}\right)$. Then $2x + 1$ will be a factor if and only if $x + \dfrac{1}{2}$ is a factor:

$$
\begin{array}{rrrr|l}
2 & 3 & -3 & -2 & \underline{-\dfrac{1}{2}} \\
\downarrow & -1 & -1 & 2 & \\
\hline
2 & 2 & -4 & 0 = P\left(-\dfrac{1}{2}\right) &
\end{array}
$$

Thus, both $2x + 1$ and $x + \dfrac{1}{2}$ are factors of $P(x)$. To find other factors the third row of the synthetic division must be divided by the 2 that was factored out of the original divisor. In factored form, $P(x) = \left(x + \dfrac{1}{2}\right)(2x^2 + 2x - 4) = (2x + 1)(x^2 + x - 2)$. This quadratic factor can itself be factored into $(x + 2)(x - 1)$. Thus, $P(x) = (2x + 1)(x + 2)(x - 1)$.

EXERCISES

1. The graph of $P(x) = 3x^5 + 5x^3 - 8x + 2$ can cross the x-axis in no more than r points. What is the value of r?

    (A) 0
    (B) 1
    (C) 2
    (D) 3
    (E) 5

2. Between which two consecutive integers is there a zero of $P(x) = 28x^3 - 11x^2 + 15x - 28$?

    (A) -2 and -1
    (B) -1 and 0
    (C) 0 and 1
    (D) 1 and 2
    (E) 2 and 3

3. $P(x) = ax^4 + x^3 - bx^2 - 4x + c$. If $P(x)$ increases without bound as x increases without bound, then, as x decreases without bound, $P(x)$

 (A) increases without bound
 (B) decreases without bound
 (C) approaches zero from above the x-axis
 (D) approaches zero from below the x-axis
 (E) cannot be determined

4. Which of the following is an odd function?

   I. $f(x) = 3x^3 + 5$
   II. $g(x) = 4x^6 + 2x^4 - 3x^2$
   III. $h(x) = 7x^5 - 8x^3 + 12x$

 (A) only I
 (B) only II
 (C) only III
 (D) only I and II
 (E) only I and III

5. Using the rational root theorem, how many possible rational roots are there for $2x^4 + 4x^3 - 6x^2 + 15x - 12 = 0$?

 (A) 8
 (B) 4
 (C) 16
 (D) 12
 (E) 6

6. How many positive real zeros would you expect to find for the polynomial function $P(x) = 3x^4 - 2x^2 - 12$?

 (A) 0
 (B) 2 or 0
 (C) 1
 (D) 3 or 1
 (E) 4 or 2 or 0

7. If both $x - 1$ and $x - 2$ are factors of $x^3 - 3x^2 + 2x - 4b$, then b must be

 (A) 0
 (B) 1
 (C) 2
 (D) 3
 (E) 4

8. If i is a root of $x^4 + 2x^3 - 3x^2 + 2x - 4 = 0$, what are the real roots?

 (A) ± 2
 (B) $1, -4$
 (C) $-1 \pm \sqrt{5}$
 (D) $1 \pm \sqrt{5}$
 (E) $-1, 4$

9. How many positive real roots does $x^4 + x^3 - 3x^2 - 3x = 0$ have?

 (A) 0
 (B) 1
 (C) 2
 (D) 3
 (E) 4

10. The sum of the zeros of $P(x) = 8x^3 - 2x^2 + 3$ is

 (A) 2
 (B) $-\dfrac{3}{8}$
 (C) $-\dfrac{1}{4}$
 (D) $\dfrac{3}{8}$
 (E) $\dfrac{1}{4}$

11. If $3x^3 - 9x^2 + Kx - 12$ is divisible by $x - 3$, then it is also divisible by

 (A) $3x^2 - x + 4$
 (B) $3x^2 - 4$
 (C) $3x^2 + 4$
 (D) $3x - 4$
 (E) $3x + 4$

12. Write the equation of lowest degree with real coefficients if two of its roots are -1 and $1 + i$.

 (A) $x^3 + x^2 + 2 = 0$
 (B) $x^3 - x^2 - 2 = 0$
 (C) $x^3 - x + 2 = 0$
 (D) $x^3 - x^2 + 2 = 0$
 (E) none of the above

2.5   INEQUALITIES

Given any algebraic expression $f(x)$, there are exactly three situations that can exist:

 (1) for some values of x, $f(x) < 0$;
 (2) for some values of x, $f(x) = 0$;
 (3) for some values of x, $f(x) > 0$.

If all three of these sets of numbers are indicated on a number line, the set of values that satisfy $f(x) < 0$ is always separated from the set of values that satisfy $f(x) > 0$ by the values of x that satisfy $f(x) = 0$.

EXAMPLE: **Find the set of values for x that satisfies $x^2 - 3x - 4 < 0$.**

Consider the associated equation:

$$x^2 - 3x - 4 = 0$$

Factoring gives

$$(x - 4)(x + 1) = 0$$
$$x - 4 = 0 \text{ or } x + 1 = 0$$

Therefore, $x = 4$ or $x = -1$.

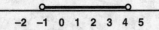

These two points, -1 and 4, separate the set of x-values that satisfy $x^2 - 3x - 4 < 0$ from the set of x-values that satisfy $x^2 - 3x - 4 > 0$. The correct region can be determined by direct substitution of a number for x from each of the three regions indicated on the number line into the original inequality.

Let $x = -2$: $x^2 - 3x - 4 = 4 + 6 - 4 > 0$.
Let $x = 0$: $x^2 - 3x - 4 = 0 + 0 - 4 < 0$.
Let $x = 5$: $x^2 - 3x - 4 = 25 - 15 - 4 > 0$.

Therefore, the region that contains zero is the only one that satisfies $x^2 - 3x - 4 < 0$, and the solution set is $\{x - 1 < x < 4\}$. The graph of this solution set on a number line is indicated below.

Although this may seem to be a rather roundabout method of solution, it does have the advantage of working for all types of inequalities. No special rules for special cases must be remembered.

EXERCISES

1. Which of the following is equivalent to $3x^2 - x < 2$?

(A) $-\dfrac{3}{2} < x < 1$

(B) $-1 < x < \dfrac{2}{3}$

(C) $-\dfrac{2}{3} < x < 1$

(D) $-1 < x < \dfrac{3}{2}$

(E) $x < -\dfrac{2}{3}$ or $x > 1$

2. If $3 < x < 7$ and $-12 < y < -6$, what are all possible values of xy?

(A) $-72 < xy < -21$
(B) $-84 < xy < -18$
(C) $-42 < xy < -36$
(D) $-84 < xy < -36$
(E) $-36 < xy < -42$

3. The number of integers that satisfy the inequality $x^2 + 48 < 16x$ is

(A) 0
(B) 4
(C) 7
(D) an infinite number
(E) none of the above

ANSWERS AND EXPLANATIONS

In these solutions the following notation is used:

a: active—Calculator use is necessary or, at a minimum, extremely helpful.

n: neutral—Answers may be found without a calculator, but a calculator may help.

i: inactive—Calculator use is not helpful and may even be a hindrance.

Part 2.2 Linear Functions

1. i **C** Slope $= \dfrac{-3 - (-2)}{-2 - 3} = \dfrac{1}{5}$.

2. i **E** $y = -\dfrac{2}{3}x - \dfrac{5}{12}$. The slope is $-\dfrac{2}{3}$.

3. i **D** $y = \dfrac{3}{5}x + \dfrac{8}{5}$. The slope of the given line is $\dfrac{3}{5}$. The slope of a perpendicular line is $-\dfrac{5}{3}$.

4. n **D** If (x, y) represents any point on the line, the slope $=$

$$\frac{y - 3}{x - 1} = \frac{-2 - 3}{5 - 1}.$$

Therefore $y = -\dfrac{5}{4}x + \dfrac{17}{4}$. The y-intercept $= \dfrac{17}{4}$.

5. i **C** The slope of the line is $\dfrac{3 - 4}{-2 - 1} = \dfrac{1}{3}$. Therefore, the slope of a perpendicular line is -3. The midpoint of the segment is $\left(\dfrac{1 - 2}{2}, \dfrac{4 + 3}{2} \right) = \left(-\dfrac{1}{2}, \dfrac{7}{2} \right)$. The equation of the line is $\dfrac{\left(y - \dfrac{7}{2} \right)}{\left(x + \dfrac{1}{2} \right)} = 3$, and so $3x + y - 2 = 0$.

6. a **B** Length $=$

$$\sqrt{(3 + 2)^2 + (-5 - 4)^2} = \sqrt{25 + 81} = \sqrt{106} \approx 10.3$$

7. i **B** $y = -\dfrac{2}{3}x + \dfrac{8}{3}$. Therefore, the slope of a parallel line $= -\dfrac{2}{3}$.

8. n E Distance $= \dfrac{|12(m) + 5(2m) - 1|}{\sqrt{144 + 25}} = 5.$

$|22m - 1| = 65$; $22m - 1 = 65$ or $-22m + 1 = 65$.

Therefore, $m = 3$ or $-\dfrac{32}{11}$.

9. a D The slope of the first line is $\dfrac{2}{3}$. The slope of the second line is 2.

$$\tan\theta = \dfrac{2 - \dfrac{2}{3}}{1 + \dfrac{4}{3}} = \dfrac{4}{7}.$$

Therefore, $\theta = \tan^{-1}\left(\dfrac{4}{7}\right) \approx 30°$.

10. a C Distance $= \dfrac{|3 \cdot 2 - 7 \cdot 4 - 8|}{\sqrt{3^2 + (-7)^2}} = \dfrac{|-30|}{\sqrt{58}} \approx 3.9.$

11. a D The slope of the first line is $-\dfrac{\pi}{\sqrt{2}}$, and the slope of the second line is $-\dfrac{a}{3}$. To be perpendicular, $-\dfrac{\pi}{\sqrt{2}} = \dfrac{3}{a}$. $a = \dfrac{-3\sqrt{2}}{\pi} \approx -1.35.$

12. a D The two lines and the x-axis form a triangle.

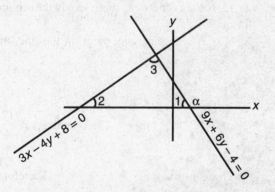

The slope of a line equals the tangent of the angle α formed by the line and the positive x-axis. Thus, $\tan\angle 2 = \dfrac{3}{4}$, which implies that $\angle 2 = \text{Tan}^{-1}\left(\dfrac{3}{4}\right) \approx 36.9°$. Since $\tan\angle\alpha = -\dfrac{3}{2}$, $\tan\angle 1 = \dfrac{3}{2}$, which implies that $\angle 1 = \text{Tan}^{-1}\left(\dfrac{3}{2}\right) \approx 56.3°$. Therefore, $\angle 3 = 180° - 36.9° - 56.3° = 86.8°$.

13. a A Multiply the second equation by -2 and add to the first equation to get $-3x + 5y = 0$. Then $3x = 5y$, $\dfrac{x}{y} = \dfrac{5}{3} \approx 1.67$. [5.9].

14. a E $d = \dfrac{|ax_0 + by_0 + c|}{\sqrt{a^2 + b^2}} = \dfrac{|11(0) + 7(0) - 5|}{\sqrt{11^2 + 7^2}}$

$= \dfrac{5}{\sqrt{121 + 49}} = \dfrac{5}{\sqrt{170}} \approx 0.38.$ [2.2].

Part 2.3 Quadratic Functions

1. n B The x coordinate of the vertex is $x = -\dfrac{b}{2a} = -\dfrac{4}{4} = -1$ and the y coordinate is $y = 2(-1)^2 + 4(-1) - 5 = -7$. Hence the vertex is the point $(-1, -7)$.

2. n C Find the vertex: $x = -\dfrac{b}{2a} = \dfrac{4}{-2} = -2$ and $y = 5 - 4(-2) - (-2)^2 = 9$. Since $a = -1 < 0$ the parabola opens down, so the range is $\{y : y \le 9\}$.

3. n B The x coordinate of the vertex is $x = -\dfrac{b}{2a} = -\dfrac{3}{4}$. Thus, the equation of the axis of symmetry is $x = -\dfrac{3}{4}$.

4. n E $2x^2 + x - 6 = (2x - 3)(x + 2) = 0$. The zeros are $\dfrac{3}{2}$ and -2.

5. n B Sum of zeros $= -\dfrac{b}{a} = -\dfrac{-6}{3} = 2$

6. n E From the discriminant $b^2 - 4ac = 4 - 4 \cdot 1 \cdot 3 = -8 < 0$.

7. a C Substitute 1.54 for x to get $f(1.54) = (1.54)^2 a + 1.54b + c = -7.3$ and substitute -1.54 for x to get $f(-1.54) = (-1.54)^2 a - 1.54b + c = 7.3$. Add these two equations to get $4.7432a + 2c = 0$. Divide each term by $4.7432c$ to get $\dfrac{a}{c} + \dfrac{2}{4.7432} = 0$. Therefore, $\dfrac{a}{c} \approx \dfrac{-0.42}{1}$.

8. a C The equation of a vertical parabola with its vertex at the origin has the form $y = ax^2$. Substitute $(7, 7)$ for x and y to find $a = \dfrac{1}{7}$. When $y = 6$, $x^2 = 42$. Therefore, $x = \pm\sqrt{42}$, and the segment $= 2\sqrt{42} \approx 13$.

Part 2.4 Higher-Degree Polynomial Functions

1. n **D** By Descartes' rule of signs, there can be two or zero positive real root and one negative real root, for a maxium of three.

2. n **C** $P(0) = -28$ $P(1) = 4$. The sign change implies Choice C is correct.

3. i **A** Since the degree of the polynomial is an even number, both ends of the graph go off in the same direction. Since $P(x)$ increases without bound as x increases without bound, $P(x)$ also increases without bound as x decreases without bound.

4. n **C** Since $h(x) = -h(-x)$, it is the only odd function.

5. i **C** Rational roots have the form $\dfrac{p}{q}$, where p is a factor of 12 and q is a factor of 2. $\dfrac{p}{q} \in \left\{ \pm 12, \pm 6, \pm 4, \pm 3, \pm 2, \pm 1, \pm\dfrac{3}{2}, \pm\dfrac{1}{2} \right\}$. The total is 16.

6. n **C** By Descartes' rule of signs, only one.

7. i **A** Sum of the roots $= 3$. Since 1 and 2 are roots, the third root is 0. Product of the roots $= 4b$. Therefore, $b = 0$.

Alternative Solution: Since $x - 1$ is a factor, $P(1) = 1^3 - 3 \cdot 1^2 + 2 \cdot 1 - 4b = 0$. Therefore, $b = 0$.

8. i **C** Using synthetic division with i and $-i$ results in a depressed equation of $x^2 + 2x - 4 = 0$ whose roots are $-1 \pm \sqrt{5}$.

9. n **B** By Descartes' rule of signs, only one.

10. i **E** Sum of roots $= -\dfrac{b}{a} = \dfrac{2}{8} = \dfrac{1}{4}$.

11. i **C** Using synthetic division with 3 gives $K = 4$ and a depressed equation of $3x^2 + 4$.

12. i **D** $1 - i$ is also a root. To find the equation, multiply $(x + 1)[x - (1 + i)][x - (1 - i)]$, which are the factors that produced the three roots.

Part 2.5 Inequalities

1. n **C** $3x^2 - x - 2 = (3x + 2)(x - 1) = 0$ when $x = -\dfrac{2}{3}$ or 1. Numbers between these satisfy the original inequality.

2. i **B** The smallest number possible is -84, and the largest is -18.

3. n **C** $x^2 - 16x + 48 = (x - 4)(x - 12) = 0$, when $x = 4$ or 12. Numbers between these satisfy the original inequality.

TRIGONOMETRIC FUNCTIONS

CHAPTER

3

3.1 DEFINITIONS

The general definitions of the six trigonometric functions are obtained from an angle placed in standard position on a rectangular coordinate system. When an angle θ is placed so that its vertex is at the origin, its initial side is along the positive x-axis, and its terminal side is anywhere on the coordinate system, it is said to be in *standard position*. The angle is given a positive value if it is measured in a counterclockwise direction from the initial side to the terminal side, and a negative value if it is measured in a clockwise direction.

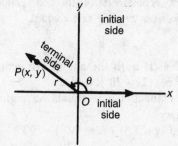

Let $P(x,y)$ be any point on the terminal side of the angle, and let r represent the distance between O and P. The six trigonometric functions are defined to be:

$$\sin\theta = \frac{\text{ordinate of } P}{OP} = \frac{y}{r}$$

$$\cos\theta = \frac{\text{abscissa of } P}{OP} = \frac{x}{r}$$

$$\tan\theta = \frac{\text{ordinate of } P}{\text{abscissa of } P} = \frac{y}{x}$$

$$\cot\theta = \frac{\text{abscissa of } P}{\text{ordinate of } P} = \frac{x}{y}$$

$$\sec\theta = \frac{OP}{\text{abscissa of } P} = \frac{r}{x}$$

$$\csc\theta = \frac{OP}{\text{ordinate of } P} = \frac{r}{y}$$

From these definitions it follows that:

$$\sin\theta \cdot \csc\theta = 1 \qquad \tan\theta = \frac{\sin\theta}{\cos\theta}$$

$$\cos\theta \cdot \sec\theta = 1 \qquad \cot\theta = \frac{\cos\theta}{\sin\theta}$$

$$\tan\theta \cdot \cot\theta = 1$$

The distance OP is always positive, and the ordinate and abscissa of P are positive or negative depending on which quadrant the terminal side of $\angle\theta$ lies in. The signs of the trigonometric functions are indicated in the following table.

45

Quadrant	I	II	III	IV
Function: sin θ, csc θ	+	+	–	–
cos θ, sec θ	+	–	–	+
tan θ, cot θ	+	–	+	–

Each angle θ whose terminal side lies in quadrant II, III, or IV has associated with it two acute angles called *reference angles*. Angle α is the acute angle formed by the x-axis and the terminal side. Angle β is the acute angle formed by the y-axis and the terminal side.

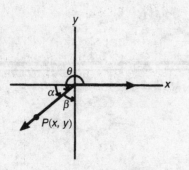

Any function of $\angle\theta = \pm$ the same function of $\angle\alpha$, and any function of $\angle\theta = \pm$ the cofunction of $\angle\beta$. The sign is determined by the quadrant in which the terminal side lies.

EXAMPLE 1: Express sin 320° in terms of $\angle\alpha$ and $\angle\beta$.

$$\alpha = 360° - 320° = 40°$$
$$\beta = 320° - 270° = 50°$$

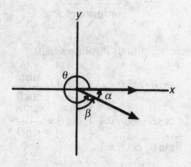

Since the sine is negative in quadrant IV, sin 320° = –sin 40° = –cos 50°.

EXAMPLE 2: Express cot 200° in terms of $\angle\alpha$ and $\angle\beta$.

$$\alpha = 200° - 180° = 20°$$
$$\beta = 270° - 200° = 70°$$

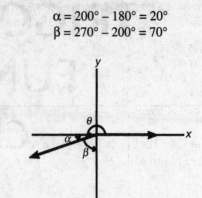

Since the cotangent is positive in quadrant III, cot 200° = cot 20° = tan 70°.

EXAMPLE 3: Express cos 130° in terms of $\angle\alpha$ and $\angle\beta$.

$$\alpha = 180° - 130° = 50°$$
$$\beta = 130° - 90° = 40°$$

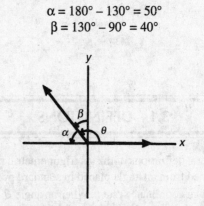

Since the cosine is negative in quadrant II, cos 130° = –cos 50° = –sin 40°.

Sine and cosine, tangent and cotangent, and secant and cosecant are *cofunctions*. These examples indicate that any function of α = the cofunction of β and that α and β are *complementary* angles because their sum is 90°. A very useful property is obtained from this observation: *Cofunctions of complementary angles are equal.*

EXAMPLE 4: If both the angles are acute and sin $(3x + 20°)$ = cos $(2x - 40°)$, find x.

Since these cofunctions are equal, the angles must be complementary.

$$\text{Therefore, } (3x + 20°) + (2x - 40°) = 90°$$
$$5x - 20° = 90°$$
$$x = 22°$$

EXERCISES

1. Express cos 320° as a function of an angle between 0° and 45°.

   (A) cos 40°
   (B) sin 40°
   (C) −cos 40°
   (D) −sin 40°
   (E) none of the above

2. If point $P(-5,12)$ lies on the terminal side of $\angle\theta$ in standard position, sin θ =

   (A) $\dfrac{-5}{12}$

   (B) $\dfrac{12}{13}$

   (C) $\dfrac{-5}{13}$

   (D) $\dfrac{12}{5}$

   (E) $-\dfrac{12}{13}$

3. If sec $\theta = -\dfrac{5}{4}$ and sin $\theta > 0$, then tan θ =

   (A) $\dfrac{4}{3}$

   (B) $\dfrac{3}{4}$

   (C) $-\dfrac{4}{3}$

   (D) $-\dfrac{3}{4}$

   (E) none of the above

4. If x is an angle in quadrant III and tan $(x - 30°) =$ cot x, find x.

   (A) 210°
   (B) 240°
   (C) 225°
   (D) 60°
   (E) none of the above

5. If $90° < \alpha < 180°$ and $270° < \beta < 360°$, then which of the following *cannot* be true?

   (A) sin α = sin β
   (B) tan α = sin β
   (C) tan α = tan β
   (D) sin α = cos β
   (E) sec α = csc β

6. Expressed as a function of an acute angle, cos 310° + cos 190° =

   (A) −cos 40°
   (B) cos 70°
   (C) −cos 50°
   (D) sin 20°
   (E) −cos 70°

3.2 ARCS AND ANGLES

Although the degree is the chief unit used to measure an angle in elementary mathematics courses, the radian has several advantages in more advanced mathematics. Degrees and radians are related by this equation: $\pi^R = 180°$.

EXAMPLE 1: In each of the following, convert the degrees to radians or the radians to degrees. (If no unit of measurement is indicated, radians are assumed.)

   (A) 30°
   (B) 270°
   (C) $\dfrac{\pi^R}{4}$
   (D) $\dfrac{17\pi}{3}$
   (E) 24

   (A) $\dfrac{x^R}{30°} = \dfrac{\pi^R}{180°}$. Solving for x, $x = \dfrac{\pi^R}{6}$. Since $\pi^R = 180°$ on the right side of the equation, $\dfrac{\pi^R}{6}$ must equal 30° on the left side of the equation.

   (B) $\dfrac{x^R}{270°} = \dfrac{\pi^R}{180°}$. $x = \dfrac{3\pi^R}{2}$

   (C) $\dfrac{\dfrac{\pi}{4}}{x°} = \dfrac{\pi^R}{180°}$. $x = 45°$

   (D) $\dfrac{\dfrac{17\pi}{3}}{x°} = \dfrac{\pi^R}{180°}$. $x = 1020°$

   (E) $\dfrac{24}{x°} = \dfrac{\pi^R}{180°}$. $x = \left(\dfrac{4320}{\pi}\right)°$

In a circle of radius r inches with an arc subtended by a central angle of θ^R, two important formulas can be derived. The length of the arc, s, is equal to $r\theta$, and the area of the sector, AOB, is equal to $\frac{1}{2}r^2\theta$.

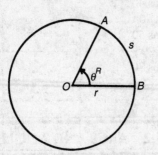

EXAMPLE 2: Find the area of the sector and the length of the arc subtended by a central angle of $\frac{2\pi}{3}$ radians in a circle whose radius is 6 inches.

$$s = r\theta \qquad\qquad A = \frac{1}{2}r^2\theta$$

$$s = 6 \cdot \frac{2\pi}{3} = 4\pi \text{ inches} \qquad A = \frac{1}{2} \cdot 36 \cdot \frac{2\pi}{3} = 12\pi \text{ square inches}$$

EXAMPLE 3: In a circle of radius 8 inches, find the area of the sector whose arc length is 6π inches.

$$s = r\theta \qquad\qquad A = \frac{1}{2}r^2\theta$$

$$6\pi = 8\theta \qquad\qquad A = \frac{1}{2} \cdot 64 \cdot \frac{3\pi}{4} = 24\pi \text{ square inches}$$

$$\theta = \frac{3\pi^R}{4}$$

EXAMPLE 4: Find the length of the radius of a circle in which a central angle of 60° subtends an arc of length 8π inches.

The 60° angle must be converted to radians:

$$\frac{x^R}{60°} = \frac{\pi^R}{180°}$$

Therefore,
$$x = \frac{\pi}{3}$$

Also,
$$s = r\theta$$
$$8\pi = r \cdot \frac{\pi}{3}$$
$$r = 24 \text{ inches}$$

EXERCISES

1. An angle of 30 radians is equal to how many degrees?

   (A) $\dfrac{\pi}{6}$

   (B) $\dfrac{30}{\pi}$

   (C) $\dfrac{5400}{\pi}$

   (D) $\dfrac{540}{\pi}$

   (E) $\dfrac{\pi}{30}$

2. If a sector of a circle has an arc length of 2π inches and an area of 6π square inches, what is the length of the radius of the circle?

   (A) 1
   (B) 2
   (C) 3
   (D) 6
   (E) 12

3. If a circle has a circumference of 16 inches, the area of a sector with a central angle of 4.7 radians is

   (A) 48
   (B) 12
   (C) 10
   (D) 15
   (E) 25

4. A central angle of 40° in a circle of radius 1 inch intercepts an arc whose length is s. Find s.

   (A) 1.4
   (B) 40
   (C) 0.7
   (D) 2.0
   (E) 3.0

5. The pendulum on a clock swings through an angle of 25°, and the tip sweeps out an arc of 12 inches. How long is the pendulum?

   (A) 43.2 inches
   (B) 27.5 inches
   (C) 86.4 inches
   (D) 1.67 inches
   (E) 13.8 inches

3.3   SPECIAL ANGLES

Although most of the values of the trigonometric functions are difficult to determine (and are usually obtained from your calculator), values of functions involving angles that are multiples of $\frac{\pi}{6}, \frac{\pi}{4}, \frac{\pi}{3}$, or $\frac{\pi}{2}$ are easy to find. Any angle that is a multiple of $\frac{\pi}{2}$ is known as a *quadrantal* angle because, in standard position, its terminal side lies on one of the axes between quadrants. Any function of a quadrantal angle is easily determined because one of the coordinates of point P on the terminal side is zero.

	0	$\frac{\pi}{2}$	π	$\frac{3\pi}{2}$	2
sine	0	1	0	−1	0
cosine	1	0	−1	0	1
tangent	0	und	0	und	0
cotangent	und	0	und	0	und
secant	1	und	−1	und	1
cosecant	und	1	und	−1	und

In the table above, "und" means that the function is undefined because the definition of the function necessitates division by zero.

Any function of an odd multiple of $\frac{\pi}{4} = \pm$ that function of any other odd multiple of $\frac{\pi}{4}$. (The sign is determined by the quadrant in which the terminal side of the angle lies.)

Any function of $k \cdot \frac{\pi}{6}$ (where k is not a multiple or factor of 6) $= \pm$ that function of any other similar multiple of $\frac{\pi}{6}$.

Any function of $k \cdot \frac{\pi}{3}$ (where k is not a multiple or factor of 3) $= \pm$ that function of any other similar multiple of $\frac{\pi}{3}$.

Because of these properties it is necessary only to learn the values of the trigonometric functions of $\frac{\pi}{4}, \frac{\pi}{6}$, and

$\frac{\pi}{3}$ and then to attach a + or −, depending on the quadrant in which the terminal side of the angle lies.

	$\frac{\pi}{6}$ or 30°	$\frac{\pi}{4}$ or 45°	$\frac{\pi}{3}$ or 60°
sine	$\frac{1}{2}$	$\frac{\sqrt{2}}{2}$	$\frac{\sqrt{3}}{2}$
cosine	$\frac{\sqrt{3}}{2}$	$\frac{\sqrt{2}}{2}$	$\frac{1}{2}$
tangent	$\frac{\sqrt{3}}{3}$	1	$\sqrt{3}$
cotangent	$\sqrt{3}$	1	$\frac{\sqrt{3}}{2}$
secant	$\frac{2\sqrt{3}}{3}$	$\sqrt{2}$	2
cosecant	2	$\sqrt{2}$	$\frac{2\sqrt{3}}{3}$

EXAMPLE: What is the exact value of each of the following?

(A) $\cos 0$

(B) $\csc \frac{3\pi}{2}$

(C) $\cos \frac{7\pi}{6}$

(D) $\tan \frac{5\pi}{4}$

(E) $\sec \frac{3\pi}{4}$

(F) $\sin 300°$

(G) $\cos (-390°)$

(H) $\cot 60°$

(I) $\tan (-45°)$

Procedures are as follows:

(A) From the first table on this page, $\cos 0 = 1$.

(B) $\frac{3\pi}{2}$ is a quadrantal angle with its terminal side along the negative y-axis. There re, $P(0, -r)$, where $OP = r$. $\csc \frac{3\pi}{2} = \frac{r}{-r} = -1$. (Or, from the first table on this page, $\csc \frac{3\pi}{2} = -1$.)

(C) $\dfrac{7\pi}{6}$ is a multiple of $\dfrac{\pi}{6}$, and so $\cos\dfrac{7\pi}{6} = \pm\ \cos\dfrac{\pi}{6} = \pm\dfrac{\sqrt{3}}{2}$. Since the terminal side of $\dfrac{7\pi}{6}$ is in quadrant III, $\cos\dfrac{7\pi}{6} = -\dfrac{\sqrt{3}}{2}$.

(D) $\dfrac{5\pi}{4}$ is a multiple of $\dfrac{\pi}{4}$, and its terminal side lies in quadrant III. Therefore, $\tan\dfrac{5\pi}{4}$ is positive, and $\tan\dfrac{5\pi}{4} = \tan\dfrac{\pi}{4} = 1$.

(E) $\dfrac{3\pi}{4}$ is a multiple of $\dfrac{\pi}{4}$, and its terminal side lies in quadrant II. Therefore, $\sec\dfrac{3\pi}{4}$ is negative, and $\sec\dfrac{3\pi}{4} = -\sec\dfrac{\pi}{4} = -\sqrt{2}$.

(F) $300° = \dfrac{5\pi}{3}$, which is a multiple of $\dfrac{\pi}{3}$, and its terminal side lies in quadrant IV. Therefore, $\sin 300°$ is negative, and $\sin 300° = \sin\dfrac{5\pi}{3} = -\sin\dfrac{\pi}{3} = -\dfrac{\sqrt{3}}{2}$.

(G) $-390° = -\dfrac{13\pi}{6}$, which is a multiple of $\dfrac{\pi}{6}$, and its terminal side lies in quadrant IV. Therefore, $\cos(-390°)$ is positive, and $\cos(-390°) = \cos\left(-\dfrac{13\pi}{6}\right) = \cos\dfrac{\pi}{6} = \dfrac{\sqrt{3}}{2}$.

(H) $60° = \dfrac{\pi}{3}$. Therefore, $\cot 60° = \cot\dfrac{\pi}{3} = \dfrac{\sqrt{3}}{3}$.

(I) $-45° = -\dfrac{\pi}{4}$, which is a multiple of $\dfrac{\pi}{4}$, and its terminal side lies in quadrant IV. Therefore, $\tan(-45°)$ is negative, and $\tan(-45°) = \tan\left(-\dfrac{\pi}{4}\right) = -\tan\dfrac{\pi}{4} = -1$.

EXERCISES

1. Tan $(-60°)$ equals

    (A) $-\tan 30°$
    (B) $\cot 30°$
    (C) $-\tan 60°$
    (D) $-\cot 60°$
    (E) $\tan 60°$

2. Tan $(-135°) + \cot 315°$ equals

    (A) 1
    (B) 2
    (C) -1
    (D) -2
    (E) 0

3. Cos $\pi - \sin 570° - \csc\left(-\dfrac{\pi}{2}\right) + \sec 0°$ equals

    (A) 1
    (B) $1\dfrac{1}{2}$
    (C) $-1\dfrac{1}{2}$
    (D) 0
    (E) $-\dfrac{1}{2}$

4. Sec $\dfrac{11\pi}{6} \cdot \tan\dfrac{2\pi}{3} \cdot \sin\dfrac{7\pi}{4}$ equals

    (A) -1
    (B) $\sqrt{2}$
    (C) $\dfrac{\sqrt{6}}{3}$
    (D) $-\sqrt{2}$
    (E) $-\dfrac{\sqrt{6}}{3}$

5. Sin $300°$ equals

    (A) $\cos 60°$
    (B) $\sin 120°$
    (C) $\cos 240°$
    (D) $\sin 240°$
    (E) $\cos 120°$

3.4   GRAPHS

Since the values of all the trigonometric functions repeat themselves at regular intervals, and, for some number p, $f(x) = f(x + p)$ for all numbers x, these functions are called *periodic functions*. The smallest positive value of p for which this property holds is called the *period* of the function.

The sine, cosine, secant, and cosecant have periods of 2π, and the tangent and cotangent have periods of π. The graphs of the six trigonometric functions, shown below, demonstrate that the tangent and cotangent repeat on intervals of length and that the others repeat on intervals of length 2π.

From the graphs or from the table of values at the right it is possible to determine the domain and range of each of the six trigonometric functions.

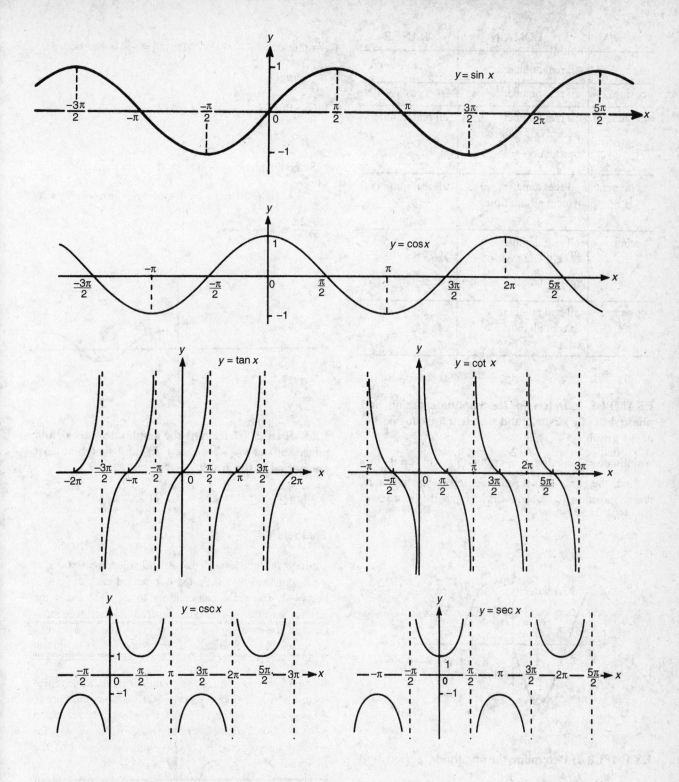

The general form of a trigonometric function, f, is given to be $y = A \cdot f(Bx + C)$, where $|A|$ is the amplitude, $\dfrac{\text{normal period of } f}{B}$ is the period of the graph, and $-\dfrac{C}{B}$ is the phase shift. Although the amplitude is associated with only the sine and cosine, the period and phase shift can be associated with any of the trigonometric functions; they are usually associated with the sine and cosine graphs. The *amplitude* is one-half the vertical distance between the lowest and highest point of the graph. The *phase shift* is the distance to the right or left that the graph is changed from its normal position. The *frequency* is equal to $\dfrac{1}{\text{period}}$.

	DOMAIN	RANGE
sine	all real numbers	$-1 \le \sin x \le 1$
cosine	all real numbers	$-1 \le \cos x \le 1$
tangent	all real numbers except odd multiples of $\frac{\pi}{2}$	all real numbers
cotangent	all real numbers except all multiples of π	all real numbers
secant	all real numbers except odd multiples of $\frac{\pi}{2}$	$\sec x \le -1$ or $\sec x \ge 1$
cosecant	all real numbers except all multiples of π	$\csc x \le -1$ or $\csc x \ge 1$

EXAMPLE 1: Determine the amplitude, period, and phase shift of $y = 2\sin 2x$ and sketch at least one period of the graph.

Amplitude $= 2$ Period $= \dfrac{2\pi}{2} = \pi$ Phase shift $= 0$

Since the phase shift is zero, the sine graph starts at its normal position, $(0,0)$, and is drawn out to the right and to the left.

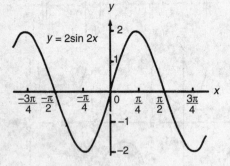

EXAMPLE 2: Determine the amplitude, period, and phase shift of $y = \dfrac{1}{2}\cos\left(\dfrac{1}{2}x - \dfrac{\pi}{3}\right)$ and sketch at least one period of the graph.

Amplitude $= \dfrac{1}{2}$ Period $= \dfrac{2\pi}{\frac{1}{2}} = 4\pi$

Phase shift $= \dfrac{\frac{\pi}{3}}{\frac{1}{2}} = \dfrac{2\pi}{3}$

Since the phase shift is $\dfrac{2\pi}{3}$, the cosine graph starts at $x = \dfrac{2\pi}{3}$ instead of $x = 0$ and one period ends at $x = \dfrac{2\pi}{3} + 4\pi$ or $\dfrac{14\pi}{3}$.

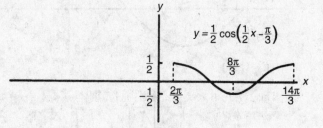

EXAMPLE 3: Determine the amplitude, period, and phase shift of $y = -2\sin(\pi x + 3\pi)$ and sketch at least one period of the graph.

Amplitude $= 2$ Period $= \dfrac{2\pi}{\pi} = 2$

Phase shift $= -\dfrac{3\pi}{\pi} = -3$

Since the phase shift is -3, the sine graph starts at $x = -3$ instead of $x = 0$, and one period ends at $-3 + 2$ or $x = -1$. The graph can continue to the right and to the left for as many periods as desired. Since the coefficient of the sine is negative, the graph starts down as x increases from -3, instead of up as a normal sine graph does.

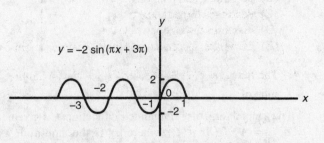

EXERCISES

1. In the figure, part of the graph of $y = \sin 2x$ is shown. What are the coordinates of point P?

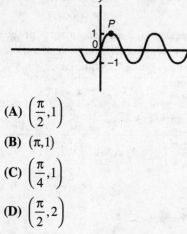

(A) $\left(\dfrac{\pi}{2}, 1\right)$

(B) $(\pi, 1)$

(C) $\left(\dfrac{\pi}{4}, 1\right)$

(D) $\left(\dfrac{\pi}{2}, 2\right)$

(E) $(\pi, 2)$

2. The figure below could be a portion of the graph whose equation is

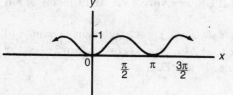

(A) $y - 1 = \sin x \cdot \cos x$
(B) $y \sec x = 1$
(C) $2y + 1 = \sin 2x$
(D) $2y + 1 = \cos 2x$
(E) $1 - 2y = \cos 2x$

3. As θ increases from $\dfrac{\pi}{4}$ to $\dfrac{5\pi}{4}$, the value of

$4\cos\dfrac{1}{2}\theta$

(A) increases, and then decreases
(B) decreases, and then increases
(C) decreases throughout
(D) increases throughout
(E) decreases, increases, and then decreases again

4. The function $f(x) = \sqrt{3}\cos x + \sin x$ has an amplitude of

(A) $\sqrt{3}$

(B) $\sqrt{3} + 1$

(C) 2

(D) $2\sqrt{3}$

(E) $\dfrac{\sqrt{3} + 1}{2}$

5. For what value of P is the period of the function $y = \dfrac{1}{3}\cos Px$ equal to $\dfrac{2\pi}{3}$?

(A) $\dfrac{1}{3}$

(B) 2

(C) 3

(D) 6

(E) $\dfrac{2}{3}$

6. If $0 \le x \le \dfrac{\pi}{2}$, what is the maximum value of the function $f(x) = \sin\dfrac{1}{3}x$?

(A) 0

(B) 1

(C) $\dfrac{1}{3}$

(D) $\dfrac{1}{2}$

(E) $\dfrac{\sqrt{3}}{2}$

7. If the graph in the figure below has an equation of the form $y = \sin(Mx + N)$, what is the value of N?

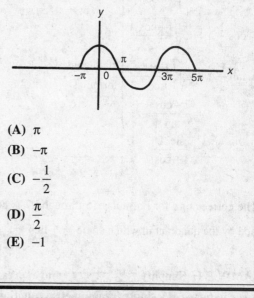

(A) π

(B) $-\pi$

(C) $-\dfrac{1}{2}$

(D) $\dfrac{\pi}{2}$

(E) -1

3.5 IDENTITIES, EQUATIONS, AND INEQUALITIES

Many of the problems involving trigonometry depend on several formulas that can be used with any angle as long as the value of the function is not undefined. A list of the most frequently used formulas follows:

1. $\sin^2 x + \cos^2 x = 1$
2. $\tan^2 x + 1 = \sec^2 x$ $\}$ Pythagorean identities
3. $\cot^2 x + 1 = \csc^2 x$

4. $\sin(A + B) = \sin A \cdot \cos B + \cos A \cdot \sin B$
5. $\sin(A - B) = \sin A \cdot \cos B - \cos A \cdot \sin B$
6. $\cos(A + B) = \cos A \cdot \cos B - \sin A \cdot \sin B$ } Sum and difference formulas
7. $\cos(A - B) = \cos A \cdot \cos B + \sin A \cdot \sin B$
8. $\tan(A + B) = \dfrac{\tan A + \tan B}{1 - \tan A \cdot \tan B}$
9. $\tan(A - B) = \dfrac{\tan A - \tan B}{1 + \tan A \cdot \tan B}$

10. $\sin 2A = 2 \sin A \cdot \cos A$
11. $\cos 2A = \cos^2 A - \sin^2 A$ } Double-angle formulas
12. $\quad\quad\ = 2 \cos^2 A - 1$
13. $\quad\quad\ = 1 - 2 \sin^2 A$
14. $\tan 2A = \dfrac{2 \tan A}{1 - \tan^2 A}$

A few other formulas are rarely used on the Math Level IIC examination but may be helpful:

15. $\sin \dfrac{1}{2} A = \pm \sqrt{\dfrac{1 - \cos A}{2}}$
16. $\cos \dfrac{1}{2} A = \pm \sqrt{\dfrac{1 + \cos A}{2}}$
17. $\tan \dfrac{1}{2} A = \pm \sqrt{\dfrac{1 - \cos A}{1 + \cos A}}$ } Half-angle formulas
18. $\quad\quad\ = \dfrac{1 - \cos A}{\sin A}$
19. $\quad\quad\ = \dfrac{\sin A}{1 + \cos}$

The correct sign for Formulas 15 through 17 is determined by the quadrant in which angle $\dfrac{1}{2} A$ lies.

EXAMPLE 1: Simplify $\dfrac{\csc A}{\cot A + \tan A}$ and express the answer in terms of a single trigonometric function.

If none of the formulas seem to be helpful, change all the functions to sines and cosines. From the basic definitions of the trigonometric functions,

$$\csc A = \frac{1}{\sin A}, \cot A = \frac{\cos A}{\sin A}, \tan A = \frac{\sin A}{\cos A}.$$

Therefore,

$$\frac{\csc A}{\cot A + \tan A} = \frac{\dfrac{1}{\sin A}}{\dfrac{\cos A}{\sin A} + \dfrac{\sin A}{\cos A}}.$$

When the numerator and the denominator of the complex fraction are multiplied by the lowest common denominator of the three "little" denominators, $\sin A \cdot \cos A$, the fraction becomes

$$\frac{\dfrac{1}{\sin A} \cdot \sin A \cdot \cos A}{\left(\dfrac{\cos A}{\sin A} + \dfrac{\sin A}{\cos A} \right) \cdot \sin A \cdot \cos A} = \frac{\cos A}{\cos^2 A + \sin^2 A}$$

$$= \frac{\cos A}{1}.$$

by Formula 1. Therefore,

$$\frac{\csc A}{\cot A + \tan A} = \cos A.$$

EXAMPLE 2: Express $\sin 4x$ in terms of $\sin x$ and $\cos x$.

$\sin 4x = \sin 2(2x)$. Let $A = 2x$ and use Formula 10.
$\sin 4x = \sin 2A = 2 \sin A \cdot \cos A = 2 \sin 2x \cdot \cos 2x$. Using Formulas 10 and 11 gives

$$\sin 4x = 2(2 \sin x \cdot \cos x) \cdot (\cos^2 x - \sin^2 x)$$
$$= 4 \sin x \cdot \cos^3 x - 4 \sin^3 x \cdot \cos x.$$

EXAMPLE 3: If $\tan A = \dfrac{5}{12}$ and $\sin B = \dfrac{3}{5}$, where A and B are acute angles, find the value of $\cos (A + B)$.

From the basic definitions of the trigonometric functions and the Pythagorean theorem, $\sin A = \dfrac{5}{13}$, $\cos A = \dfrac{12}{13}$, and $\cos B = \dfrac{4}{5}$.

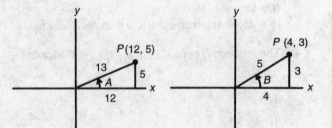

Therefore, using Formula 6 gives

$$\cos(A + B) = \cos A \cdot \cos B - \sin A \cdot \sin B$$
$$= \frac{12}{13} \cdot \frac{4}{5} - \frac{5}{13} \cdot \frac{3}{5}$$
$$= \frac{48}{65} - \frac{15}{65}$$
$$= \frac{33}{65}.$$

EXAMPLE 4: If A is an angle in the third quadrant, B is an angle in the second quadrant, $\tan A = \dfrac{3}{4}$, and $\tan B = -\dfrac{1}{2}$, in which quadrant does angle $(A + B)$ lie?

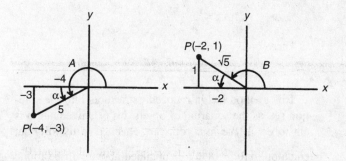

Since $180° < A < 270°$ and $90° < B < 180°$, therefore $270° < A + B < 450°$, meaning that angle $(A + B)$ must lie in quadrant I or IV. Since the sine is positive in quadrant I and negative in quadrant IV, the value of $\sin(A + B)$ determines the correct quadrant. From the basic definitions of the trigonometric functions and the Pythagorean theorem, $\sin A = -\dfrac{3}{5}$, $\cos A = -\dfrac{4}{5}$, $\sin B = \dfrac{\sqrt{5}}{5}$, and $\cos B = \dfrac{-2\sqrt{5}}{5}$. From Formula 4,

$$\sin(A + B) = \sin A \cdot \cos B + \cos A \cdot \sin B$$
$$= \frac{-3}{5} \cdot \frac{-2\sqrt{5}}{5} + \frac{-4}{5} \cdot \frac{\sqrt{5}}{5}$$
$$= \frac{6\sqrt{5}}{25} - \frac{4\sqrt{5}}{25}$$
$$= \frac{2\sqrt{5}}{25} \qquad > 0$$

Therefore, angle $(A + B)$ lies in quadrant I.

EXAMPLE 5: Find the value of $\cos 2\theta - \sin(90° + \theta)$ if $\tan \theta = -\dfrac{3}{4}$ and $\sin \theta$ is positive.

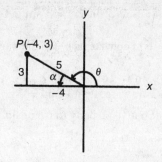

Since $\tan \theta < 0$ and $\sin \theta > 0$, the terminal side of angle θ must lie in quadrant II. From the basic definitions of the trigonometric functions and the Pythagorean theorem,

$$\sin \theta = \frac{3}{5} \quad \text{and} \quad \cos \theta = \frac{-4}{5}.$$

From Formula 11,

$$\cos 2\theta = \cos^2 \theta - \sin^2 \theta = \frac{16}{25} - \frac{9}{25} = \frac{7}{25}.$$

From Formula 4,

$$\sin(90° + \theta) = \sin 90° \cdot \cos \theta + \cos 90° \cdot \sin \theta$$
$$= 1 \cdot \frac{-4}{5} + 0 \cdot -\frac{4}{5}$$

Therefore,

$$\cos 2\theta - \sin(90° + \theta) = \frac{7}{25} - \left(\frac{-4}{5}\right) = \frac{7}{25} + \frac{4}{5} = \frac{27}{25}.$$

EXAMPLE 6: If $\sin \theta = a$, find the value of $\sin 2\theta$ in terms of a.

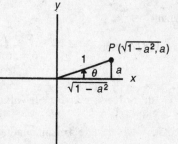

Choose θ to be an angle in any quadrant. (It can be chosen in quadrant I for convenience and without any loss of generality.) From the basic definitions of the trigonometric functions and the Pythagorean theorem, the coordinates of P are $\left(\sqrt{1 - a^2}, a\right)$. From Formula 10, $\sin 2\theta = 2 \sin \theta \cdot \cos \theta = 2 \cdot a \cdot \sqrt{1 - a^2}$.

EXAMPLE 7: If $\cos 23° = z$, find the value of $\cos 46°$ in terms of z.

Since $46 = 2(23)$, Formula 12 can be used: $\cos 2A = 2 \cos^2 A - 1$. $\cos 46° = \cos 2(23°) = 2 \cos^2 23° - 1 = 2(\cos 23°)^2 - 1 = 2z^2 - 1$.

EXAMPLE 8: Solve **sin 2x = 3 sin x for x, where 0 ≤ x < 2π.**
Use Formula 10 to convert each term into a function of x only.

$$2 \sin x \cdot \cos x = 3 \sin x$$
$$2 \sin x \cdot \cos x - 3 \sin x = 0$$

Use the distributive property to factor $\sin x$ out of each term:

$$\sin x(2 \cos x - 3) = 0$$

Therefore, $\sin x = 0$ or $2 \cos x - 3 = 0$. The second equation implies that $\cos x = \dfrac{3}{2}$, which is impossible because the range of cosine = $\{y\colon -1 \le y \le 1\}$. Therefore, $\sin x = 0$ contributes the only roots of the original equation. Sin $x = 0$, when $x = 0, \pi$.

EXAMPLE 9: Solve **2sin x + cos 2x = 2 sin² x − 1 for all values of x such that 0 ≤ x < 2π.**
Use Formula 13 to convert every term to $\sin x$. The equation becomes $2 \sin x + 1 - 2 \sin^2 x = 2 \sin^2 x - 1$. Then

$$4 \sin^2 x - 2 \sin x - 2 = 0$$
$$2 \sin^2 x - \sin x - 1 = 0$$
$$(2 \sin x + 1)(\sin x - 1) = 0$$
$$2 \sin x + 1 = 0 \text{ or } \sin x - 1 = 0$$
$$\sin x = -\frac{1}{2} \text{ or } \sin x = 1.$$

Since sine is negative in quadrants III and IV and since $\sin \dfrac{\pi}{6} = \dfrac{1}{2}$, $\sin x = -\dfrac{1}{2}$ when $x = \dfrac{7\pi}{6}$ and $\dfrac{11\pi}{6}$. Also, $\sin x = 1$ when $x = \dfrac{\pi}{2}$.

Therefore, the solution set $= \left\{\dfrac{\pi}{2}, \dfrac{7\pi}{6}, \dfrac{11\pi}{6}\right\}$.

EXAMPLE 10: Solve **tan 3x for all values of x such that 0° ≤ x < 360° = 1.**
Tan $3x$ is positive when angle $3x$ is in quadrants I and III.

Therefore, $3x = \dfrac{\pi}{4} = 45°$ or $3x = \dfrac{5\pi}{4} = 225°$. However, to find all values of x such that $0° \le x < 360°$, two angles coterminal with $45°$ and $225°$ must also be found for values of $3x$. Since $3x = 45°, 45° + 360°, 45° + 720°$, or $3x = 225°, 225° + 360°, 225° + 720°$, therefore, if $3x = 45°, 225°, 405°, 585°, 765°$, or $945°$, then $x = 15°, 75°, 135°, 195°, 255°$, or $315°$.

In this type of problem make sure that enough coterminal angles are chosen to determine all values of x. If n is the coefficient of x (in this problem it is 3), $n - 1$ angles coterminal to the principal values are necessary.

EXAMPLE 11: Solve **sin x = √3 cos x for all values of x such that 0 ≤ x < 2π.**

Method 1: Since a solution is not obtained when $\cos x = 0$, it is possible to divide both sides of the equation by $\cos x$.

$\dfrac{\sin x}{\cos x} = \sqrt{3}$, which becomes $\tan x = \sqrt{3}$

$\tan x = \sqrt{3}$ when $x = \dfrac{\pi}{3}$ or $\dfrac{4\pi}{3}$

Method 2: Square both sides of the equation and use Formula 1.

$$\sin^2 x = 3 \cos^2 x$$
$$\sin^2 x = 3(1 - \sin^2 x) = 3 - 3 \sin^2 x$$
$$4 \sin^2 x = 3$$
$$\sin^2 x = \frac{3}{4}$$
$$\sin x = \pm \frac{\sqrt{3}}{2}$$

Therefore,

$$x = \frac{\pi}{3}, \frac{2\pi}{3}, \frac{4\pi}{3}, \frac{5\pi}{3}$$

This method has introduced extra roots into the solution set, as the squaring of both sides of any equation is apt to do. If the four roots are checked in the original equation, only $\dfrac{\pi}{3}$ and $\dfrac{4\pi}{3}$ obtained above will be found to be solutions.

Thus, the solution set obtained by either method is $\left\{\dfrac{\pi}{4}, \dfrac{4\pi}{3}\right\}$.

EXAMPLE 12: Solve **sin x · sin 40° − cos x · cos 40° = $\dfrac{1}{2}$ for all values of x such that 0° ≤ x < 360°.**
When this equation is multiplied through by -1, the left side is similar to the right side of Formula 6. From Formula 6, $\cos (x + 40°) = \cos x \cdot \cos 40° - \sin x \cdot \sin 40°$.

Thus, this equation becomes $\cos (x + 40°) = -\dfrac{1}{2}$. Since $\cos \dfrac{2\pi}{3} = \cos \dfrac{4\pi}{3} = -\dfrac{1}{2}, \dfrac{2\pi}{3} = 120°$, and $\dfrac{4\pi}{3} = 240°$, $x + 40° = 120°$ or $x + 40° = 240°$. Solving for x, the solution set is $\{80°, 200°\}$.

EXAMPLE 13: Solve **√3 cos x − sin x = 2 for all values of x such that 0 ≤ x < 2π.**

Method 1: Divide the equation through by 2, obtaining $\dfrac{\sqrt{3}}{2} \cos x - \dfrac{1}{2} \sin x = 1$. Since $\sin \dfrac{\pi}{3} = \dfrac{\sqrt{3}}{2}$ and

$\cos\dfrac{\pi}{3} = \dfrac{1}{2}$, the equation becomes $\sin\dfrac{\pi}{3} \cdot \cos x - \cos\dfrac{\pi}{3} \cdot$ $\sin x = 1.$ Using Formula 5, the equation becomes

$$\sin\left(\dfrac{\pi}{3} - x\right) = 1.$$

Therefore, $\dfrac{\pi}{3} - x = \dfrac{\pi}{2}$ and $x = \dfrac{\pi}{3} - \dfrac{\pi}{2} = -\dfrac{\pi}{6},$ which is not in the range $0 \le x < 2\pi$. Add 2π to this angle to find a coterminal angle within this range:

$-\dfrac{\pi}{6} + 2\pi = \dfrac{11\pi}{6}.$

Therefore, the solution set has only one member:

$\left\{\dfrac{11\pi}{6}\right\}.$

Method 2: Add $\sin x$ to both sides of the equation and then square each side, obtaining:

$$(\sqrt{3}\cos x)^2 = (2 + \sin x)^2$$

$$3\cos^2 x = 4 + 4\sin x + \sin^2 x$$

Using a form of Formula 1 ($\cos^2 x = 1 - \sin^2 x$) and setting everything equal to zero gives

$$3(1 - \sin^2 x) = 4 + 4\sin x + \sin^2 x$$

$$3 - 3\sin^2 x = 4 + 4\sin x + \sin^2 x$$

$$4\sin^2 x + 4\sin x + 1 = 0$$

$$(2\sin x + 1)^2 = 0$$

$$\sin x = -\dfrac{1}{2}$$

EXAMPLE 14: For what values of x between 0 and 2π is $\sin x < \cos x$?

Sketch the graphs of $y = \sin x$ and $y = \cos x$ on the same set of axes between 0 and 2π. Observe where the graph of $y = \sin x$ is below the graph of $y = \cos x$ (the sections indicated with xxxxx).

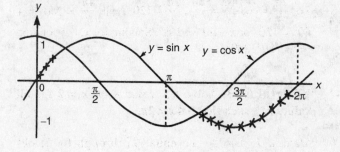

To find the points of intersection of the two graphs solve the equation

$$\sin x = \cos x$$

$$\dfrac{\sin x}{\cos x} = 1$$

$$\tan x = 1$$

$$x = \dfrac{\pi}{4} \text{ and } \dfrac{5\pi}{4}$$

Therefore, the set of x values that solve the inequality is $0 < x < \dfrac{\pi}{4}$ or $\dfrac{5\pi}{4} < x < 2\pi$.

EXERCISES

1. Solve $\cos 2x + \cos x + 1 = 0$ when $0 \le x < 2\pi$. The solutions set is

    (A) $\left\{\dfrac{\pi}{3}, \dfrac{\pi}{2}, \dfrac{3\pi}{2}, \dfrac{5\pi}{3}\right\}$

    (B) $\left\{\dfrac{2\pi}{3}, \dfrac{4\pi}{3}\right\}$

    (C) $\left\{\dfrac{\pi}{2}, \dfrac{3\pi}{2}\right\}$

    (D) $\left\{\dfrac{\pi}{2}, \dfrac{2\pi}{3}, \dfrac{4\pi}{3}, \dfrac{3\pi}{2}\right\}$

    (E) $\left\{\dfrac{\pi}{3}, \dfrac{5\pi}{3}\right\}$

2. Which of the following is the solution set of $\sin\theta \cdot \cos\theta = \dfrac{1}{4}$ when $0° \le \theta < 360°$?

    (A) $\{30°, 150°\}$
    (B) $\{30°, 150°, 210°, 330°\}$
    (C) $\{15°, 75°\}$
    (D) $\{15°, 75°, 195°, 225°\}$
    (E) $\{60°, 300°\}$

3. How many positive values of $x \le 2\pi$ make $\tan 4x = \sqrt{3}$?

    (A) 0
    (B) 1
    (C) 2
    (D) 4
    (E) 8

4. If $\cos\theta = \dfrac{3}{5}$, $\tan 2\theta$ equals

(A) $\dfrac{24}{7}$

(B) $-\dfrac{24}{7}$

(C) $\dfrac{24}{5}$

(D) $-\dfrac{24}{5}$

(E) $\pm\dfrac{24}{7}$

5. Solve the equation $\cos 2\theta = -\sin\theta$ for all positive values of $\theta < 2\pi$. Here, θ equals

(A) $\dfrac{\pi}{2}, \dfrac{7\pi}{6}, \dfrac{11\pi}{6}$

(B) $\dfrac{\pi}{2}, \dfrac{7\pi}{6}, \dfrac{3\pi}{2}, \dfrac{11\pi}{6}$

(C) $\dfrac{\pi}{6}, \dfrac{\pi}{2}, \dfrac{5\pi}{6}$

(D) $\dfrac{\pi}{6}, \dfrac{\pi}{2}, \dfrac{5\pi}{6}, \dfrac{3\pi}{2}$

(E) $\dfrac{\pi}{2}$

6. If $\sin 37° = z$, then $\sin 74°$ equals

(A) $2z\sqrt{1-z^2}$

(B) $2z^2 + 1$

(C) $2z$

(D) $2z^2 - 1$

(E) $\dfrac{z}{\sqrt{1-z^2}}$

7. $\text{Cot}(A + B)$ equals

(A) $\dfrac{\cot A \cdot \cos B + 1}{\cot A - \cot B}$

(B) $\dfrac{\cot A \cdot \cot B - 1}{\cot A + \cot B}$

(C) $\dfrac{\tan A - \tan B}{1 - \tan A \cdot \tan B}$

(D) $\dfrac{1 + \tan A \cdot \tan B}{\tan A - \tan B}$

(E) $\dfrac{\cot A \cdot \cot B + 1}{\cot B - \cot A}$

8. $\dfrac{\tan 140° + \tan 70°}{1 - \tan 140° \cdot \tan 70°}$ equals

(A) $-\sqrt{3}$

(B) $\dfrac{\sqrt{3}}{3}$

(C) $\dfrac{\sqrt{3}}{1-\sqrt{3}}$

(D) $\sqrt{3}$

(E) $\dfrac{3-\sqrt{3}}{3}$

9. Which of the following is equal to $\cos^4 40° - \sin^4 40°$?

(A) $\cos 80°$
(B) $\sin 80°$
(C) 0
(D) 1
(E) none of the above

10. If $\sin A = \dfrac{2}{3}$ and $\cos A < 0$, find the value of $\tan 2A$.

(A) $\dfrac{4\sqrt{5}}{5}$

(B) $-4\sqrt{5}$

(C) $-\dfrac{4\sqrt{5}}{9}$

(D) $\dfrac{4\sqrt{5}}{9}$

(E) $4\sqrt{5}$

11. What values of x satisfy the inequality $\sin 2x < \sin x$ so that $0 \le x \le 2\pi$?

(A) $\dfrac{\pi}{3} < x < \dfrac{5\pi}{3}$

(B) $0 < x < \dfrac{\pi}{3}$ or $\dfrac{5\pi}{3} < x < 2\pi$

(C) $\dfrac{2\pi}{3} < x < 2\pi$

(D) $\dfrac{\pi}{3} < x < \pi$ or $\dfrac{5\pi}{3} < x < 2\pi$

(E) $0 < x < \dfrac{\pi}{3}$ or $\pi < x < \dfrac{5\pi}{3}$

3.6   INVERSE FUNCTIONS

If the graph of any trigonometric function $f(x)$ is reflected about the line $y = x$ (see Section 1.3), the graph of the inverse of that trigonometric function, $f^{-1}(x)$, results. These inverse trigonometric functions are called Sin^{-1} or Arcsin, Cos^{-1} or Arccos, and so on. In every case the resulting graph is *not* the graph of a function. To obtain a function, the range of the inverse relation must be severely limited. The particular range of each inverse trigonometric function is accepted by convention. The ranges of the inverse functions are as follows:

$$-\frac{\pi}{2} \le \text{Sin}^{-1}x,\ x \le \frac{\pi}{2}$$

$$0 \le \text{Cos}^{-1}x,\ x \le \pi$$

$$-\frac{\pi}{2} < \text{Tan}^{-1}x,\ x < \frac{\pi}{2}$$

$$0 < \text{Cot}^{-1}x,\ x < \pi$$

$$0 \le \text{Sec}^{-1}x,\ x \le \pi \text{ and } \text{Sec}^{-1}x \ne \frac{\pi}{2}$$

$$-\frac{\pi}{2} \le \text{Csc}^{-1}x,\ x \le \frac{\pi}{2} \text{ and } \text{Csc}^{-1}x \ne 0$$

The last three inverse functions are rarely used. The inverse trigonometric functions are used to represent angles that cannot be expressed in any other way without the help of tables or a calculator. It is important to understand that each inverse trigonometric function represents an angle.

EXAMPLE 1: Express each of the following in simpler form:

(A) $\text{Sin}^{-1}\dfrac{\sqrt{3}}{2}$

(B) $\text{Arccos}\left(-\dfrac{\sqrt{2}}{2}\right)$

(C) $\text{Arctan}(-1)$

(D) $\text{Arccot}\sqrt{3}$

(E) $\text{Sec}^{-1}\sqrt{2}$

(F) $\text{Csc}^{-1}(-2)$

Referring to the ranges of the inverse trigonometric functions:

(A) $\text{Sin}^{-1}\dfrac{\sqrt{3}}{2} = \theta$. Find the principal value of θ (the angle that is in the range of $\text{Sin}^{-1}\ x$) such that $\sin\theta = \dfrac{\sqrt{3}}{2}$. Since $\sin\theta > 0$, θ must be in quadrant I, $\theta = \dfrac{\pi}{3}$.

(B) $\text{Arccos}\left(-\dfrac{\sqrt{2}}{2}\right) = \theta$. Find the principal value of θ such that $\cos\theta = -\dfrac{\sqrt{2}}{2}$. Since $\cos\theta < 0$, θ must be in quadrant II, $\theta = \dfrac{3\pi}{4}$.

(C) $\text{Arctan}(-1) = \theta$. Find the principal value of θ such that $\tan\theta = -1$. Since $\tan\theta < 0$, θ must be in quadrant IV and $-\dfrac{\pi}{2} < \text{Arctan}\ x < \dfrac{\pi}{2}$; $\theta = -\dfrac{\pi}{4}$.

(D) $\text{Arccot}\sqrt{3} = \theta$. Find the principal value of θ such that $\cot\theta = \sqrt{3}$. Since $\cot\theta > 0$, θ must be in quadrant I; $\theta = \dfrac{\pi}{6}$.

(E) $\text{Sec}^{-1}\sqrt{2} = \theta$. Find the principal value of θ such that $\sec\theta = \sqrt{2}$. Since $\sec\theta > 0$, θ must be in quadrant I; $\theta = \dfrac{\pi}{4}$.

(F) $\text{Csc}^{-1}(-2) = \theta$. Find the principal value of θ such that $\csc\theta = -2$. Since $\csc\theta < 0$, θ must be in quadrant IV and $-\dfrac{\pi}{2} \le \text{Csc}^{-1}x \le \dfrac{\pi}{2}$; $\theta = -\dfrac{\pi}{6}$.

EXAMPLE 2: Evaluate $\sin\left(\text{Arccos}\dfrac{3}{5}\right)$.

Let $\text{Arccos}\dfrac{3}{5} = \theta$. The problem now becomes "Evaluate $\sin\theta$ when $\cos\theta = \dfrac{3}{5}$ and θ is in quadrant I." From the basic definitions of the trigonometric functions and the Pythagorean theorem, $\sin\theta = \dfrac{4}{5}$.

Therefore, $\sin\left(\text{Arccos}\dfrac{3}{5}\right) = \dfrac{4}{5}$.

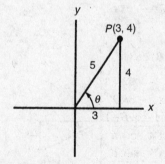

EXAMPLE 3: Evaluate

$$\cos\left[\text{Arcsin}\left(-\frac{3}{5}\right) + \text{Arccos}\frac{5}{13}\right].$$

Let $\text{Arcsin}\left(-\dfrac{3}{5}\right) = \alpha$ and $\text{Arccos}\dfrac{5}{13} = \beta$. The problem now becomes "Evaluate $\cos(\alpha + \beta)$ when $\sin\alpha = -\dfrac{3}{5}$

and α is in quadrant IV, and $\cos\beta = \dfrac{5}{13}$ and β is in quadrant I." From the basic definitions of sine and cosine and the Pythagorean theorem, $\cos\alpha = \dfrac{4}{5}$ and $\sin\beta = \dfrac{12}{13}$.

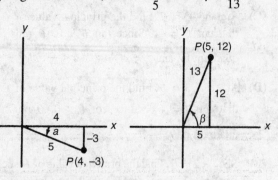

From Formula 6,

$$\cos(\alpha+\beta) = \cos\alpha \cdot \cos\beta - \sin\alpha \cdot \sin\beta$$

$$= \frac{4}{5} \cdot \frac{5}{13} - \left(-\frac{3}{5}\right) \cdot \frac{12}{13}$$

$$= \frac{20}{65} + \frac{36}{65} = \frac{56}{65}$$

Therefore, $\cos\left[\text{Arcsin}\left(-\dfrac{3}{5}\right) + \text{Arccos}\dfrac{5}{13}\right] = \dfrac{56}{65}$.

EXAMPLE 4: Evaluate $\sin\left[2\,\text{Arctan}\left(-\dfrac{8}{15}\right)\right]$.

Let $\theta = \text{Arctan}\left(-\dfrac{8}{15}\right)$. The problem now becomes "Evaluate $\sin 2\theta$ when $\tan\theta = -\dfrac{8}{15}$ and θ is in quadrant IV." From the basic definitions of the trigonometric functions and the Pythagorean theorem, $\sin\theta = -\dfrac{8}{17}$ and $\cos\theta = \dfrac{15}{17}$.

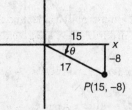

From Formula 10,

$$\sin 2\theta = 2\sin\theta \cdot \cos\theta = 2\left(-\frac{8}{17}\right)\cdot\left(\frac{15}{17}\right) = -\frac{240}{289}.$$

Therefore, $\sin\left[2\,\text{Arctan}\left(-\dfrac{8}{15}\right)\right] = \dfrac{240}{289}$.

EXAMPLE 5: Solve $3\sin^2\theta + 10\sin\theta - 8 = 0$ **for all values of** θ **such that** $0 \le \theta < 2\pi$.

The equation $3\sin^2\theta + 10\sin\theta - 8 = 0$ factors into $(3\sin\theta - 2)(\sin\theta + 4) = 0$, which leads to $3\sin\theta - 2 = 0$ or $\sin\theta + 4 = 0$. Thus, $\sin\theta = \dfrac{2}{3}$ or $\sin\theta = -4$. Since -4 is not in the range of the sine function, the second factor gives no solutions. $\sin\theta = \dfrac{2}{3}$ implies that $\theta = \text{Sin}^{-1}\left(\dfrac{2}{3}\right) \approx 41.81°$. Since $\sin\theta > 0$, it is necessary to find angles in quadrants I and II. In quadrant I, $\theta \approx 41.81°$. The quadrant II angle that has the same reference angles as $41.81°$ is $180° - 41.81° \approx 138.19°$. Therefore, the solution set $= \{41.81°, 138.19°\}$.

EXERCISES

1. Solve for x: $\text{Arccos}(2x^2 - 2x) = \dfrac{2\pi}{3}$.

   (A) $\pm\dfrac{1}{2}$

   (B) $\dfrac{1}{2}$

   (C) $-\dfrac{1}{2}$

   (D) $\dfrac{\pi}{3}$

   (E) 0

2. $\text{Arcsin}\left(\sin\dfrac{7\pi}{6}\right) =$

   (A) $-\dfrac{1}{2}$

   (B) $\dfrac{\pi}{6}$

   (C) $\dfrac{7\pi}{6}$

   (D) $-\dfrac{\pi}{6}$

   (E) $\dfrac{1}{2}$

3. Which of the following is (are) true?
   I. $\text{Arcsin}\,1 + \text{Arcsin}(-1) = 0$
   II. $\text{Cos}^{-1}(1) + \text{Cos}^{-1}(-1) = 0$
   III. $\text{Arccos}\,x = \text{Arccos}(-x)$ for all values of x in the domain of Arccos.

   (A) only I
   (B) only II
   (C) only III
   (D) only I and II
   (E) only II and III

4. Express in terms of an inverse trigonometric function the angle that a diagonal of a cube makes with the base of the cube.

(A) $\text{Arcsin}\dfrac{\sqrt{2}}{2}$

(B) $\text{Arcsin}\dfrac{\sqrt{3}}{3}$

(C) $\text{Arccos}\dfrac{\sqrt{6}}{2}$

(D) $\text{Arccos}\dfrac{\sqrt{3}}{3}$

(E) $\text{Arctan}\sqrt{2}$

5. $\text{Tan}\left[\text{Arcsin}\left(-\dfrac{3}{5}\right)\right]$ equals

(A) $\dfrac{3}{4}$

(B) $-\dfrac{3}{4}$

(C) $\dfrac{4}{3}$

(D) $-\dfrac{4}{3}$

(E) $-\dfrac{4}{5}$

6. When only principal values are used, $\text{Arcsin}\left(-\dfrac{5}{13}\right)+\text{Arccos}\left(-\dfrac{3}{5}\right)$ represents an angle lying in which quadrant?

(A) I
(B) II
(C) III
(D) IV
(E) I or II

7. Which of the following is not defined?

(A) $\text{Arcsin}\dfrac{1}{9}$

(B) $\text{Arccos}\left(-\dfrac{4}{3}\right)$

(C) $\text{Arctan}\dfrac{11}{12}$

(D) $\text{Arccot}(-4)$

(E) $\text{Arcsec}\,3\pi$

8. Which of the following is a solution of $\cos 3x = \dfrac{1}{2}$?

(A) $60°$

(B) $\dfrac{5\pi}{3}$

(C) $\text{Arccos}\dfrac{1}{6}$

(D) $\text{Arccos}\dfrac{\sqrt{3}}{2}$

(E) $\dfrac{1}{3}\text{Arccos}\dfrac{1}{2}$

3.7 TRIANGLES

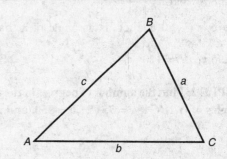

The final topic in trigonometry concerns the relationship between the angles and sides of a triangle that is *not* a right triangle. Depending on which of the sides and angles of the triangle are supplied, the following formulas are helpful. In $\triangle ABC$

Law of sines: $\dfrac{\sin A}{a}=\dfrac{\sin B}{b}=\dfrac{\sin C}{c}$

used when the lengths of two sides and the value of the angle opposite one, or two angles and the length of one side are given.

Law of cosines: $a^2 = b^2 + c^2 - 2bc \cdot \cos A$

$$b^2 = a^2 + c^2 - 2ac \cdot \cos B$$

$$c^2 = a^2 + b^2 - 2ab \cdot \cos C$$

used when the lengths of two sides and the included angle, or the lengths of three sides, are given.

Area of a $\triangle$: $\text{Area} = \dfrac{1}{2}bc \cdot \sin A$

$$\text{Area} = \dfrac{1}{2}ac \cdot \sin B$$

$$\text{Area} = \dfrac{1}{2}ab \cdot \sin C$$

used when two sides and the included angle are given.

EXAMPLE 1: Find the number of degrees in the largest angle of a triangle whose sides are 3, 5, and 7.

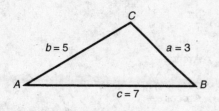

The largest angle is opposite the longest side. Use the law of cosines:

$$c^2 = a^2 + b^2 - 2ab \cdot \cos C$$
$$49 = 9 + 25 - 30 \cdot \cos C$$

Therefore, $\cos C = -\dfrac{15}{30} = -\dfrac{1}{2}$.

Since $\cos C < 0$ and $\angle C$ is an angle of a triangle, $90° < \angle C < 180°$.

Therefore, $\angle C = 120°$.

EXAMPLE 2: Find the number of degrees in the other two angles of $\triangle ABC$ if $c = 75\sqrt{2}$, $b = 150$, and $\angle C = 30°$.

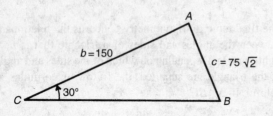

Use the law of sines:

$$\frac{75\sqrt{2}}{\sin 30°} = \frac{150}{\sin B}$$
$$75\sqrt{2} \cdot \sin B = 150 \cdot \sin 30°$$

$$\sin B = \frac{150 \cdot \dfrac{1}{2}}{75\sqrt{2}} = \frac{75}{75\sqrt{2}} = \frac{1}{\sqrt{2}} = \frac{\sqrt{2}}{2}$$

Therefore, $\angle B = 45°$ or $135°$; $\angle A = 105°$ or $15°$ since there are $180°$ in the sum of the three angles of a triangle.

EXAMPLE 3: Find the area of $\triangle ABC$ if $a = 180$ inches, $b = 150$ inches, and $\angle C = 30°$.

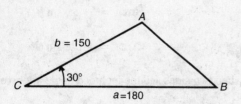

$$\text{Area} = \frac{1}{2}ab \cdot \sin C$$
$$= \frac{1}{2} \cdot 180 \cdot 150 \cdot \sin 30°$$
$$= \frac{1}{2} \cdot 180 \cdot 150 \cdot \frac{1}{2}$$
$$= 90 \cdot 75 = 675 \text{ square inches.}$$

If the length of two sides of a triangle and the angle opposite one of those sides are given, it is possible that two triangles, one triangle, or no triangle can be constructed with the data. This is called the *ambiguous* case. If the length of sides a and b and the value of $\angle A$ are given, the length of side b determines the number of triangles that can be constructed.

Case 1: If $\angle A > 90°$ and $a \leq b$, no triangle can be formed because side a would not reach the base line. If $a > b$, one obtuse triangle can be drawn.

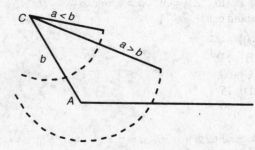

Let the length of the altitude from C to the base line be h. From the basic definition of sine, $\sin A = \dfrac{h}{b}$, and thus, $h = b \cdot \sin A$.

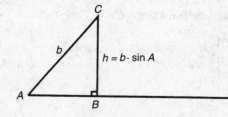

Case 2: If $\angle A < 90°$ and side $a < b \cdot \sin A$, no triangle can be formed. If $a = b \cdot \sin A$, one triangle can be formed. If $a > b$, there also will be only one triangle. If, on the other hand, $b \cdot \sin A < a < b$, two triangles can be formed.

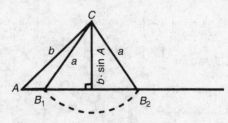

If a compass is opened the length of side a and a circle is drawn with center at C, the circle will cut the baseline at two points, B_1 and B_2. Thus, $\triangle AB_1C$ satisfies the conditions of the problem, as does $\triangle AB_2C$.

EXAMPLE 1: How many triangles can be formed if $a = 24$, $b = 31$, and $\angle A = 30°$?

Because $\angle A < 90°$, $b \cdot \sin A = 31 \cdot \sin 30° = 31 \cdot \dfrac{1}{2} = 15\dfrac{1}{2}$. Since $b \cdot \sin A < a < b$, there are two triangles.

EXAMPLE 2: How many triangles can be formed if $a = 24$, $b = 32$, and $\angle A = 150°$?

Since $\angle A > 90°$ and $a < b$, no triangle can be formed.

EXERCISES

1. In $\triangle ABC$, $\angle A = 30°$, $b = 8$, and $a = 4\sqrt{2}$. Angle C could equal

    (A) 45°
    (B) 135°
    (C) 60°
    (D) 15°
    (E) 90°

2. In $\triangle ABC$, $\angle A = 30°$, $a = 6$, and $c = 8$. Which of the following must be true?

    (A) $0° < \angle C < 90°$
    (B) $90° < \angle C < 180°$
    (C) $45° < \angle C < 135°$
    (D) $0° < \angle C < 45°$ or $90° < \angle C < 135°$
    (E) $0° < \angle C < 45°$ or $130° < \angle C < 180°$

3. The angles of a triangle are in a ratio of $8 : 3 : 1$. The ratio of the longest side of the triangle to the next longest side is

    (A) $\sqrt{6} : 2$
    (B) $8 : 3$
    (C) $\sqrt{3} : 1$
    (D) $8 : 5$
    (E) $2\sqrt{2} : \sqrt{3}$

4. The sides of a triangle are in a ratio of $4 : 5 : 6$. The smallest angle is

    (A) 82°
    (B) 69°
    (C) 56°
    (D) 41°
    (E) 27°

5. Find the length of the longer diagonal of a parallelogram if the sides are 6 inches and 8 inches and the smaller angle is 60°.

    (A) 8
    (B) 11
    (C) 12
    (D) 7
    (E) 17

6. What are all values of side a in the figure below such that two triangles can be constructed?

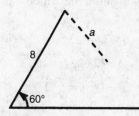

    (A) $a > 4\sqrt{3}$
    (B) $a > 8$
    (C) $a = 4\sqrt{3}$
    (D) $4\sqrt{3} < a < 8$
    (E) $8 < a < 8\sqrt{3}$

7. In $\triangle ABC$, $\angle B = 30°$, $\angle C = 105°$, and $b = 10$. The length of side a equals

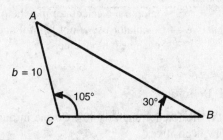

    (A) 7
    (B) 17
    (C) 9
    (D) 10
    (E) 14

8. The area of $\triangle ABC$, $= 24\sqrt{3}$, side $a = 6$, and side $b = 16$. The value of $\angle C$ is

    (A) 30°
    (B) 30° or 150°
    (C) 60°
    (D) 60° or 120°
    (E) none of the above

9. The area of $\triangle ABC = 12\sqrt{3}$, side $a = 6$, and side $b = 8$. Side $c =$

(A) $2\sqrt{37}$

(B) $2\sqrt{13}$

(C) $2\sqrt{37}$ or $2\sqrt{13}$

(D) 10

(E) 10 or 12

10. Given the following data, which can form two triangles?

I. $\angle C = 30°$, $c = 8$, $b = 12$

II. $\angle B = 45°$, $a = 12\sqrt{2}$, $b = 15\sqrt{2}$

III. $\angle C = 60°$, $b = 12$, $c = 5\sqrt{3}$

(A) only I

(B) only II

(C) only III

(D) only I and II

(E) only I and III

ANSWERS AND EXPLANATIONS

In these solutions the following notation is used:

a: active — Calculator use is necessary or, at a minimum, extremely helpful.

n: neutral — Answers may be found without a calculator, but a calculator may help.

i: inactive — Calculator use is not helpful and may even be a hindrance.

Part 3.1 Definitions

1. n A Reference angle is 40°. Cosine in quadrant IV is positive.

2. i B See corresponding figure. Therefore, $\sin \theta = \dfrac{12}{13}$.

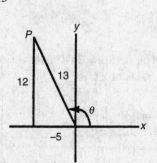

3. i D Angle θ is in quadrant II since sec < 0 and sin > 0. Therefore, $\tan\theta = \dfrac{3}{-4} = \dfrac{-3}{4}$.

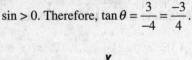
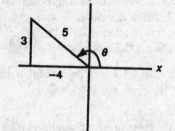

4. i B Cofunctions of complementary angles are equal. $x - 30 + x = 90$ finds a reference angle of 60° for x. The angle in quadrant III that has a reference angle of 60° is 240°.

5. i A Angle α is in quadrant II, and $\sin \alpha$ is positive. Angle β is in quadrant IV, and $\sin \beta$ is negative.

6. a E Put your calculator in degree mode, $\cos 310° + \cos 190° \approx 0.643 + (-0.985) \approx -0.342$. Checking the answer choices shows that $-\cos 70° \approx -0.342$.

Part 3.2 Arcs and Angles

1. n C $\dfrac{30}{\pi} = \dfrac{x}{180°}$. Therefore, $x = \dfrac{5400°}{\pi}$.

2. i D $s = r\theta$. $2\pi = r\theta$. $A = \dfrac{1}{2}r^2\theta$.

$6\pi = \dfrac{1}{2}r^2\theta = \dfrac{1}{2}r(r\theta) = \dfrac{1}{2}r(2\pi)$. $r = 6$.

3. a D $C = 2\pi r = 16$. $r = \dfrac{8}{\pi} \approx 2.55$.

$A = \dfrac{1}{2}r^2\theta \approx \dfrac{1}{2}(2.55)^2 \cdot 4.7 \approx 15$.

4. a C $40° = \dfrac{2\pi^R}{9} \approx 0.7$ radians.

$s = r\theta \approx 1 \cdot 0.7 \approx 0.7$.

5. a B Change 25° to 0.436 radian $\left(0.436 = \dfrac{25}{180}\pi \right)$.

$s = r\theta$, and so $12 = r(0.436)$ and $r = 27.5$ inches

Part 3.3 Special Angles

1. n   C $-60°$ is in quadrant IV, where tan < 0.

2. n   E $-135°$ is in quadrant III with reference angle 45°. Tan > 0 is in quadrant III. Therefore $\tan(-135°) = 1$. Reference angle for 315° is 45°. Cot < 0 is in quadrant IV. Therefore, cot 315° = -1. $1 + (-1) = 0$.

3. n   B   The terms equal, in order, $-1, -\dfrac{1}{2}, -1, 1$.

$$-1 - \left(-\dfrac{1}{2}\right) - (-1) + 1 = \dfrac{3}{2}.$$

4. n   B   The terms equal, in order, $\dfrac{2\sqrt{3}}{3}, \dfrac{-\sqrt{3}}{1}, \dfrac{-\sqrt{2}}{2}$.

5. n   D $\sin 300° = \dfrac{-\sqrt{3}}{2}$. Choices A and B are ruled out because they are positive. Choices C and E are ruled out because they are cosines with the same reference angle (60°) as sin 300°.

Part 3.4 Graphs

1. n   C   Period = $\dfrac{2\pi}{2} = \pi$. Point P is $\dfrac{1}{4}$ of the way through the period. Amplitude is 1 because the coefficient of sin is 1. Therefore, point P is at $\left(\dfrac{1}{4}\pi, 1\right)$.

2. n   E   Amplitude = $\dfrac{1}{2}$. Period = π. Graph shifted $\dfrac{1}{2}$ unit up. Graph looks like a cosine graph reflected about x-axis and shifted up $\dfrac{1}{2}$ unit.

3. n   C   Graph has amplitude of 4 and period of 4π.

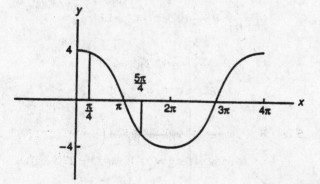

4. n   C   Multiply and divide by 2.

$$f(x) = 2\left(\dfrac{\sqrt{3}}{2}\cos x + \dfrac{1}{2}\sin x\right)$$
$$= 2(\sin 60° \cos x + \cos 60° \sin x)$$
$$= 2\sin(60° + x).$$

Therefore, the amplitude is 2 because the coefficient is 2.

5. i   C   Period = $\dfrac{2\pi}{P} = \dfrac{2\pi}{3}$.

6. n   D   Period = 6π. Curve reaches its maximum when $x = \dfrac{6\pi}{4}$, which is not in the allowable set of x values. Therefore, maximum is reached when $x = \dfrac{\pi}{2}$. Maximum = $\sin\dfrac{1}{3} \cdot \dfrac{\pi}{2} = \dfrac{1}{2}$.

7. i   D   Period = $\dfrac{2\pi}{M} = 4$ (from the figure). $M = \dfrac{1}{2}$. Phase shift for a sine curve in the figure is $-\pi$. Therefore, $\dfrac{1}{2}x + N = 0$ when $x = -\pi$. Therefore, $N = \dfrac{\pi}{2}$.

Part 3.5 Identities, Equations, and Inequalities

1. n   D $\cos 2x = 2\cos^2 x - 1$. The equation becomes
$$2\cos^2 x + \cos x = 0.$$
$$(2\cos x + 1)\cos x = 0.$$

2. n   D $2\sin\theta \cdot \cos\theta = \dfrac{1}{2}$ becomes $\sin 2\theta = \dfrac{1}{2}$. $2\theta = 30°, 150°, 390°, 510°$. $\theta = 15°, 75°, 195°, 225°$.

3. n   E $4x = 60°, 240°, 420°, 600°, 780°, 960°, 1140°, 1320°$. $x = 15°, 60°, 105°, 150°, 195°, 240°, 285°, 330°$.

4. n   E $\tan 2\theta = \dfrac{2\tan\theta}{1 - \tan^2\theta} = \dfrac{2\left(\dfrac{4}{3}\right)}{1 - \left(\dfrac{16}{9}\right)}$. Since θ can be in quadrant I or IV, $\tan 2\theta = \pm\dfrac{24}{7}$.

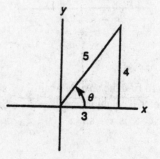

5. n A $1 - 2\sin^2\theta = -\sin\theta$. $2\sin^2\theta - \sin\theta - 1 = 0$. $(2\sin\theta + 1)(\sin\theta - 1) = 0$. $\sin\theta = -\dfrac{1}{2}$ or $\sin\theta = 1$.

6. i A Since $\sin 74° = \sin 2(37°)$, Formula 10 can be used: $\sin 2A = 2\sin A \cdot \cos A$. $\sin 74° = 2\sin 37° \cdot \cos 37°$. From the figure, $\cos 37° = \sqrt{1 - z^2}$. Therefore, $\sin 74° = 2z\sqrt{1 - z^2}$.

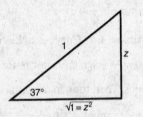

7. i B $\text{Cot}(A + B) = \dfrac{1}{\tan(A + B)}$

$= \dfrac{1 - \tan A \cdot \tan B}{\tan A + \tan B}$

$= \dfrac{1 - \dfrac{1}{\cot A \cdot \cot B}}{\dfrac{1}{\cot A} + \dfrac{1}{\cot B}}$

8. n B This is part of the formula for $\tan(A + B) =$ $\tan(140° + 70°) = \tan 210° = \dfrac{\sqrt{3}}{3}$.

9. n A $\cos^4 40° - \sin^4 40°$
$= (\cos^2 40° + \sin^2 40°)(\cos^2 40° - \sin^2 40°)$
$= 1[\cos 2(40°)] = \cos 80°$.

10. n B A is in quadrant II.

$\tan 2A = \dfrac{2\tan A}{1 - \tan^2 A} = \dfrac{2\left(-\dfrac{2}{\sqrt{5}}\right)}{1 - \dfrac{4}{5}} = -4\sqrt{5}.$

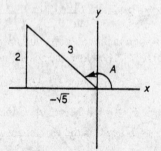

11. n D To find the answer to $\sin 2x < \sin x$, solve the associated equation, $\sin 2x = \sin x$. $\sin 2x - \sin x = 2\sin x \cdot \cos x - \sin x = \sin x (2\cos x - 1) = 0$.

$\sin x = 0$ or $\cos x = \dfrac{1}{2}$. So $x = 0, \dfrac{\pi}{3}, \pi, \dfrac{5\pi}{3}, 2\pi$. The only regions where inequality is satisfied are $\dfrac{\pi}{3} < x < \pi$ or $\dfrac{5\pi}{3} < x < 2\pi$.

Part 3.6 Inverse Functions

1. i B $2x^2 - 2x = \cos\dfrac{2\pi}{3} = -\dfrac{1}{2}$. $4x^2 - 4x + 1 = 0$. $(2x - 1)^2 = 0$. Therefore, $x = \dfrac{1}{2}$.

2. n D $\sin\dfrac{7\pi}{6} = -\dfrac{1}{2}$. $\text{Arcsin}\left(-\dfrac{1}{2}\right)$ is an angle in quadrant IV.

Note: The answer is not the "obvious" choice, $\dfrac{7\pi}{6}$, because $-\dfrac{\pi}{2} \le \text{Sin}^{-1}x \le \dfrac{\pi}{2}$.

3. n A $\text{Arcsin } 1 = \dfrac{\pi}{2}$ and $\text{Arcsin}(-1) = -\dfrac{\pi}{2}$; I is true. $\text{Arccos } 1 = 0$ and $\text{Arccos}(-1) = \pi$; II is false. Since Arccos is defined in quadrants I and II, III is false.

4. i B If the side of a cube is 1, its diagonal $= \sqrt{3}$ and the diagonal of the base is $\sqrt{2}$. Therefore, $\theta = \text{Arcsin}\dfrac{\sqrt{3}}{3}$.

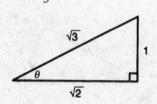

5. n B Let $\theta = \text{Arcsin}\left(-\dfrac{3}{5}\right)$. $\sin\theta = -\dfrac{3}{5}$. Therefore, $\tan\theta = -\dfrac{3}{4}$.

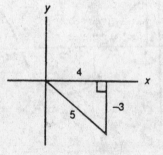

6. n B Let $A = \text{Arcsin}\left(-\dfrac{5}{13}\right)$ and $B = \text{Arccos}\left(-\dfrac{3}{5}\right)$. Since $-90° \le A \le 0°$ and $90° \le B \le 180°$, $0° \le A + B \le 180°$. Since cosine has different signs in quadrants I and II, evaluate:

$$\cos(A + B) = \frac{12}{13} \cdot \frac{-3}{5} - \frac{-5}{13} \cdot \frac{4}{5} = \frac{-16}{65} < 0.$$

Therefore, $A + B$ is in quadrant II.

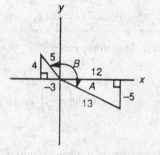

7. n B $-1 \le \cos\theta \le 1$. Thus, Choice B is not defined.

8. i E $3x = \text{Arccos}\left(\dfrac{1}{2}\right)$, and so $x = \dfrac{1}{3}\text{Arccos}\left(\dfrac{1}{2}\right)$.

Part 3.7 Triangles

1. i D Law of sines: $\dfrac{\sin B}{8} = \dfrac{\frac{1}{2}}{4\sqrt{2}}$. $\sin B = \dfrac{\sqrt{2}}{2}$. $B = 45°$ or $135°$. Therefore, $C = 105°$ or $15°$.

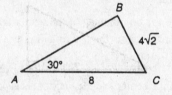

2. n E Law of sines: $\dfrac{\sin C}{8} = \dfrac{\frac{1}{2}}{6}$. $\sin C = \dfrac{2}{3} \approx 0.67$.

$\sin 45° = \dfrac{\sqrt{2}}{2} \approx 0.7$. Since sine is an increasing function in quadrant I, $0° \le C \le 45°$. Angles in quadrant II greater than $135°$ use these values of C as reference angles.

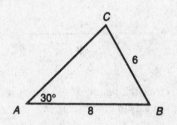

3. n A The angles are $15°$, $45°$, and $120°$. Let c be the longest side and b the next longest. $\dfrac{\sin 120°}{c} = \dfrac{\sin 45°}{b}$.

$$\frac{c}{b} = \frac{\sin 120°}{\sin 45°} = \frac{\frac{\sqrt{3}}{2}}{\frac{\sqrt{2}}{2}} = \frac{\sqrt{6}}{2}.$$

4. a D Use the law of cosines. Let the sides be 4, 5, and 6. $16 = 25 + 36 - 60\cos A$. $\cos A = \dfrac{45}{60} = \dfrac{3}{4}$, which implies that $A = \text{Cos}^{-1}(0.75) \approx 41°$.

5. a C Law of cosines: $d^2 = 36 + 64 - 96\cos 120°$. $d^2 = 148$. Therefore, $d \approx 12$.

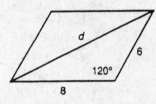

6. i D Altitude to base $= 8\sin 60° = 4\sqrt{3}$. Therefore, $4\sqrt{3} < a < 8$.

7. a E $A = 45°$. Law of sines: $\dfrac{\sin 45°}{a} = \dfrac{\sin 30°}{10}$. Therefore, $a = 10\sqrt{2} \approx 14$.

8. n D Area $= \dfrac{1}{2}ab\sin C$. $24\sqrt{3} = \dfrac{1}{2} \cdot 6 \cdot 16\sin C$. $\sin C = \dfrac{\sqrt{3}}{2}$. Therefore, $C = 60°$ or $120°$.

9. n C Area $= \dfrac{1}{2}ab\sin C$. $12\sqrt{3} = \dfrac{1}{2} \cdot 6 \cdot 8\sin C$. $\sin C = \dfrac{\sqrt{3}}{2}$. $C = 60°$ or $120°$. Use law of cosines with $60°$ and then with $120°$.

Note: At this point in the solution you know there have to be two values for C. Therefore, the answer must be Choice C or E. If $C = 10$ (from Choice E), ABC is a right triangle with area $= \dfrac{1}{2} \cdot 6 \cdot 8 = 24$. Therefore, Choice E is not the answer, and so Choice C is the correct answer.

10. n A In I the altitude $= 12 \cdot \dfrac{1}{2} = 6$, $6 < c < 12$, and so 2 triangles. In II $b > 12\sqrt{2}$, so only 1 triangle. In III the altitude $= 12 \cdot \dfrac{\sqrt{3}}{2} > 5\sqrt{3}$, so no triangle.

MISCELLANEOUS RELATIONS AND FUNCTIONS

CHAPTER

4

4.1 CONIC SECTIONS

The general quadratic equation in two variables has the form $Ax^2 + Bxy + Cy^2 + Dx + Ey + F = 0$, where A, B, and C are not all zero. Depending on the values of the coefficients A, B, and C, the equation represents a circle, a parabola, a hyperbola, an ellipse, or a degenerate case of one of these (e.g., a point, a line, two parallel lines, two intersecting lines, or no graph at all).

If the graph is not a degenerate case, the following indicates which conic section the equation represents:

If $B^2 - 4AC < 0$ and $A = C$, the graph is a circle.
If $B^2 - 4AC < 0$ and $A \neq C$, the graph is an ellipse.
If $B^2 - 4AC = 0$, the graph is a parabola.
If $B^2 - 4AC > 0$, the graph is a hyperbola.

Most of the conic sections encountered on the Level II examination will not have an xy-term (i.e., $B = 0$). This term causes the graph to be rotated so that a major axis of symmetry is not parallel to either the x- or the y-axis. The general quadratic equation in two variables can be changed into a more useful form by completing the square separately on the x's and separately on the y's. After the square has been completed and, for convenience, the letters used as constants have been changed, the equation becomes:

(a) for a circle, $(x - h)^2 + (y - k)^2 = r^2$, where (h,k) are the coordinates of the center and r is the radius of the circle.

(b) for an ellipse, if $C > A$, $\dfrac{(x-h)^2}{a^2} + \dfrac{(y-k)^2}{b^2} = 1$, where (h,k) are the coordinates of the center and the major axis is parallel to the x-axis (Figure a), If $C < A$, $\dfrac{(x-h)^2}{b^2} + \dfrac{(y-k)^2}{a^2} = 1$, where (h,k) are the coordinates of the center and the major axis is parallel to the y-axis (Figure b).

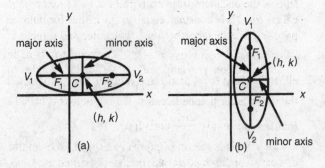

In both cases the length of the *major axis* is $2a$ and the length of the *minor axis* is $2b$.

(c) for a parabola, if $C = 0$, $(x - h)^2 = 4p(y - k)$ (Figure a). If $A = 0$, $(y - k)^2 = 4p(x - h)$ (Figure b). In both cases (h,k) are the coordinates of the vertex and p is the directed distance from the vertex to the focus.

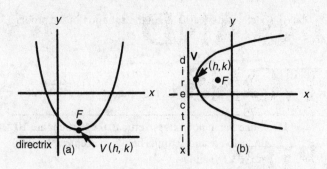

(d) for a hyperbola, $\dfrac{(x - h)^2}{a^2} - \dfrac{(y - k)^2}{b^2} = 1$, where (h,k) are the coordinates of the center, and the graph opens to the side (Figure a). If the graph opens up and down, the equation is $\dfrac{(y - k)^2}{a^2} - \dfrac{(y - h)^2}{b^2} = 1$, where (h,k) are the coordinates of the center (Figure b).

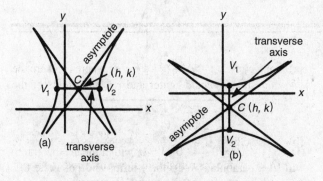

In both cases the length of the *transverse axis* is $2a$, and the length of the *conjugate axis* is $2b$.

An *ellipse* is the set of points in a plane such that the sum of the distances from each point to two fixed points, called *foci,* is a constant equal to $2a$ in the formula on page 68. The distance between the center and a focus is called c and is equal to $\sqrt{a^2 - b^2}$. The *eccentricity* of an ellipse is always less than 1 and is equal to $\dfrac{c}{a}$. The chord through a focus perpendicular to the major axis is called a *latus rectum* and is $\dfrac{2b^2}{a}$ units long.

A *parabola* is the set of points in a plane such that the distance from each point to a fixed point, called the *focus,* is equal to the distance to a fixed line, called the *directrix.* The *eccentricity* is always equal to 1. The chord through the focus perpendicular to the axis of symmetry is called the *latus rectum* and is $4p$ units long.

A *hyperbola* is the set of points in a plane such that the absolute value of the difference of the distances from each point to two fixed points, called *foci*, is a constant equal to $2a$ in the formula previously stated. The distance between the center and a focus is called c and is equal to $\sqrt{a^2 + b^2}$. The *eccentricity* of a hyperbola is always greater than 1 and is equal to $\dfrac{c}{a}$. Every hyperbola has associated with it two lines, called *asymptotes,* that intersect at the center. In general, an asymptote is a line that a curve approaches, but never quite touches, as one or both variables become increasingly larger or smaller. If the hyperbola opens to the side, the slopes of the two asymptotes are $\pm\dfrac{b}{a}$. If the hyperbola opens up and down, the slopes of the two asymptotes are $\pm\dfrac{a}{b}$. The chord through a focus perpendicular to the transverse axis is called a *latus rectum* and is equal to $\dfrac{2b^2}{a}$ units.

The equation $xy = k$, where k is a constant, is the equation of a *rectangular hyperbola* whose asymptotes are the x- and y-axes. If $k > 0$, the branches of the hyperbola lie in quadrants I and III. If $k < 0$, the branches lie in quadrants II and IV.

EXAMPLE 1: Each of the following is an equation of a conic section. State which one and find, if they exist: (I) the coordinates of the center, (II) the coordinates of the vertices, (III) the coordinates of the foci, (IV) the eccentricity, (V) the equations of the asymptotes. Also, sketch the graph.

(A) $9x^2 - 16y^2 - 18x + 96y + 9 = 0$
(B) $4x^2 + 4y^2 - 12x - 20y - 2 = 0$
(C) $4x^2 + y^2 + 24x - 16y = 0$
(D) $y^2 + 6x - 8y + 4 = 0$

In all cases $B = 0$.

(A) $B^2 - 4AC = 0 - 4 \cdot 9 \cdot (-16) > 0$, which means the graph will be a hyperbola. To complete the square, group the x-terms and the y-terms. Factor out of each group the coefficient of the quadratic term:

$$9(x^2 - 2x \quad\;) - 16(y^2 - 6y \quad\;) + 9 = 0$$

A perfect trinomial square is obtained within both parentheses by taking one-half of the coefficient of the linear term, squaring it, and adding it to the existing polynomial. A similar amount must be added to the right side of the equation.

$$9(x^2 - 2x + 1) - 16(y^2 - 6y + 9) + 9 = 9 \cdot 1 - 16 \cdot 9$$
$$9(x - 1)^2 - 16(y - 3)^2 = -144$$

Divide each term by -144 to get the equation in the proper form: $\dfrac{(y-3)^2}{9} - \dfrac{(x-1)^2}{16} = 1.$

  I.   Center $(1,3)$. $a = 3$, $b = 4$, and so $c = \sqrt{9+16} = 5$.

  II.   The vertices are 3 units above and 3 units below the center. Vertices are $(1,6)$ and $(1,0)$.

  III.   The foci are 5 units above and 5 units below the center. Foci are $(1,8)$ and $(1,-2)$.

  IV.   Eccentricity $= \dfrac{c}{a} = \dfrac{5}{3}$.

  V.   The slopes of the asymptotes are $\pm\dfrac{a}{b} = \pm\dfrac{3}{4}$. The asymptotes pass through the center $(1,3)$. Therefore, the equations of the asymptotes are $y - 3 = \pm\dfrac{3}{4}(x-1)$.

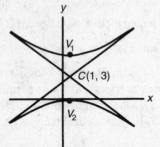

(B) $B^2 - 4AC = 0 - 4 \cdot 4 \cdot 4 < 0$ and $A = C = 4$, which means the graph will be a circle. Divide through by 4, group the x-terms together, group the y-terms together, and complete the square.

$$\left(x^2 - 3x + \frac{9}{4}\right) + \left(y^2 - 5y + \frac{25}{4}\right) - \frac{2}{4} = \frac{9}{4} + \frac{25}{4}$$

$$\left(x - \frac{3}{2}\right)^2 + \left(y - \frac{5}{2}\right)^2 = 9$$

  I.   Center $\left(\dfrac{3}{2}, \dfrac{5}{2}\right)$. Radius $= 3$. None of the other items is defined.

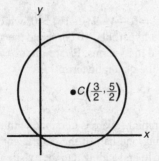

(C) $B^2 - 4AC = 0 - 4 \cdot 4 \cdot 1 < 0$ and $A \neq C$, which means the graph will be an ellipse. Group the x-terms, group the y-terms, and factor out of each group the coefficient of the quadratic term:

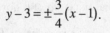

$$4(x^2 + 6x \quad) + 1(y^2 - 16y \quad) = 0$$

Complete the square and add similar amounts to both sides of the equation:

$$4(x^2 + 6x + 9) + 1(y^2 - 16y + 64) = 4 \cdot 9 + 1 \cdot 64$$
$$4(x+3)^2 + (y-8)^2 = 100$$

Divide each term by 100 to get the equation in the proper form:

$$\frac{(x+3)^2}{25} + \frac{(y-8)^2}{100} = 1$$

  I.   Center $(-3,8)$. $a = 10$, $b = 5$, and so $c = \sqrt{100-25} = \sqrt{75} = 5\sqrt{3}$.

  II.   Since the major axis is vertical, the vertices are 10 units above and 10 units below the center. Vertices are $(-3,18)$ and $(-3,-2)$.

  III.   The foci are $5\sqrt{3}$ units above and $5\sqrt{3}$ units below the center. Foci are $(-3, 8 + 5\sqrt{3})$ and $(-3, 8 - 5\sqrt{3})$.

  IV.   Eccentricity $= \dfrac{c}{a} = \dfrac{5\sqrt{3}}{10} = \dfrac{\sqrt{3}}{2}$.

  V.   There are no asymptotes.

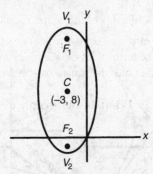

(D) $B^2 - 4AC = 0 - 4 \cdot 0 \cdot 1 = 0$, which means the graph will be a parabola. Group the y-terms and complete the square.

$$(y^2 - 8y + 16) = -6x - 4 + 16$$
$$(y - 4)^2 = -6(x - 2)$$

  I.   No center.

  II.   Vertex is $(2,4)$.

  III. $4p = -6$, and so $p = -\dfrac{3}{2}$. Since the y-term is squared, the parabola opens to the side. Thus, the focus is $\dfrac{3}{2}$ units to the left of the vertex. Focus $\left(\dfrac{1}{2}, 4\right)$.

  IV.   Eccentricity $= 1$.

  V.   No asymptotes.

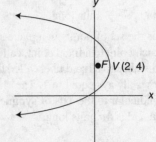

EXAMPLE 2: Find the equation of the hyperbola with center at (3,–4), eccentricity 4, and conjugate axis of length 6.

Two cases are possible: $\dfrac{(x-3)^2}{a^2} - \dfrac{(y+4)^2}{b^2} = 1$ and $\dfrac{(y+4)^2}{a^2} - \dfrac{(x-3)^2}{b^2} = 1$.

Eccentricity $= \dfrac{c}{a} = 4$. Thus, $c = 4a$.

Conjugate axis $= 2b = 6$. Thus, $b = 3$.

In the case of a hyperbola,

$$c^2 = a^2 + b^2$$
$$(4a)^2 = a^2 + 9$$
$$16a^2 = a^2 + 9$$
$$15a^2 = 9$$
$$a^2 = \frac{9}{15}$$

Therefore, the equation becomes:

$$\frac{15(x-3)^2}{9} - \frac{(y+4)^2}{9} = 1 \text{ or } \frac{15(y+4)^2}{9} - \frac{(x-3)^2}{9} = 1.$$

The *degenerate case* occurs when one of the variables drops out, or the terms with the variables equal zero, or a negative number after the square has been completed.

EXAMPLE 3: What is the graph of each of the following?

(A) $x^2 + y^2 - 4x + 2y + 5 = 0$
(B) $xy = 0$
(C) $2x^2 - 3y^2 + 8x + 6y + 5 = 0$
(D) $3x^2 + 4y^2 - 6x - 16y + 19 = 0$
(E) $x^2 + y^2 + 5 = 0$

(A) $B^2 - 4AC = 0 - 4 \cdot 1 \cdot 1 < 0$ and $A = C$, which indicates the graph should be a circle. Completing the square on the x's and y's separately gives:

$$(x^2 - 4x + 4) + (y^2 + 2y + 1) + 5 = 0 + 4 + 1$$
$$(x - 2)^2 + (y + 1)^2 = 0$$

This is a circle with center (2,–1) and radius zero; thus, a degenerate case. The graph is the single point (2,–1).

(B) $xy = 0$ if and only if $x = 0$ or $y = 0$. Thus, the graph is two intersecting lines: the x-axis ($y = 0$) and the y-axis ($x = 0$).

(C) $B^2 - 4AC = 0 - 4 \cdot 2 \cdot (-3) > 0$, which indicates the graph should be a hyperbola. Factoring and completing the square on the x's and y's separately gives:

$$2(x^2 + 4x + 4) - 3(y^2 - 2y + 1) + 5 = 0 + 2 \cdot 4 - 3 \cdot 1$$
$$2(x + 2)^2 - 3(y - 1)^2 = 0$$

Since the right side of the equation is zero and the left side is the difference between two perfect squares, the left side of the equation can be factored and simplified.

$$\left[\sqrt{2}(x+2) + \sqrt{3}(y-1)\right] \cdot \left[\sqrt{2}(x+2) - \sqrt{3}(y-1)\right] = 0$$
$$\left[\sqrt{2}x + 2\sqrt{2} + \sqrt{3}y - \sqrt{3}\right] = 0$$

or

$$\left[\sqrt{2}x + 2\sqrt{2} - \sqrt{3}y + \sqrt{3}\right] = 0$$
$$y = -\frac{\sqrt{2}}{\sqrt{3}}x + \frac{\sqrt{3} - 2\sqrt{2}}{\sqrt{3}} \approx -0.82x - 0.63$$

or

$$y = \frac{\sqrt{2}}{\sqrt{3}}x + \frac{\sqrt{3} + 2\sqrt{2}}{\sqrt{3}} \approx 0.82x + 2.63$$

Thus, the graph is two lines intersecting at (–2,1), one with slope $-\dfrac{\sqrt{2}}{\sqrt{3}} \approx -0.82$ and y-intercept $\dfrac{\sqrt{3} - 2\sqrt{2}}{\sqrt{3}} \approx$ –0.63 and the other with slope $\dfrac{\sqrt{2}}{\sqrt{3}} \approx 0.82$ and y-intercept $\dfrac{\sqrt{3} + 2\sqrt{2}}{\sqrt{3}} \approx 2.63$.

(D) $B^2 - 4AC = 0 - 4 \cdot 3 \cdot 4 < 0$ and $A \neq C$, which indicates the graph should be an ellipse. Factoring and completing the square on the x's and y's separately gives:

$$3(x^2 - 2x + 1) + 4(y^2 - 4y + 4) + 19 = 0 + 3 \cdot 1 + 4 \cdot 4$$
$$3(x - 1)^2 + 4(y - 2)^2 = 0$$

Since $3(x - 1)^2 \geq 0$ and $4(y - 2)^2 \geq 0$, the only point that satisfies the equation is (1,2), which makes each term on the left side of the equation equal to zero. The graph is the one point (1,2).

(E) $x^2 + y^2 = -5$. The graph does not exist since $x^2 \geq 0$ and $y^2 \geq 0$ and there is no way that their sum can equal –5.

EXERCISES

1.   Which of the following are the coordinates of a focus of $5x^2 + 4y^2 - 20x + 8y + 4 = 0$?

(A) (1, –1)
(B) (2, –1)
(C) (3, –1)
(D) (2, –2)
(E) (–2, 1)

2. If the graphs of $x^2 + y^2 = 4$ and $xy = 1$ are drawn on the same set of axes, how many points will they have in common?

(A) 0
(B) 2
(C) 4
(D) 1
(E) 3

3. The graph of $(x - 2)^2 = 4y$ has a

(A) vertex at (4,2)
(B) focus at (2,0)
(C) directrix $y = -1$
(D) latus rectum 2 units in length
(E) none of these

4. Which of the following is an asymptote of $3x^2 - 4y^2 - 12 = 0$?

(A) $y = \dfrac{4}{3}x$

(B) $y = -\dfrac{2}{\sqrt{3}}$

(C) $y = -\dfrac{3}{4}x$

(D) $y = \dfrac{\sqrt{3}}{2}x$

(E) $y = \dfrac{2\sqrt{3}}{3}x$

5. The graph of $x^2 = (2y + 3)^2$ is

(A) a circle
(B) an ellipse
(C) a hyperbola
(D) a point
(E) two intersecting lines

6. The area bounded by the curve $y = \sqrt{4 - x^2}$ and the x-axis is

(A) 4π
(B) 8π
(C) 16π
(D) 2π
(E) π

7. Which of the following is the equation of the circle with center at the origin and tangent to the line with equation $3x - 7y = 29$?

(A) $x^2 + y^2 = 12$
(B) $2x^2 + 2y^2 = 29$
(C) $x^2 + y^2 = 15$
(D) $3x^2 + 3y^2 = 40$
(E) $x^2 + y^2 = 10$

8. An equilateral triangle is inscribed in the circle whose equation is $x^2 + 2x + y^2 - 4y = 0$. The length of the side of the triangle is

(A) 5
(B) 1.9
(C) 2.2
(D) 3.9
(E) 4.5

4.2 EXPONENTIAL AND LOGARITHMIC FUNCTIONS

The basic properties of exponents and logarithms and the fact that the exponential function and the logarithmic function are inverses lead to many interesting problems.

The basic exponential properties:
For all positive real numbers x and y, and all real numbers a and b:

$$x^a \cdot x^b = x^{a+b} \qquad x^0 = 1$$

$$\frac{x^a}{x^b} = x^{a-b} \qquad x^{-a} = \frac{1}{x^a}$$

$$\left(x^a\right)^b = x^{ab} \qquad x^a \cdot y^a = (xy)^a$$

The basic logarithmic properties:
For all positive real numbers a, b, p, and q, and all real numbers x, where $a \neq 1$ and $b \neq 1$:

$$\log_b(p \cdot q) = \log_b p + \log_b q \qquad \log_b 1 = 0 \qquad b^{\log_b p} = p$$

$$\log_b\left(\frac{p}{q}\right) = \log_b p - \log_b q \qquad \log_b b = 1$$

$$\log_b\left(p^x\right) = x \cdot \log_b p \qquad \log_b p = \frac{\log_a p}{\log_a b}$$

The basic property that relates the exponential and logarithmic functions is:
For all real numbers x, and all positive real numbers b and N,

$$\log_b N = x \text{ is equivalent to } b^x = N.$$

EXAMPLE 1: Simplify $x^{n-1} \cdot x^{2n} \cdot \left(x^{2-n}\right)^2$

This is equal to $x^{n-1} \cdot x^{2n} \cdot x^{4-2n} = x^{n-1+2n+4-2n} = x^{n+3}$.

EXAMPLE 2: Simplify $\dfrac{3^{n-2} \cdot 9^{2-n}}{3^{2-n}}$.

In order to combine exponents using the properties above, the base of each factor must be the same.

$$\frac{3^{n-2} \cdot 9^{2-n}}{3^{2-n}} = \frac{3^{n-2} \cdot (3^2)^{2-n}}{3^{2-n}} = \frac{3^{n-2} \cdot 3^{4-2n}}{3^{2-n}}$$

$$= 3^{n-2+4-2n-(2-n)} = 3^0 = 1$$

Note: Examples 3 and 4 can be easily evaluated with a calculator.

EXAMPLE 3: If $\log_{10} 23 = z$, what does $\log_{10} 2300$ equal?

$\log_{10} 2300 = \log_{10} 23 \cdot 100 = \log_{10} 23 + \log_{10} 100 = z + \log_{10} 10^2 =$

$$z + 2 \cdot \log_{10} 10 = z + 2$$

EXAMPLE 4: If $\log_{10} 2 = a$ and $\log_{10} 3 = b$, find the value of $\log_{10} 15$.

$\log_{10} 15 = \log_{10} (3 \cdot 5) = \log_{10} 3 + \log_{10} \left(\dfrac{10}{2}\right) = \log_{10} 3 + \log_{10} 10 - \log_{10} 2 = b + 1 - a$.

EXAMPLE 5: Solve for x: $\log(x + 5) = \log x + \log 5$.

(When no base is indicated, any arbitrary base can be used. If you plan to use your calculator, you should use base 10.)

$$\log x + \log 5 = \log 5 \cdot x$$

Therefore, $\log(x + 5) = \log(5x)$, which is true only when:

$$x + 5 = 5x$$
$$5 = 4x$$
$$x = \frac{5}{4}$$

EXAMPLE 6: Evaluate $\log_{27} \sqrt{54} - \log_{27} \sqrt{6}$.

$\log_{27} \sqrt{54} - \log_{27} \sqrt{6} = \log_{27}\left(\dfrac{\sqrt{54}}{\sqrt{6}}\right)$

$$= \log_{27} \sqrt{9} = \log_{27} 3 = x$$

The last equality implies that

$$27^x = 3$$
$$(3^3)^x = 3$$
$$3^{3x} = 3^1$$

Therefore, $3x = 1$ and $x = \dfrac{1}{3}$.

Thus, $\log_{27} \sqrt{54} - \log_{27} \sqrt{6} = \dfrac{1}{3}$.

<u>Alternative Solution:</u>

$$\log_{27} \sqrt{54} = \frac{\log_{10} \sqrt{54}}{\log_{10} 27} \approx \frac{0.866}{1.431} \approx 0.605$$

$$\log_{27} \sqrt{6} = \frac{\log_{10} \sqrt{6}}{\log_{10} 27} \approx \frac{0.389}{1.431} \approx 0.272$$

Therefore, $\log_{27} \sqrt{54} - \log_{27} \sqrt{6} \approx 0.333 \approx \dfrac{1}{3}$.

EXAMPLE 7: Evaluate $\log_8 2 - \log_6 216 + \log_{81} 3 - \log_5 (625)^{1/3}$.

Since each term has a different base, the terms must be evaluated separately.

Let $\log_8 2 = x$. This implies that

$$8^x = 2$$
$$(2^3)^x = 2$$
$$2^{3x} = 2^1$$

Therefore, $3x = 1$ and $x = \dfrac{1}{3}$.

Let $\log_6 216 = y$. This implies that

$$6^y = 216$$
$$6^y = 6^3$$

Therefore, $y = 3$.

Let $\log_3 81 = z$. This implies that

$$3^z = 81$$
$$3^z = 3^4$$

Therefore, $z = 4$.

$\text{Log}_5 (625)^{1/3} = \dfrac{1}{3} \cdot \log_5 625$. Let $\log_5 625 = w$. This implies that

$$5^w = 625$$
$$5^w = 5^4$$

Therefore, $w = 4$. Thus, $\log_5 (625)^{1/3} = \dfrac{1}{3} \cdot 4 = \dfrac{4}{3}$. Putting the four parts together, we find that the original expression is equal to $\dfrac{1}{3} - 3 + 4 - \dfrac{4}{3} = 0$.

The graphs of all exponential functions $y = b^x$ have roughly the same shape and pass through point (0,1). If $b > 1$, the graph increases as x increases and approaches the x-axis as an asymptote as x decreases. The amount of curvature becomes greater as the value of b is made greater. If $0 < b < 1$, the graph increases as x decreases and approaches the x-axis as an asymptote as x increases. The amount of curvature becomes greater as the value of b is made smaller.

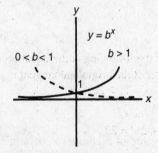

The graphs of all logarithmic functions $y = \log_b x$ have roughly the same shape and pass through point (1,0). If

$b > 1$, the graph increases as x increases and approaches the y-axis as an asymptote as x approaches zero. The amount of curvature becomes greater as the value of b is made greater. If $0 < b < 1$, the graph decreases as x increases and approaches the y-axis as an asymptote as x approaches zero. The amount of curvature becomes greater as the value of b is made smaller.

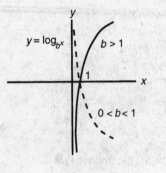

EXERCISES

1. If $x^a \cdot (x^{a+1})^a \cdot (x^a)^{1-a} = x^k$, then $k =$

(A) $2a + 1$
(B) $a + a^2$
(C) $3a$
(D) $3a + 1$
(E) $a^3 + a$

2. If $\log_8 3 = x \cdot \log_2 3$, then $x =$

(A) 4
(B) $\log_4 3$
(C) 3
(D) $\dfrac{1}{3}$
(E) $\log_8 9$

3. If $\log_{10} m = \dfrac{1}{2}$, then $\log_{10} 10m^2 =$

(A) 2.5
(B) 2
(C) 100
(D) 10.25
(E) 3

4. If $\log_b 5 = a$, $\log_b 2.5 = c$, and $5^x = 2.5$, then $x =$

(A) ac
(B) $\dfrac{c}{a}$
(C) $a + c$
(D) $c - a$
(E) The value of x cannot be determined from the information given.

5. If $f(x) = \log_2 x$, then $f\!\left(\dfrac{2}{x}\right) + f(x) =$

(A) $\log\!\left(\dfrac{2}{x}\right) + \log_2 x$
(B) 1
(C) $\log_2\!\left(\dfrac{2 + x^2}{x}\right)$
(D) $\log_2\!\left(\dfrac{2}{x}\right) \cdot \log_2 x$
(E) 0

6. If $\log_b (xy) < 0$, which of the following must be true?

(A) $xy < 0$
(B) $xy < 1$
(C) $xy > 1$
(D) $xy > 0$
(E) none of the above

7. If $\log_2 m = \sqrt{7}$ and $\log_7 n = \sqrt{2}$, $mn =$

(A) 98
(B) 2
(C) 1
(D) 96
(E) 103

8. If $f(x, y) = \dfrac{\log x}{\log y}$, $f(e, \pi) =$

(A) 2.01
(B) 0.50
(C) -1.73
(D) 0.87
(E) -0.37

9. $\mathrm{Log}_7\, 5 =$

(A) 1.2
(B) 1.1
(C) 0.9
(D) 0.8
(E) -0.7

10. $\left(\sqrt[3]{2}\right)\left(\sqrt[5]{4}\right)\left(\sqrt[9]{8}\right) =$

(A) 1.9
(B) 2.0
(C) 2.1
(D) 2.3
(E) 2.5

11. If \$300 is invested at 3%, compounded continuously, how long (to the nearest year) will it take for the money to double? (If P is the amount invested, the formula for the amount, A, that is available after t years is $A = Pe^{0.03t}$.)

(A) 26
(B) 25
(C) 24
(D) 23
(E) 22

4.3 ABSOLUTE VALUE

The absolute value of x (written as $|x|$) is defined as follows:
if $x \geq 0$, then $|x| = x$;
if $x < 0$, then $|x| = -x$ (note that $-x$ is a positive number);
$|x| \geq 0$ for all values of x.

EXAMPLE 1: If $|x - 3| = 2$, find x.
Absolute value problems can always be solved by using the definition to restate the problem in two parts.

Part 1: If $x - 3 \geq 0$, then $|x - 3| = x - 3$. Thus,
$$x - 3 = 2$$
$$x = 5$$

Notice that both conditions, $x - 3 \geq 0$ and $x = 5$, must be satisfied. They are in this case.

Part 2: If $x - 3 < 0$, then $|x - 3| = -(x - 3) = -x + 3$. Thus,
$$-x + 3 = 2$$
$$x = 1$$

Both conditions, $x - 3 < 0$ and $x = 1$, are satisfied. The solution set is $\{1, 5\}$.

EXAMPLE 2: If $|3x + 5| = 2x + 3$, find x.

Part 1: If $3x + 5 \geq 0$, then $|3x + 5| = 3x + 5$. Thus,
$$3x + 5 = 2x + 3$$
$$x = -2$$

Both conditions, $x \geq -\dfrac{5}{3}$ and $x = -2$, are not satisfied, and so no solution is supplied by this part.

Part 2: If $3x + 5 < 0$, then $|3x + 5| = -(3x + 5) = -3x - 5$. Thus,
$$-3x - 5 = 2x + 3$$
$$-5x = 8$$
$$x = -\frac{8}{5}$$

Both conditions, $x < -\dfrac{5}{3}$ and $x = -\dfrac{8}{5}$, are not satisfied, so no solution is supplied by this part either. Therefore the solution set is the empty set, $\emptyset$.

EXAMPLE 3: Find all values of x such that $|2x + 3| \geq 5$.
Consider the equality $|2x + 3| = 5$, which is associated with this inequality. Using the definition of absolute value gives the following:

If $2x + 3 \geq 0$, then $2x + 3 = 5$.

Thus, $x \geq -\dfrac{3}{2}$ and $x = 1$.

If $2x + 3 < 0$, then $-(2x + 3) = 5$.

Thus, $x < -\dfrac{3}{2}$ and $x = -4$.

The solution set of the equality is $\{1, -4\}$. Consider the regions of a number line indicated by these two numbers.

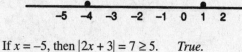

If $x = -5$, then $|2x + 3| = 7 \geq 5$.   *True.*
If $x = 0$, then $|2x + 3| = 3 \leq 5$.   *True.*
If $x = 2$, then $|2x + 3| = 7 \geq 5$.   *False.*

Therefore, the set of numbers that satisfies the original inequality, $|2x + 3| \geq 5$, is $\{x : x \leq -4 \text{ or } x \geq 1\}$. The graph of this solution set on a number line is indicated below.

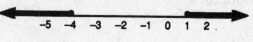

EXAMPLE 4: If the graph of $f(x)$ is shown below, sketch the graph of (A) $|f(x)|$ (B) $f(|x|)$.

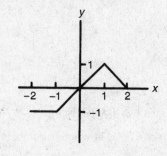

(A) Since $|f(x)| \geq 0$, by the definition of absolute value, the graph cannot have any points below the x-axis. If $f(x) < 0$, then $|f(x)| = -f(x)$. Thus, all points below the x-axis are reflected about the x-axis, and all points above the x-axis remain unchanged.

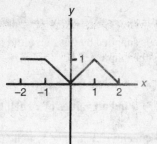

(B) Since the absolute value of x is taken before the function value is found, and since $|x| = -x$ when $x < 0$, any negative value of x will graph the same y-values as the corresponding positive values of x. Thus, the graph to the left of the y-axis will be a reflection of the graph to the right of the y-axis.

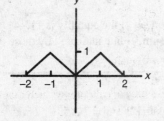

EXAMPLE 5: If $f(x) = |x + 1| - 1$, what is the minimum value of $f(x)$?
Since $|x + 1| \geq 0$, its smallest value is 0. Therefore, the smallest value of $f(x)$ is $0 - 1 = -1$. The graph of $f(x)$ is indicated below.

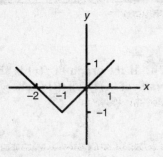

EXERCISES

1. $|2x - 1| = 4x + 5$ has how many numbers in its solution set?

(A) 0
(B) 1
(C) 2
(D) an infinite number
(E) none of the above

2. Which of the following is equivalent to $1 \leq |x - 2| \leq 4$?

(A) $3 \leq x \leq 6$
(B) $x \leq 1$ or $x \geq 3$
(C) $1 \leq x \leq 3$
(D) $x \leq -2$ or $x \geq 6$
(E) $-2 \leq x \leq 1$ or $3 \leq x \leq 6$

3. The area bound by the relation $|x| + |y| = 2$ is

(A) 8
(B) 1
(C) 2
(D) 4
(E) There is no finite area.

4. Given a function, $f(x)$, such that $f(x) = f(|x|)$. Which one of the following could be the graph of $f(x)$?

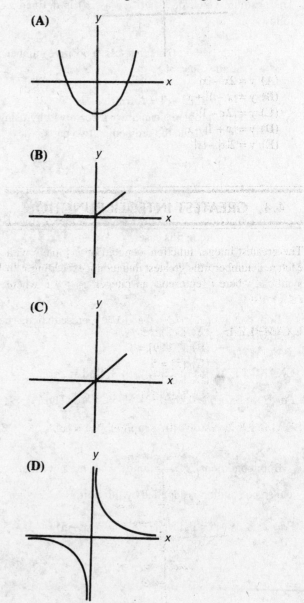

(E)

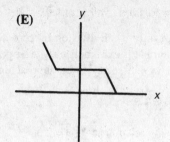

5. The figure shows the graph of which one of the following?

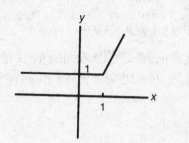

(A) $y = 2x - |x|$
(B) $y = |x - 1| + x$
(C) $y = |2x - 1|$
(D) $y = |x + 1| - x$
(E) $y = 2|x| - |x|$

4.4 GREATEST INTEGER FUNCTION

The greatest integer function, denoted by $[x]$, pairs with each real number x the greatest integer not exceeding x. In symbols, where i represents an integer: $f(x) = i$, where $i \le x < i + 1$.

EXAMPLE 1:   (A) $[3.2] = 3$
(B) $[1.999] = 1$
(C) $[5] = 5$
(D) $[-3.12] = -4$
(E) $[-0.123] = -1$.

EXAMPLE 2: Sketch the graph of $f(x) = [x]$.

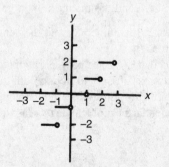

EXAMPLE 3: What is the range of $f(x) = \left[\dfrac{[x]}{x}\right]$.

If x is not an integer, $[x] < x$. Therefore, $\dfrac{[x]}{x}$ represents a decimal between 0 and 1, and $\left[\dfrac{[x]}{x}\right] = 0$. If x is an integer, $[x] = x$, and $\left[\dfrac{[x]}{x}\right] = \left[\dfrac{x}{x}\right] = [1] = 1$. Therefore, the range is $\{0, 1\}$.

EXERCISES

1. If the postal rate for first-class mail is 32 cents for the first ounce or portion thereof and 23 cents for each additional ounce or portion thereof, then the cost in cents of first-class postage for a letter weighing N ounces is always

(A) $32 + [N - 1] \cdot 23$
(B) $[N - 32] \cdot 23$
(C) $32 + [N] \cdot 23$
(D) $1 + [N] \cdot 23$
(E) none of the above

2. If $f(x) = i$, where i is an integer such that $i \le x < i + 1$, the range of $f(x)$ is

(A) the set of all real numbers
(B) the set of all positive integers
(C) the set of all integers
(D) the set of all negative integers
(E) the set of all nonnegative real numbers

3. If $f(x) = [2x] - 4x$ with domain $0 \le x \le 2$, then $f(x)$ can also be written as

(A) $2x$
(B) $-x$
(C) $-2x$
(D) $x^2 - 4x$
(E) none of the above

4.5 RATIONAL FUNCTIONS AND LIMITS

F is a rational function if and only if $F(x) = \dfrac{p(x)}{q(x)}$, where $p(x)$ and $q(x)$ are both polynomial functions and $q(x)$ is not zero. As a general rule, the graphs of rational functions are not continuous (i.e., they have holes, or sections of the graphs are separated from other sections by asymptotes). A point of discontinuity occurs at any value of x that would cause $q(x)$ to become zero.

If $p(x)$ and $q(x)$ can be factored so that $F(x)$ can be reduced, removing the factors that caused the discontinuities, the graph will contain only holes. If the factors that caused the discontinuities cannot be removed, asymptotes will occur.

EXAMPLE 1: Sketch the graph of $F(x) = \dfrac{x^2 - 1}{x + 1}$.

There is a discontinuity at $x = -1$ since this value would cause division by zero. The fraction $\dfrac{x^2 - 1}{x + 1} = \dfrac{(x-1)(x+1)}{(x+1)} = (x - 1)$, and so the graph of $F(x)$ is the same as the graph of $y = x - 1$ except for a hole at $x = -1$.

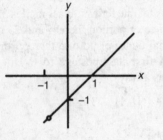

EXAMPLE 2: Sketch the graph of $F(x) = \dfrac{1}{x - 2}$.

Since this fraction cannot be reduced, and $x = 2$ would cause division by zero, a vertical asymptote occurs when $x = 2$. This is true because, as x approaches very close to 2, $F(x)$ gets either extremely large or extremely small. As x becomes extremely large or extremely small, $f(x)$ gets closer and closer to zero. This means that a horizontal asymptote occurs when $y = 0$. Plotting a few points indicates that the graph looks like the figure below.

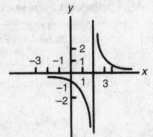

A more compact way of expressing the preceding statements about x approaching 2 or x getting extremely large follows: The statement $\lim\limits_{x \to \infty} f(x) = 0$ is read as "the limit of $f(x)$ is zero as x increases without bound (or as x approaches infinity)." $\lim\limits_{x \to 2^-} f(x) = -\infty$ is read as "the limit of $f(x)$ decreases without bound (or tends to negative infinity) as x approaches 2 from the negative side."

EXAMPLE 3: What does $\lim\limits_{x \to 1} \dfrac{x^2 - 1}{x + 1}$ equal?

Since $\dfrac{x^2 - 1}{x + 1}$ reduces to $x - 1$,

$$\lim_{x \to 1} \frac{x^2 - 1}{x + 1} = \lim_{x \to 1} x - 1 = 0.$$

EXAMPLE 4: What does $\lim\limits_{x \to 2^+} 3x + 5$ equal?

Since "problems" occur only when division by zero appears imminent, this example is extremely easy. As x gets closer and closer to 2, $3x + 5$ seems to be approaching closer and closer to 11. Therefore, $\lim\limits_{x \to 2^+} 3x + 5 = 11$.

EXAMPLE 5: What does $\lim\limits_{x \to 2} \left(\dfrac{3x^2 + 5}{x - 2} \right)$ equal?

The denominator does not factor out, and so the graph of this rational function has a vertical asymptote. As x approaches 2 from above (i.e., 2.1, 2.01, 2.001, . . .), the numerator and denominator both remain positive and so $\dfrac{3x^2 + 5}{x - 2}$ gets larger and larger and approaches positive infinity. As x approaches 2 from below (i.e., 1.9, 1.99, 1.999, . . .), the numerator is still positive, but the denominator is negative and so $\dfrac{3x^2 + 5}{x - 2}$ gets smaller and smaller and approaches negative infinity. Thus, $\lim\limits_{x \to 2^-} \dfrac{3x^2 + 5}{x - 2}$ does not exist since $\lim\limits_{x \to 2^-} \dfrac{3x^2 + 5}{x - 2}$ and $\lim\limits_{x \to 2^+} \dfrac{3x^2 + 5}{x - 2} = +\infty$, which are not the same.

EXAMPLE 6: If $f(x) = \begin{cases} 3x + 2 & \text{when } x \neq 0 \\ 0 & \text{when } x = 0 \end{cases}$, what does $\lim\limits_{x \to 0} f(x)$ equal?

As x approaches zero, $3x + 2$ approaches 2, in spite of the fact that $f(x) = 0$ when $x = 0$. Therefore, $\lim\limits_{x \to 0} f(x) = 2$.

EXAMPLE 7: What does $\lim\limits_{x \to \infty} \left(\dfrac{3x^2 + 4x + 2}{2x^2 + x - 5} \right)$ equal?

At first glance it appears that either the answer is obvious or the problem is impossible. The answer could be $\dfrac{3}{2}$ because, if both the numerator and the denominator are increasing without bound, their quotient must approach $\dfrac{3}{2}$. On the other hand, no matter how large a number is substituted for x, the numerator and the denominator are never the same.

This is, in fact, a rather easy problem to solve. The method used most often is to divide both the numerator and the denominator through by the variable raised to its highest exponent in the problem. In this case divide through by x^2.

$$\lim_{x \to \infty} \left(\frac{3x^2 + 4x + 2}{2x^2 + x - 5} \right) = \lim_{x \to \infty} \left(\frac{3 + \dfrac{4}{x} + \dfrac{2}{x^2}}{2 + \dfrac{1}{x} - \dfrac{5}{x^2}} \right).$$

Now, as $x \to \infty$, $\dfrac{4}{x}, \dfrac{2}{x^2}, \dfrac{1}{x}$, and $\dfrac{5}{x^2}$, each approaches

zero. Thus, the entire fraction approaches $\dfrac{3 + 0 + 0}{2 + 0 - 0} = \dfrac{3}{2}$.

Therefore, $\lim_{x \to \infty} \left(\dfrac{3x^2 + 4x + 2}{2x^2 + x - 5} \right) = \dfrac{3}{2}$.

EXERCISES

1. To be continuous at $x = 1$, the value of $\dfrac{x^4 - 1}{x^3 - 1}$ must be defined to be equal to

(A) $\dfrac{4}{3}$

(B) 0

(C) 1

(D) 4

(E) -1

2. If $f(x) = \begin{cases} \dfrac{3x^2 + 2x}{x} & \text{when } x \neq 0 \\ k & \text{when } x = 0 \end{cases}$, what must the

value of k be equal to in order for $f(x)$ to be a continuous function?

(A) 0

(B) 2

(C) $-\dfrac{2}{3}$

(D) $-\dfrac{3}{2}$

(E) No value of k can make $f(x)$ a continuous function.

3. $\lim_{x \to 2} \left(\dfrac{x^3 - 8}{x^4 - 16} \right) =$

(A) $\dfrac{3}{8}$

(B) $\dfrac{1}{2}$

(C) $\dfrac{4}{7}$

(D) 0

(E) This expression is undefined.

4. $\lim_{x \to \infty} \left(\dfrac{5x^2 - 2}{3x^2 + 8} \right) =$

(A) ∞

(B) $\dfrac{3}{11}$

(C) $\dfrac{5}{3}$

(D) $-\dfrac{1}{4}$

(E) 0

5. Which of the following is the equation of an asymptote of $y = \dfrac{3x^2 - 2x - 1}{9x^2 - 1}$?

(A) $x = -\dfrac{1}{3}$

(B) $y = \dfrac{1}{3}$

(C) $x = 1$

(D) $y = 1$

(E) $y = -\dfrac{1}{3}$

4.6 PARAMETRIC EQUATIONS

At times it is convenient to express a relationship between x and y in terms of a third variable. This third variable is called a *parameter*, and the equations are called *parametric equations*.

EXAMPLE 1: $\begin{cases} x = 3t + 4 \\ y = t - 5 \end{cases}$ **are parametric equations with a parameter t.**

As different values are substituted for the parameter, ordered pairs of points, (x, y), represent points on the graph. At times it is possible to eliminate the parameter and to rewrite the equation in familiar xy-form. When eliminating the parameter, it is necessary to be aware of the fact that the resulting equation may consist of points not on the graph of the original set of equations.

EXAMPLE 2: $\begin{cases} x = t^2 \\ y = 3t^2 + 1 \end{cases}$. **Eliminate the parameter and sketch the graph.**

Substituting x for t^2 in the second equation results in $y = 3x + 1$, which is the equation of a straight line with a slope of 3 and a y-intercept of 1. However, the original parametric equations indicate that $x \geq 0$ and $y \geq 1$ since t^2 cannot be negative. Thus, the proper way to indicate this set of points without the parameter is as follows: $y = 3x + 1$ and $x \geq 0$. The graph is the ray indicated in the figure.

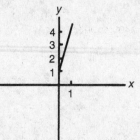

EXAMPLE 3: Sketch the graph of the parametric equations $\begin{cases} x = 4\cos\theta \\ y = 3\sin\theta \end{cases}$.

It is possible to eliminate the parameter, θ, by dividing the first equation by 4 and the second equation by 3, squaring each, and then adding the equations together.

$$\left(\frac{x}{4}\right)^2 = \cos^2\theta \quad \text{and} \quad \left(\frac{y}{3}\right)^2 = \sin^2\theta$$

$$\frac{x^2}{16} + \frac{y^2}{9} = \cos^2\theta + \sin^2\theta = 1$$

Here, $\frac{x^2}{16} + \frac{y^2}{9} = 1$ is the equation of an ellipse with its center at the origin, $a = 4$, and $b = 3$. Since $-1 \leq \cos\theta \leq 1$ and $-1 \leq \sin\theta \leq 1$, $-4 \leq x \leq 4$ and $-3 \leq y \leq 3$ from the two parametric equations. In this case the parametric equations do not limit the graph obtained by removing the parameter.

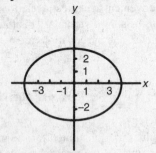

EXERCISES

1. The domain of the function defined by the parametric equations $\begin{cases} x = t^2 + t \\ y = t^2 - t \end{cases}$ is

    (A) $\{x : x \geq 0\}$

    (B) $\left\{ x : x \geq -\frac{1}{4} \right\}$

    (C) all real numbers

    (D) $\{x : x \geq -1\}$

    (E) $\{x : x \leq 1\}$

2. The graph of $\begin{cases} x = \sin^2 t \\ y = 2\cos t \end{cases}$ is a

    (A) straight line
    (B) line segment
    (C) parabola
    (D) portion of a parabola
    (E) semicircle

3. Which of the following is (are) a pair of parametric equations that represent a circle?

    I. $\begin{cases} x = \sin\theta \\ y = \cos\theta \end{cases}$

    II. $\begin{cases} x = t \\ y = \sqrt{1 - t^2} \end{cases}$

    III. $\begin{cases} x = \sqrt{s} \\ y = \sqrt{1 - s} \end{cases}$

    (A) only I
    (B) only II
    (C) only III
    (D) only II and III
    (E) I, II, and III

4.7   POLAR COORDINATES

Although the most common way to represent a point in a plane is in terms of its distance from two perpendicular axes, there are several other ways. One such way is in terms of the distance of the point from the origin and the angle between the positive x-axis and the ray emanating from the origin going through the point.

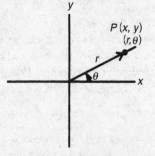

In the figure, the regular rectangular coordinates of P are (x, y) and the *polar coordinates* are (r, θ). Note that $r > 0$ because it represents a distance.

Since $\sin\theta = \frac{y}{r}$ and $\cos\theta = \frac{x}{r}$, there is an easy relationship between rectangular and polar coordinates:

$x = r \cdot \cos\theta$
$y = r \cdot \sin\theta$
$x^2 + y^2 = r^2$, using the Pythagorean theorem.

Unlike the case involving rectangular coordinates, each point in the plane can be named by an infinite number of polar coordinates.

EXAMPLE 1: $(2,30°)$, $(2,390°)$, $(2,-330°)$, $(-2,210°)$, $(-2,-150°)$ all name the same point.

In general, a point in the plane represented by (r,θ) can also be represented by $(r, \theta + 2\pi n)$ or $(-r, \theta + (2n-1)\pi)$, where n is an integer.

EXAMPLE 2: Express point P, whose rectangular coordinates are $\left(3, 3\sqrt{3}\right)$, in terms of polar coordinates.

$$r^2 = x^2 + y^2 = 9 + 27 = 36$$
$$r = 6$$
$$r \cdot \cos\theta = x$$
$$\cos\theta = \frac{3}{6} = \frac{1}{2}$$

Therefore, $\theta = 60°$, and the coordinates of P are $(6, 60°)$.

EXAMPLE 3: Without sketching, describe the graphs

of (A) $r = 2$ and (B) $r = \dfrac{1}{\sin\theta}$.

(A) $r^2 = x^2 + y^2$
$$r = 2$$

Therefore, $x^2 + y^2 = 4$, which is the equation of a circle whose center is at the origin and whose radius is 2.

(B) $r \cdot \sin\theta = x$
$$r = \frac{1}{\sin\theta}$$
Therefore, $x = 1$.

Thus, $r = \dfrac{1}{\sin\theta}$ is the equation of a vertical line one unit to the right of the y-axis.

Since the complex number $a + bi$ can be represented by point $P(a,b)$ on a coordinate plane, it can also be represented in terms of polar coordinates. Thus, $a = r \cdot \cos\theta$ and $b = r \cdot \sin\theta$. Therefore,

$$a + bi = r \cdot \cos\theta + i \cdot r \cdot \sin\theta = r(\cos\theta + i \cdot \sin\theta).$$

This last statement is often abbreviated as $r \cdot$ cis θ.

There are several useful formulas for manipulating complex numbers when they are written in polar form, which is sometimes called "trigonometric form." If $z_1 = r_1(\cos\theta_1 + i \cdot \sin\theta_1)$ and $z_2 = r_2(\cos\theta_2 + i \cdot \sin\theta_2)$, then

1. $z_1 \cdot z_2 = r_1 r_2\left[\cos(\theta_1 + \theta_2) + i \cdot \sin(\theta_1 + \theta_2)\right]$ or
$$z_1 \cdot z_2 = r_1 r_2 \text{ cis}(\theta_1 + \theta_2)$$

2. $\dfrac{z_1}{z_2} = \dfrac{r_1}{r_2}\left[\cos(\theta_1 - \theta_2) + i \cdot \sin(\theta_1 - \theta_2)\right]$ or
$$\frac{z_1}{z_2} = \frac{r_1}{r_2} \text{ cis}(\theta_1 - \theta_2)$$

De Moivre's theorem states that

3. $(z_1)^n = (r_1)^n(\cos n\theta_1 + i \cdot \sin n\theta_1)$ or
$$(z_1)^n = (r_1)^n \text{ cis } n\theta_1, \text{ where } n \text{ is any positive integer.}$$

The periodicity of the trigonometric functions allows the use of De Moivre's theorem to find the nth roots of any complex number.

4. $(z_1)^{1/n} = (r_1)^{1/n}\left(\cos \dfrac{\theta_1 + 2\pi k}{n} + i \cdot \sin \dfrac{\theta_1 + 2\pi k}{n}\right)$ or
$$(z_1)^{1/n} = (r_1)^{1/n} \text{ cis } \frac{\theta_1 + 2\pi k}{n}, \text{ where } k \text{ is an integer}$$
taking on values from 0 to $n - 1$.

EXAMPLE 4: Using the polar form, find the value of $\dfrac{2 + 2i}{\sqrt{3} - 3i}$.

For the numerator, $r^2 = x^2 + y^2 = 4 + 4 = 8$. Therefore, $r = 2\sqrt{2}$. $r \cdot \cos\theta = x$. $\cos\theta = \dfrac{2}{2\sqrt{2}} = \dfrac{\sqrt{2}}{2}$. Therefore, $\theta = 45°$.

Thus, $2 + 2i = 2\sqrt{2}$ cis $45°$.

For the denominator, $r^2 = x^2 + y^2 = 3 + 9 = 12$. Therefore, $r = 2\sqrt{3}$. $r \cdot \cos\theta = x$.

$$\cos\theta = \frac{\sqrt{3}}{2\sqrt{3}} = \frac{1}{2} \quad \text{and} \quad \sin\theta = \frac{-3}{2\sqrt{3}} = \frac{-\sqrt{3}}{2}$$

Therefore, θ is in quadrant IV and $\theta = -60°$ or $300°$.

Thus, $\sqrt{3} + 3i = 2\sqrt{3}$ cis $(-60°)$.

$$\frac{2 + 2i}{\sqrt{3} - 3i} = \frac{2\sqrt{2} \text{ cis } 45°}{2\sqrt{3} \text{ cis } (-60°)}$$
$$= \frac{\sqrt{2}}{\sqrt{3}} \text{ cis}[45 - (-60°)]° = \frac{\sqrt{6}}{3} \text{ cis}(105°)$$

EXAMPLE 5: If $z = 2$ cis $\dfrac{\pi}{9}$, what does z^5 equal?

$$z^5 = 2^5 \text{ cis}\left(5 \cdot \frac{\pi}{9}\right) = 32 \text{ cis } \frac{5\pi}{9}, \quad \text{using De Moivre's}$$
theorem.

EXAMPLE 6: Find the three cube roots of i.

The complex number i is represented on the rectangular coordinate system by the point with coordinates $(0,1)$. Therefore, the polar form of i is $1\left(\cos\dfrac{\pi}{2}+i\cdot\sin\dfrac{\pi}{2}\right)$ or $\text{cis}\,\dfrac{\pi}{2}$. Since the cube roots are desired, i must be represented by $\text{cis}\left(\dfrac{\pi}{2}+2k\pi\right)$, where $k=0,1,2$.

$$\sqrt[3]{i}=1^{1/3}\,\text{cis}\,\frac{1}{3}\left(\frac{\pi}{2}+2k\pi\right),\text{ where }k=0,1,2$$

$$\sqrt[3]{i}=\text{cis}\left(\frac{\pi}{6}+\frac{2k\pi}{3}\right),\text{ where }k=0,1,2$$

$$\sqrt[3]{i}=\text{cis}\,\frac{\pi}{6}$$

or

$$\sqrt[3]{i}=\text{cis}\left(\frac{\pi}{6}+\frac{2k\pi}{3}\right)=\text{cis}\,\frac{5\pi}{6}$$

or

$$\sqrt[3]{i}=\text{cis}\left(\frac{\pi}{6}+\frac{4\pi}{3}\right)=\text{cis}\,\frac{9\pi}{6}=\text{cis}\,\frac{3\pi}{2}.$$

EXERCISES

1. A point has polar coordinate $(2,60°)$. The same point can be represented by

   (A) $(-2,240°)$
   (B) $(2,240°)$
   (C) $(-2,60°)$
   (D) $(2,-60°)$
   (E) $(2,-240°)$

2. The graph of $r=\cos\,\theta$ intersects the graph of $r=\sin 2\theta$ at points

   (A) $\left(\dfrac{\sqrt{3}}{2},\dfrac{\pi}{6}\right),\left(\dfrac{\sqrt{3}}{2},\dfrac{5\pi}{6}\right),$ and $\left(0,\dfrac{\pi}{2}\right)$

   (B) $\left(\dfrac{\sqrt{3}}{2},\dfrac{\pi}{6}\right)$ and $\left(\dfrac{\sqrt{3}}{2},\dfrac{11\pi}{6}\right)$

   (C) $\left(\dfrac{\sqrt{3}}{2},\dfrac{\pi}{3}\right)$ and $\left(-\dfrac{\sqrt{3}}{2},\dfrac{\pi}{3}\right)$

   (D) $\left(\dfrac{\sqrt{3}}{2},\dfrac{\pi}{6}\right),\left(0,\dfrac{\pi}{2}\right),$ and $\left(-\dfrac{\sqrt{3}}{2},\dfrac{5\pi}{6}\right)$

   (E) only $(0,0)$

3. An equation in rectangular form equivalent to $r^2=36\sec 2\theta$ is

   (A) $x^2-y^2=36$
   (B) $(x-y)^2=36$
   (C) $x^4-y^4=36$
   (D) $xy=36$
   (E) $y=6$

4. If $-2-2i\sqrt{3}$ is divided by $-1+i\sqrt{3}$, the quotient, in trigonometric form is

   (A) $1(\cos 240°+i\cdot\sin 240°)$
   (B) $2(\cos 120°+i\cdot\sin 120°)$
   (C) $2(\cos 135°+i\cdot\sin 135°)$
   (D) $2(\cos 150°+i\cdot\sin 150°)$
   (E) $2(\cos 210°+i\cdot\sin 210°)$

5. If $A=2(\cos 20°+i\cdot\sin 20°)$ and $B=3(\cos 40°+i\cdot\sin 40°)$, product $AB=$

   (A) $6\,\text{cis}\,20°$
   (B) $6\,\text{cis}\,60°$
   (C) $5\,\text{cis}\,20°$
   (D) $5\,\text{cis}\,60°$
   (E) $5(\cos 80°+i\cdot\sin 80°)$

6. $\sqrt[6]{4\sqrt{3}+4i}=$

   (A) $\sqrt{2}\left(\cos\dfrac{\pi}{36}-i\cdot\sin\dfrac{\pi}{36}\right)$

   (B) $\sqrt{2}\,\text{cis}\,\dfrac{25\pi}{36}$

   (C) $\sqrt{8}\,\text{cis}\,\dfrac{25\pi}{36}$

   (D) $\sqrt{2}\left(\cos\dfrac{13\pi}{36}-i\cdot\sin\dfrac{13\pi}{36}\right)$

   (E) $\sqrt{8}\,\text{cis}\,\dfrac{13\pi}{36}$

7. The polar coordinates of a point P are $(2,200°)$. The Cartesian (rectangular) coordinates of P are

   (A) $(-1.88,-0.68)$
   (B) $(-0.68,-1.88)$
   (C) $(-0.34,-0.94)$
   (D) $(-0.94,-0.34)$
   (E) $(-0.47,-0.17)$

ANSWERS AND EXPLANATIONS

In these solutions the following notation is used:

a: active—Calculator use is necessary or, at a minimum, extremely helpful.

n: neutral—Answers can be found without a calculator, but a calculator may help.

i: inactive—Calculator use is not helpful and may even be a hindrance.

Part 4.1 Conic Sections

1. i **D** Complete the squares: $5(x^2 - 4x + 4) + 4(y^2 + 2y + 1) = -4 + 20 + 4$. $5(x-2)^2 + 4(y+1)^2 = 20$. $\dfrac{(x-2)^2}{4} + \dfrac{(y+1)^2}{5} = 1$. $a^2 = 5$, $b^2 = 4$, and so $c^2 = 1$. Therefore, the foci are 1 unit above and below the center, which is at $(2,-1)$.

2. n **C** See the figure below.

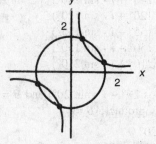

3. n **C** Vertex is at $(2,0)$, focus is at $(2,1)$, directrix is $y = -1$.

4. i **D** $\dfrac{x^2}{4} - \dfrac{y^2}{3} = 1$. Asymptotes $y = \pm\dfrac{\sqrt{3}}{2}x$.

5. i **E** $x^2 - (2y + 3)^2 = 0$ factors into $(x - 2y - 3)(x + 2y + 3) = 0$, which breaks into $x - 2y - 3 = 0$ or $x + 2y + 3 = 0$. These are the equations of two intersecting lines.

6. i **D** This is the equation of a semicircle with radius 2. $A = \dfrac{1}{2}\pi r^2 = 2\pi$.

7. a **B** Distance $= \dfrac{|3\cdot 0 - 7\cdot 0 - 29|}{\sqrt{3^2 + 7^2}} = \dfrac{29}{\sqrt{58}} \approx 3.808$. Therefore, $r^2 = 14.5$. Choice B can be written as $x^2 + y^2 = 14.5$.

8. a **D** $x^2 + 2x + 1 + y^2 - 4y + 4 = 0 + 1 + 4$. $(x + 1)^2 + (y - 2)^2 = 5$. Therefore, radius $= \sqrt{5}$. The center is at the centroid of the triangle, which is two-thirds of the distance from the vertex to the midpoint of the opposite side. Therefore, the radius is two-thirds of

the altitude (h): $\sqrt{5} = \dfrac{2}{3}h$. $h = 1.5\sqrt{5}$. Triangle AMC is a 30-60-90 triangle so $AC = 2\left(1.5\sqrt{5}\right) \div \sqrt{3} \approx \dfrac{6.708}{1.732} \approx 3.9$.

Alternative Solution: Since $\triangle ABC$ is an equilateral triangle, arc $AB = 120°$. Central angle $\angle AOB = 120°$, OM is an altitude, and $\angle AOM = 60°$. Thus, $\triangle AOM$ is a 30-60-90 triangle. Therefore, $AM = \dfrac{\sqrt{5}}{2}\sqrt{3}$ and $AB = \sqrt{15} \approx 3.9$.

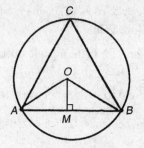

Part 4.2 Exponential and Logarithmic Functions

1. i **C** $x^a \cdot x^{a^2 + a} \cdot x^{a - a^2} = x^{a + a^2 + a + a - a^2} = x^{3a}$.

2. n **D** Let $\log_8 3 = y$. Then $\log_2 3^x = y$. $8^y = 3$. $2^{3y} = 3$. $2^y = 3^{1/3}$. From the second equation $2^y = 3x$. Therefore, $3^x = 3^{1/3}$, and so $x = \dfrac{1}{3}$.

3. n **B** $\text{Log}_{10}(10m^2) = \log_{10} 10 + 2\log_{10} m = 1 + 2\cdot\dfrac{1}{2} = 2$.

4. i **B** $b^a = 5$, $b^c = 2.5 = 5^x$, using the relationship between logs and exponents: $b^{ax} = 5^x = b^c$. Therefore, $ax = c$ and $x = \dfrac{c}{a}$.

5. i **B** $f\left(\dfrac{2}{x}\right) + f(x) = \log_2\left(\dfrac{2}{x}\right) + \log_2 x$
$= \log_2 2 - \log_2 x + \log_2 x = 1$.

6. i **E** If $b > 1$, Choice B is the answer. If $b < 1$, Choice C is the answer. Since no restriction was put on b, however, the correct answer is Choice E.

7. i **A** Converting the log expressions to exponential expressions gives $m = 2^{\sqrt{7}}$ and $n = 7^{\sqrt{2}}$. Therefore, $mn = 2^{\sqrt{7}} \cdot 7^{\sqrt{2}} \approx 6.2582 \cdot 15.673 \approx 98$.

8. a **D** $f(e,\pi) = \dfrac{\log e}{\log \pi} \approx \dfrac{0.434}{0.497} \approx 0.87$.

9. a **D** $\log_7 5 = \dfrac{\log_{10} 5}{\log_{10} 7} \approx \dfrac{0.699}{0.845} \approx 0.8$.

10. a **C** $\sqrt[3]{2}\sqrt[5]{4}\sqrt[9]{8} = 2^{1/3}4^{1/5}8^{1/9} \approx 2.1$.

11. a   D   Substitute in $A = Pe^{0.03t}$ to get $600 = 300e^{0.03t}$. Simplify to get $2 = e^{0.03t}$. Then take ln of both sides to get $ln\,2 = 0.03t$ and $t = \dfrac{ln\,2}{0.03}$. Use your calculator to find that t is approximately 23.

Part 4.3 Absolute Value

1. i   B   If $2x - 1 \ge 0$, the equation becomes $2x - 1 = 4x + 5$ and $x = -3$. However, $2x - 1 \ge 0$ implies $x \ge \dfrac{1}{2}$, and so -3 does not work. If $2x - 1 < 0$, the equation becomes $-2x + 1 = 4x + 5$ and $x = -\dfrac{2}{3}$. Since $2x - 1 < 0$ implies that $x < \dfrac{1}{2}$, $x = -\dfrac{2}{3}$ is the only root.

2. i   E   Here, x must be more than 1 unit from 2 but less than 4 units from 2 (including 1 and 4).

```
 ●        ●    ●        ●
-3 -2 -1  0  1  2  3  4  5  6  7
```

3. i   A   The figure is a square $2\sqrt{2}$ on a side. The area is 8.

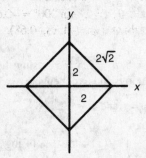

4. i   A   Since $f(x)$ must $= f(|x|)$, the graph must be symmetric about the y-axis. The only graph meeting this requirement is Choice A.

5. i   B   Since the point where a major change takes place is at $(1,1)$, the statement in the absolute value should equal zero when $x = 1$. This occurs only in Choice B.

Part 4.4 Greatest Integer Function

1. i   C   If $N < 1$, the value should be 32; and if $1 \le N < 2$, the value should be 55. The only answer that satisfies this is Choice C.

2. i   C   Since $f(x) =$ an integer by definition, the answer is Choice C.

3. n   E   Plotting a few numbers between 0 and 2 leads to a rather steep line going down to the right.

x	0	0.4	0.5	0.6	0.9	1	1.1
$f(x)$	0	–1.6	–1	–1.4	–2.6	–2	–2.4

Part 4.5 Rational Functions and Limits

1. i   A   Factor and reduce: $\dfrac{(x-1)(x+1)(x^2+1)}{(x-1)(x^2+x+1)}$. Substitute 1 for x and the fraction equals $\dfrac{4}{3}$.

2. i   B   Factor and reduce the fraction, which becomes $3x + 2$. As x approaches zero, this approaches 2.

3. i   A   Factor and reduce $\dfrac{(x-2)(x^2+2x+4)}{(x-2)(x+2)(x^2+4)}$. Substitute 2 for x and the fraction equals $\dfrac{3}{8}$.

4. i   C   Divide numerator and denominator through by x^2. As $x \to \infty$, the fraction approaches $\dfrac{5}{3}$.

5. i   B   Factor and reduce $\dfrac{(3x+1)(x-1)}{(3x+1)(3x-1)}$. Therefore a vertical asymptote occurs when $3x - 1 = 0$ or $x = \dfrac{1}{3}$, but this is not an answer choice. As $x \to \infty$, $y \to \dfrac{1}{3}$. Therefore, $y = \dfrac{1}{3}$ is the correct answer choice.

Part 4.6 Parametric Equations

1. i   B   Complete the square on the x equation: $x = \left(t^2 + t + \dfrac{1}{4}\right) - \dfrac{1}{4} = \left(t + \dfrac{1}{2}\right)^2 - \dfrac{1}{4}$. This represents a parabola that opens up with vertex at $\left(-\dfrac{1}{2}, -\dfrac{1}{4}\right)$. Therefore, $x \ge -\dfrac{1}{4}$.

2. i   D $\dfrac{y}{2} = \cos t$. So $\dfrac{y^2}{4} = \cos^2 t$. Adding this to $x = \sin^2 t$ gives $\dfrac{y^2}{4} + x = \cos^2 t + \sin^2 t = 1$. Since $0 \le x \le 1$ because $0 \le \sin^2 t \le 1$, this can only be a portion of the parabola given by the equation $y^2 + 4x = 4$.

3. i A Removing the parameter in I by squaring and adding gives $x^2 + y^2 = 1$, which is a circle of radius 1. Substituting x for t in the y equation of II and squaring gives $x^2 + y^2 = 1$, but $y \geq 0$ so this is only a semicircle. Squaring and substituting x^2 for s in the y equation of III gives $x^2 + y^2 = 1$, but $x \geq 0$ and so this is only a semicircle.

Part 4.7 Polar Coordinates

1. i A The angle must be either coterminal with 60° or 180° away. If coterminal, r must equal 2. No such points are given. 180° away gives +240° or –60°. In either case r must equal –2.

2. i D Set the two equations equal and use the double-angle formula:
$\cos \theta = 2 \sin \theta \cdot \cos \theta$.
$2 \sin \theta \cdot \cos \theta - \cos \theta = 0$.

Then $\cos \theta (2 \sin \theta - 1) = 0$, and so $\cos \theta = 0$ or $\sin \theta = \frac{1}{2}$.

3. n A $\operatorname{Sec} 2\theta = \dfrac{1}{\cos 2\theta} = \dfrac{1}{2\cos^2 \theta - 1}$. Substituting in $r^2 = \dfrac{36}{2\cos^2 \theta - 1}$ gives $r^2 = \dfrac{36}{2 \cdot \dfrac{x^2}{r^2} - 1}$.

$r^2 = \dfrac{36r^2}{2x^2 - r^2}$. Dividing by r^2 gives $1 = \dfrac{36}{2x^2 - x^2 - y^2}$. Thus, $x^2 - y^2 = 36$.

4. i B $-2 - 2\sqrt{3}\,i = 4 \operatorname{cis} 240°$
$-1 + \sqrt{3}\,i = 2 \operatorname{cis} 120°$
$\dfrac{-2 - 2\sqrt{3}\,i}{-1 + \sqrt{3}\,i} = \dfrac{4}{2} \operatorname{cis}(240° - 120°)$
$= 2 \operatorname{cis} 120°$.

5. i B $AB = 2 \cdot 3 \operatorname{cis}(20° + 40°) = 6 \operatorname{cis} 60°$.

6. n B $4\sqrt{3} + 4i = 8 \operatorname{cis} \dfrac{\pi}{6}$. The sixth root of this $= \sqrt{2} \operatorname{cis} \dfrac{1}{6}\left(\dfrac{\pi}{6} + 2\pi k\right)$, where $k = 0, 1, 2, 3, 4, 5$.

7. a A Put your calculator in degree mode. Since $x = r \cdot \cos \theta$, $x = 2 \cdot \cos 200° \approx -1.88$, and since $y = r \cdot \sin \theta$, $x = 2 \cdot \sin 200° \approx -0.68$. Therefore, the coordinates are (–1.88,–0.68).

MISCELLANEOUS TOPICS

CHAPTER

5.1 PERMUTATIONS AND COMBINATIONS

Any arrangement of the elements of a set in a definite order is called a *permutation*. If all n elements of a set are to be arranged, there are $n!$ (read as "n factorial") ways to arrange them; $n! = n(n-1)(n-2)\ldots 3 \cdot 2 \cdot 1$.

EXAMPLE 1: (A) $5! = 5 \cdot 4 \cdot 3 \cdot 2 \cdot 1 = 120$
(B) $3! = 3 \cdot 2 \cdot 1 = 6$

If only r elements of a set that contains n elements are to be arranged, there are $\dfrac{n!}{(n-r)!}$ arrangements. $_nP_r = P(n,r) =$ the number of permutations of n elements taken r at a time. Therefore, $_nP_r = \dfrac{n!}{(n-r)!}$.

EXAMPLE 2: Evaluate each of the following:
(A) $_4P_4$ (B) $_7P_2$ (C) $_7P_5$.

(A) $_4P_4$ means "the number of permutations of four elements taken four at a time." From the original definition, this is equal to $4! = 4 \cdot 3 \cdot 2 \cdot 1 = 24$. From the definition of $_nP_r$, $P = \dfrac{4!}{(4-4)!} = \dfrac{4!}{0!}$. This will equal 24 if and only if 0! is defined to equal 1. In fact, 0! is *always* defined to equal 1. Therefore, using either definition, $_4P_4 = 24$.

(B) $_7P_2 = \dfrac{7!}{(7-2)!} = \dfrac{7 \cdot 6 \cdot 5 \cdot 4 \cdot 3 \cdot 2 \cdot 1}{5 \cdot 4 \cdot 3 \cdot 2 \cdot 1} = 7 \cdot 6 = 42$

(C) $_7P_5 = \dfrac{7!}{(7-5)!} = \dfrac{7 \cdot 6 \cdot 5 \cdot 4 \cdot 3 \cdot 2 \cdot 1}{2 \cdot 1}$
$= 7 \cdot 6 \cdot 5 \cdot 4 \cdot 3 = 2520$

EXAMPLE 3: How many four-digit lottery numbers can be drawn with no repetition of digits?

Method 1: This is a permutation problem:

$_{10}P_4 = \dfrac{10!}{(10-4)!}$
$= \dfrac{10!}{6!} = \dfrac{10 \cdot 9 \cdot 8 \cdot 7 \cdot 6 \cdot 5 \cdot 4 \cdot 3 \cdot 2 \cdot 1}{6 \cdot 5 \cdot 4 \cdot 3 \cdot 2 \cdot 1}$
$= 10 \cdot 9 \cdot 8 \cdot 7 = 5040$

Method 2: Consider putting the digits in a grid.

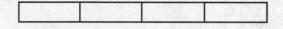

Any one of the 10 digits can be put in the first space. That leaves only 9 digits for the second space, only 8 for the third, and only 7 for the fourth. Thus, there are $10 \cdot 9 \cdot 8 \cdot 7 = 5040$ different lottery numbers.

EXAMPLE 4: In a club with 15 members, one member is to be chosen president, one secretary, and one treasurer. How many different slates of candidates can be chosen?

Method 1: The president can be any one of 15 people. After he or she is chosen, there are 14 remaining members from whom to choose the secretary and then only 13 members from whom to choose the treasurer. Therefore, there can be $15 \cdot 14 \cdot 13 = 2730$ different slates of candidates.

Method 2: This is a permutation problem:

$$_{15}P_3 = \frac{15!}{(15-3)!} = \frac{15!}{12!} = 15 \cdot 14 \cdot 13 = 2730$$

different slates of candidates.

In all these problems, the answer was obtained by taking the largest r factors of $n!$

EXAMPLE 5: Evaluate: (A) $_8P_3$, (B) $_5P_5$, (C) $_{12}P_2$.

(A) $_8P_3 = 8 \cdot 7 \cdot 6 = 336$
(B) $_5P_5 = 5 \cdot 4 \cdot 3 \cdot 2 \cdot 1 = 120$
(C) $_{12}P_2 = 12 \cdot 11 = 132$

If the n elements of a set are to be arranged in a circle, they can be so arranged in $\frac{n!}{n}$ or $(n-1)!$ ways. There are fewer such permutations because there is no beginning to a circle as there is with a line.

EXAMPLE 6: How many ways can six people be seated at a round table?
This is a circular permutation, and so there are $(6-1)! = 5! = 120$ ways.

If the circular arrangement can be viewed from either side or turned over like a bracelet, there are one-half as many permutations. Therefore, the number of permutations is $\frac{(n-1)!}{2}$.

If the n elements of a set are not all different, several permutations will appear to be the same. To take care of this repetition, the formula for a permutation of n things where there are a of one kind and b of another kind is

$$\frac{n!}{a!b!}$$

EXAMPLE 7: How many permutations of the letters in the word TATTLETALE are there?
Of the 10 letters, there are 4 T's, 2 E's, 2 L's, and 2 A's. Therefore, the number of permutations is

$$\frac{10!}{4!2!2!2!} = \frac{10 \cdot 9 \cdot 8 \cdot 7 \cdot 6 \cdot 5 \cdot 4 \cdot 3 \cdot 2 \cdot 1}{4 \cdot 3 \cdot 2 \cdot 1 \cdot 2 \cdot 1 \cdot 2 \cdot 1 \cdot 2 \cdot 1} = 6300.$$

If a definite order in the grouping of the elements of a set is *not* necessary, the grouping is called a *combination*. The number of combinations of n things taken r at a time is denoted by $_nC_r$ or $C(n,r)$ or $\binom{n}{r}$.

$$\binom{n}{r} = \frac{_nP_r}{r!} = \frac{\text{the product of the largest } r \text{ factors of } n!}{r!}$$

EXAMPLE 8: Set $A = \{a,b,c,d\}$. (A) List all the permutations of the elements of A taken two at a time. (B) List all the combinations of the elements of A taken two at a time.

(A) $_4P_2 = 4 \cdot 3 = 12$. There should be 12 items in the permutation.

$$\left.\begin{matrix} ab & ba & ca & da \\ ac & bc & cb & db \\ ad & bd & cd & dc \end{matrix}\right\}\begin{matrix}\text{The permutations of } A \\ \text{taken 2 at a time.}\end{matrix}$$

(B) $\binom{4}{2} = \frac{4 \cdot 3}{2 \cdot 1} = 6$. There should be 6 items in the combination.

$$\left.\begin{matrix} ab & & \\ ac & bc & \\ ad & bd & cd \end{matrix}\right\}\begin{matrix}\text{The combinations of } A \\ \text{taken 2 at a time.}\end{matrix}$$

EXAMPLE 9: Evaluate (A) $\binom{9}{3}$, (B) $\binom{9}{2}$, (C) $\binom{8}{2}$, (D) $\binom{8}{6}$.

(A) $\binom{9}{3} = \frac{9 \cdot 8 \cdot 7}{3 \cdot 2 \cdot 1} = 84$

(B) $\binom{9}{6} = \frac{9 \cdot 8 \cdot 7 \cdot 6 \cdot 5 \cdot 4}{6 \cdot 5 \cdot 4 \cdot 3 \cdot 2 \cdot 1} = 84$

(C) $\binom{8}{2} = \frac{8 \cdot 7}{2 \cdot 1} = 28$

(D) $\binom{8}{6} = \frac{8 \cdot 7 \cdot 6 \cdot 5 \cdot 4 \cdot 3}{6 \cdot 5 \cdot 4 \cdot 3 \cdot 2 \cdot 1} = 28$

In this example, the answers to (A) and (B) are the same, and the answers to (C) and (D) are the same. The bottom numbers in (A) and (B), 3 and 6, add up to the top number, 9, and the bottom numbers in (C) and (D), 2 and 6, add up to the top number, 8. There was much less work to do when the bottom number was the smaller of the two numbers. In general, $\binom{n}{r} = \binom{n}{n-r}$.

EXAMPLE 10: From a deck of 52 cards, how many different piles of 50 cards can be selected?

This is a combination problem. Evaluate $\binom{52}{50}$. Since, $\binom{52}{50} = \binom{52}{2}$, it is easier to evaluate $\binom{52}{2}$, which is equal to $\frac{52 \cdot 51}{2 \cdot 1} = 1326$ different piles.

EXAMPLE 11: From a deck of 52 cards, in how many ways can a hand of 13 cards be dealt so that it contains 4 hearts and 9 spades?

Since there are 13 cards of each suit, 4 of 13 hearts and 9 of 13 spades must be selected.

$$\binom{13}{4} = \frac{13 \cdot 12 \cdot 11 \cdot 10}{4 \cdot 3 \cdot 2 \cdot 1} = 715 \quad \text{and} \quad \binom{13}{9} = \binom{13}{4} = 715.$$

Therefore, the number of ways this hand can be chosen is the product $(715)(715) = 511,225$.

All scientific calculators have commands for computing factorials, permutations, and combinations. The difficult part of a problem requiring permutations and combinations is figuring out how to set them up. Once this is done, you can easily use your calculator to do the computations. Explanations of solutions for the following exercises show the setup and indicate the calculation required.

EXERCISES

1. $\frac{(5+3)!}{5! + 3!} =$

   (A) 1
   (B) 56
   (C) 320
   (D) 8
   (E) 5040

2. Frisbees come in 5 models, 8 colors, and 3 sizes. How many Frisbees must the local dealer have on hand in order to have one of each kind available?

   (A) 24
   (B) 120
   (C) 16
   (D) 39
   (E) 55

3. A craftsperson has six different kinds of seashells. How many different bracelets can be constructed if only four shells are to be used in any one bracelet?

   (A) 90
   (B) 45
   (C) 60
   (D) 360
   (E) 180

4. How many different arrangements of the letters in the word RADAR are possible?

   (A) 120
   (B) 6
   (C) 30
   (D) 60
   (E) 20

5. If a person is dealt two cards from a 52-card deck, how many different hands are possible if order is not important?

   (A) 104
   (B) 13
   (C) 1326
   (D) 103
   (E) 2652

6. If $\binom{6}{x} = \binom{4}{x}$, then $x =$

   (A) 5
   (B) 4
   (C) 11
   (D) 0
   (E) 1

7. How many odd numbers of three digits each can be formed from the digits 2, 4, 6, and 7 if repetition of digits is permitted?

   (A) 6
   (B) 27
   (C) 24
   (D) 16
   (E) 256

8. Given eight points in a plane, no three of which are colinear. How many lines do the points determine?

   (A) 16
   (B) 64
   (C) 28
   (D) 7
   (E) 36

5.2   BINOMIAL THEOREM

Expanding a binomial, $(a + b)^n$, where n is a natural number, is a tedious operation for large values of n. The binomial theorem simplifies the work. For reference, consider the expansion for the first few values of n:

$$(a+b)^1 = a+b$$
$$(a+b)^2 = a^2 + 2ab + b^2$$
$$(a+b)^3 = a^3 + 3a^2b + 3ab^2 + b^3$$
$$(a+b)^4 = a^4 + 4a^3b + 6a^2b^2 + 4ab^3 + b^4$$
$$\vdots \qquad \vdots$$

Observations: There are $n + 1$ terms in each expansion.
The sum of the exponents in each term equals n.
The exponent of b is 1 less than the number of the term.
The coefficient of each term equals
$$\binom{n}{\text{either exponent}}$$

EXAMPLE 1:   What is the coefficient of the second term of $(a + b)^4$?
The coefficient of the second term of $(a + b)^4$ is equal to $\binom{4}{1}$ or $\binom{4}{3}$, where 1 is the exponent of b and 3 is the exponent of a. The first exponent in each case is 1. Using the fourth observation above, we can express the first coefficient as $\binom{n}{n}$ or $\binom{n}{0}$.

Therefore, $\binom{n}{n} = \binom{n}{0} = 1$.

EXAMPLE 2:   What is the third term of $(a + b)^{10}$?
The exponent of b is $3 - 1$ or 2.

The exponent of a must be 8 because the sum of the exponents, 8 and 2, must equal 10.

The coefficient is $\binom{10}{8} = \binom{10}{2} = \dfrac{10 \cdot 9}{2 \cdot 1} = 45$.

Therefore, the third term is $45a^8b^2$.

EXAMPLE 3:   What is the middle term of $(a - b)^8$?
$(a - b)^8 = [a + (-b)]^8$. There are nine terms. The middle term is the fifth term. The exponent of $(-b)$ is $5 - 1$ or 4. The exponent of a is 4. The coefficient of the middle term is $\binom{8}{4} = \dfrac{8 \cdot 7 \cdot 6 \cdot 5}{4 \cdot 3 \cdot 2 \cdot 1} = 70$.

Therefore, the middle term is $70a^4b^4$.

EXAMPLE 4:   What is the fourth term of $(a - 2b)^{12}$?
$(a - 2b)^{12} = [a + (-2b)]^{12}$. The exponent of $(-2b)$ is 3. The exponent of a is 9. The coefficient is $\binom{12}{3} = \dfrac{12 \cdot 11 \cdot 10}{3 \cdot 2 \cdot 1} = 220$.

The fourth term is $220a^9(-2b)^3 = 220a^9(-8b^3) = -1760a^9b^3$.

The binomial theorem method of expanding $(a + b)^n$ can be enlarged to include any real-number exponent, n, since the exponent of b is always a nonnegative integer. Although the terminology of combinations is no longer meaningful, the symbolism is still useful. When n is a nonnegative integer (greater than or equal to 3),
$$\binom{n}{3} = \frac{n(n-1)(n-2)}{3 \cdot 2 \cdot 1}.$$

Extending this process to any real number, n, gives the correct result.

EXAMPLE 5:   Give the first three terms of $(a + b)^{-3}$.
The exponent of b must be 1 less than the number of the term, and the sum of the exponents must equal -3.

Thus, the expansion is
$$\binom{-3}{0}a^{-3} + \binom{-3}{1}a^{-4}b + \binom{-3}{2}a^{-5}b^2 + \cdots$$
$$= 1a^{-3} + \frac{-3}{1}a^{-4}b + \frac{(-3)(-4)}{2 \cdot 1}a^{-5}b^2 + \cdots$$
$$+ a^{-3} - 3a^{-4}b + 6a^{-5}b^2 - \cdots$$

EXAMPLE 6:   Give the first four terms of $(a + b)^{1/2}$.
The exponent of b must be 1 less than the number of the term, and the sum of the exponents must equal $\dfrac{1}{2}$.

Thus, the expansion is
$$\binom{1/2}{0}a^{1/2} + \binom{1/2}{1}a^{-1/2}b + \binom{1/2}{2}a^{-3/2}b^2$$
$$+ \binom{1/2}{3}a^{-5/2}b^3 + \cdots$$
$$= 1a^{1/2} + \frac{\frac{1}{2}}{1}a^{-1/2}b + \frac{\frac{1}{2}\left(-\frac{1}{2}\right)}{2 \cdot 1}a^{-3/2}b^2$$
$$+ \frac{\frac{1}{2}\left(-\frac{1}{2}\right)\left(-\frac{3}{2}\right)}{3 \cdot 2 \cdot 1}a^{-5/2}b^3 + \cdots$$
$$= a^{1/2} + \frac{1}{2}a^{-1/2}b - \frac{1}{8}a^{-3/2}b^2 + \frac{1}{8}a^{-5/2}b^3 + \cdots$$

EXAMPLE 7:   In the expansion $(a + b)^n$, the rth term is equal to the $(r + 1)$st term when a is 3 times as large as b. What is the relationship between n and r?

The r th term $= \binom{n}{r-1} \cdot a^{n-r+1} \cdot b^{r-1}$
$$= \binom{n}{r-1} \cdot (3b)^{n-r+1} \cdot b^{r-1}$$
$$= \binom{n}{r-1} \cdot 3^{n-r+1} \cdot b^n.$$

The $(r + 1)$st term $= \binom{n}{r} \cdot a^{n-r} \cdot b^r$
$$= \binom{n}{r} \cdot (3b)^{n-r} \cdot b^r$$
$$= \binom{n}{r} \cdot 3^{n-r} \cdot b^n.$$

Therefore,

$$\binom{n}{r-1} \cdot 3 \cdot 3^{n-r} \cdot b^n = \binom{n}{r} \cdot 3^{n-r} \cdot b^n$$

which, when simplified, is

$$\binom{n}{r-1} \cdot 3 = \binom{n}{r}.$$

Expanding gives

$$3 \cdot \frac{n!}{(r-1)!(n-r+1)!} = \frac{n!}{r!(n-r)!}.$$

Dividing out factors and simplifying, we have

$$3 \cdot \frac{1}{n-r+1} = \frac{1}{r}$$

which gives

$$3r = n - r + 1 \quad \text{or} \quad n = 4r - 1.$$

EXERCISES

1. The middle term of $\left(\dfrac{1}{x} - x\right)^{10}$ is
   (A) –126
   (B) 126
   (C) –252
   (D) $252x^5$
   (E) none of these

2. The seventh term of $\left(a^3 - \dfrac{1}{a}\right)^8$ is
   (A) $28a^6$
   (B) $28a^{-6}$
   (C) 28
   (D) $8a^5$
   (E) $8a^{-5}$

3. What is the coefficient of x^{17} in the expansion of $x^5(1-x^2)^{12}$?
   (A) 12,376
   (B) –924
   (C) –12,376
   (D) –6188
   (E) 924

4. The coefficient of the third term of $(8x-y)^{1/3}$, after simplification, is
   (A) $-\dfrac{1}{9}$
   (B) $-\dfrac{1}{288}$
   (C) $\dfrac{5}{81}$
   (D) $\dfrac{5}{2592}$
   (E) $-\dfrac{1}{576}$

5.3 PROBABILITY

The probability of an event happening is a number defined to be the number of ways the event can happen successfully divided by the total number of ways the event can happen.

EXAMPLE 1: What is the probability of getting a head when a coin is flipped?
A coin can fall in one of two ways, heads or tails, and each is equally likely.

$$P(\text{head}) = \frac{\text{number of ways a head can come up}}{\text{total number of ways the coin can fall}}$$
$$= \frac{1}{2}.$$

EXAMPLE 2: What is the probability of getting a 3 when one die is thrown?
A die can fall with any one of six different numbers showing, and there is only one way a 3 can show.

$$P(3) = \frac{\text{number of ways a 3 can come up}}{\text{total number of ways the die can fall}} = \frac{1}{6}.$$

EXAMPLE 3: What is the probability of getting a sum of 7 when two dice are thrown?
Since it is not obvious how many different throws will produce a sum of 7, or how many different ways the two dice will land, it will be useful to consider all the possible outcomes. The set of all outcomes of an experiment is called the *sample space* of the experiment. In order to keep track of the elements of the sample space in this experiment, let the first die be green and the second die be red. Since the green die can fall in one of six ways, and the red die can fall in one of six ways, there should be $6 \cdot 6$ or 36 elements in the sample space. The elements of the sample space are as follows:

green	red	green	red	green	red	green	red	green	red	green	red
1	1	2	1	3	1	4	1	5	1	6	1
1	2	2	2	3	2	4	2	5	2	6	2
1	3	2	3	3	3	4	3	5	3	6	3
1	4	2	4	3	4	4	4	5	4	6	4
1	5	2	5	3	5	4	5	5	5	6	5
1	6	2	6	3	6	4	6	5	6	6	6

The circled elements of the sample space are those whose sum is 7.

$$P(7) = \frac{\text{number of successes}}{\text{total number}} = \frac{6}{36} = \frac{1}{6}.$$

The probability, p, of any event is a number such that $0 \le p \le 1$. If $p = 0$, the event cannot happen. If $p = 1$, the event is sure to happen.

EXAMPLE 4: (A) What is the probability of getting a 7 when one die is thrown? (B) What is the probability of getting a number less than 12 when one die is thrown?

(A) $P(7) = 0$ since a single die has only numbers 1 through 6 on its faces.

(B) $P(\# < 12) = 1$ since any face number is less than 12.

The *odds* in favor of an event happening are defined to be the probability of the event happening successfully divided by the probability of the event not happening successfully.

EXAMPLE 5: What are the odds in favor of getting a number greater than 2 when one die is thrown?

$P(\# > 2) = \dfrac{4}{6} = \dfrac{2}{3}$ and $P(\# \not> 2) = \dfrac{2}{6} = \dfrac{1}{3}$. Therefore, the odds in favor of a number greater than 2 =

$$\frac{P(\# > 2)}{P(\# \not> 2)} = \frac{\dfrac{2}{3}}{\dfrac{1}{3}} = \frac{2}{1} \quad \text{or} \quad 2{:}1.$$

Independent events are events that have no effect on one another. Two events are defined to be independent if and only if $P(A \cap B) = P(A) \cdot P(B)$, where $A \cap B$ means the intersection of the two sets A and B. If two events are not independent, they are said to be *dependent*.

EXAMPLE 6: If two fair coins are flipped, what is the probability of getting two heads?

Since the flip of each coin has no effect on the outcome of any other coin, these are independent events.

$$P(\text{HH}) = P(\text{H}) \cdot P(\text{H}) = \frac{1}{2} \cdot \frac{1}{2} = \frac{1}{4}.$$

EXAMPLE 7: When two dice are thrown, what is the probability of getting two 5s?

These are independent events because the result of one die does not affect the result of the other.

$$P(\text{two 5s}) = P(5) \cdot P(5) = \frac{1}{6} \cdot \frac{1}{6} = \frac{1}{36}$$

EXAMPLE 8: Two dice are thrown. Event A is "the sum of 7." Event B is "at least one die is a 6." Are A and B independent?

From the chart in Example 3, $A = \{(1,6), (6,1), (2,5), (5,2), (3,4), (4,3)\}$ and $B = \{(1,6), (2,6), (3,6), (4,6), (5,6), (6,6), (6,1), (6,2), (6,3), (6,4), (6,5)\}$. Therefore, $P(A) = \dfrac{1}{6}$ and $P(B) = \dfrac{11}{36}$. $A \cap B = \{(1,6)(6,1)\}$. Therefore, $P(A \cap B) = \dfrac{2}{36} = \dfrac{1}{18}$. $P(A) \cdot P(B) = \dfrac{11}{216} \ne \dfrac{1}{18}$. Therefore, $P(A \cap B) \ne P(A) \cdot P(B)$, and so events A and B are dependent.

EXAMPLE 9: If the probability that John will buy a certain product is $\dfrac{3}{5}$, that Bill will buy that product is $\dfrac{2}{3}$, and that Sue will buy that product is $\dfrac{1}{4}$, what is the probability that at least one of them will buy the product?

Since the purchase by any one of the people does not affect the purchase by anyone else, these events are independent. The best way to approach this problem is to consider the probability that none of them buys the product.

Let A = the event "John does not buy the product."
Let B = the event "Bill does not buy the product."
Let C = the event "Sue does not buy the product."

$$P(A) = 1 - \frac{3}{5} = \frac{2}{5}; \quad P(B) = 1 - \frac{2}{3} = \frac{1}{3}$$
$$P(C) = 1 - \frac{1}{4} = \frac{3}{4}$$

The probability that none of them buys the product =
$$P(A \cap B \cap C) = P(A) \cdot P(B) \cdot P(C) = \frac{2}{5} \cdot \frac{1}{3} \cdot \frac{3}{4} = \frac{1}{10}.$$
Therefore, the probability that at least one of them buys the product is $1 - \dfrac{1}{10} = \dfrac{9}{10}$.

In general, the probability of event A happening or event B happening or both happening is equal to the sum of $P(A)$ and $P(B)$ less the probability of both happening. In symbols, $P(A \cup B) = P(A) + P(B) - P(A \cap B)$, where $A \cup B$ means the union of sets A and B. If $P(A \cap B) = 0$, the events are said to be *mutually exclusive*.

EXAMPLE 10: What is the probability of drawing a spade or a king from a deck of 52 cards?

Let A = the event "drawing a spade."
Let B = the event "drawing a king."
Since there are 13 spades and 4 kings in a deck of cards,

$$P(A) = \frac{13}{52} = \frac{1}{4}; \quad P(B) = \frac{4}{52} = \frac{1}{13}$$

$P(A \cap B) = P(\text{drawing the king of spades}) = \dfrac{1}{52}$
$P(A \cup B) = P(A) + P(B) - P(A \cap B)$
$$= \frac{13}{52} + \frac{4}{52} - \frac{1}{52} = \frac{16}{52} = \frac{4}{13}.$$

These events are *not* mutually exclusive.

EXAMPLE 11: In a throw of two dice, what is the probability of getting a sum of 7 or 11?

Let A = the event "throwing a sum of 7."
Let B = the event "throwing a sum of 11."

$P(A \cap B) = 0$, and so these events *are* mutually exclusive.

$P(A \cup B) = P(A) + P(B)$. From the chart in Example 3,
$$P(A) = \frac{6}{36} \quad \text{and} \quad P(B) = \frac{2}{36}.$$

Therefore,

$$P(A \cup B) = \frac{6}{36} + \frac{2}{36} = \frac{8}{36} = \frac{2}{9}.$$

EXERCISES

1. With the throw of two dice, what is the probability that the sum will be a prime number?

(A) $\frac{7}{18}$

(B) $\frac{5}{11}$

(C) $\frac{5}{12}$

(D) $\frac{4}{11}$

(E) $\frac{1}{2}$

2. If a coin is flipped and one die is thrown, what is the probability of getting a head or a 4?

(A) $\frac{2}{3}$

(B) $\frac{5}{12}$

(C) $\frac{1}{12}$

(D) $\frac{7}{12}$

(E) $\frac{1}{3}$

3. Three cards are drawn from an ordinary deck of 52 cards. Each card is replaced in the deck before the next card is drawn. What is the probability that at least one of the cards will be a spade?

(A) $\frac{3}{8}$

(B) $\frac{3}{4}$

(C) $\frac{3}{52}$

(D) $\frac{9}{64}$

(E) $\frac{37}{64}$

4. A coin is tossed three times. Let A = {three heads occur} and B = {at least one head occurs}. What is $P(A \cup B)$?

(A) $\frac{1}{8}$

(B) $\frac{1}{4}$

(C) $\frac{1}{2}$

(D) $\frac{3}{4}$

(E) $\frac{7}{8}$

5. A class has 12 boys and 4 girls. If three students are selected at random from the class, what is the probability that all will be boys?

(A) $\frac{1}{55}$

(B) $\frac{1}{3}$

(C) $\frac{1}{4}$

(D) $\frac{11}{15}$

(E) $\frac{11}{28}$

6. A red box contains eight items, of which three are defective, and a blue box contains five items, of which two are defective. An item is drawn at random from each box. What is the probability that both items will be nondefective?

(A) $\frac{17}{20}$

(B) $\frac{3}{8}$

(C) $\frac{8}{13}$

(D) $\frac{3}{20}$

(E) $\frac{5}{13}$

7. A hotel has five single rooms available, for which six men and three women apply. What is the probability that the rooms will be rented to three men and two women?

(A) $\dfrac{5}{8}$

(B) $\dfrac{10}{21}$

(C) $\dfrac{23}{112}$

(D) $\dfrac{5}{9}$

(E) $\dfrac{97}{251}$

8. Of all the articles in a box, 80% are satisfactory, while 20% are not. The probability of obtaining exactly five good items out of eight randomly selected articles is

(A) 0.800
(B) 0.003
(C) 0.147
(D) 0.013
(E) 0.132

5.4 SEQUENCES AND SERIES

A *sequence* is a function with a domain consisting of the natural numbers in order. A *series* is the sum of the terms of a sequence.

EXAMPLE 1: Give an example of (A) an infinite sequence of numbers, (B) a finite sequence of numbers, (C) an infinite series of numbers.

(A) $\dfrac{1}{2}, \dfrac{1}{3}, \dfrac{1}{4}, \dfrac{1}{5}, \ldots, \dfrac{1}{n+1}, \ldots$, is an *infinite* sequence of numbers with $t_1 = \dfrac{1}{2}$, $t_2 = \dfrac{1}{3}$, $t_3 = \dfrac{1}{4}$, $t_4 = \dfrac{1}{5}$, and $t_n = \dfrac{1}{n+1}$.

(B) $2, 4, 6, \ldots, 20$ is a *finite* sequence of numbers with $t_1 = 2, t_2 = 4, t_3 = 6, t_{10} = 20$.

(C) $\dfrac{1}{2} + \dfrac{1}{4} + \dfrac{1}{8} + \dfrac{1}{16} + \cdots + \dfrac{1}{2^n} + \cdots$ is an infinite series of numbers.

EXAMPLE 2: If $t_n = \dfrac{2n}{n+1}$, find the first five terms of the sequence.

When 1, 2, 3, 4, and 5 are substituted for n, $t_1 = \dfrac{2}{2} = 1, t_2 = \dfrac{4}{3}, \; t_3 = \dfrac{6}{4} = \dfrac{3}{2}, t_4 = \dfrac{8}{5},$ and $t_5 = \dfrac{10}{6} = \dfrac{5}{3}.$

The first five terms are $1, \dfrac{4}{3}, \dfrac{3}{2}, \dfrac{8}{5}, \dfrac{5}{3}.$

EXAMPLE 3: If $a_1 = 2$ and $a_n = \dfrac{a_{n-1}}{2}$, find the first five terms of the sequence.

Since every term is expressed with respect to the immediately preceding term, this is called a *recursion formula*.

$$a_1 = 2, a_2 = \dfrac{a_1}{2} = \dfrac{2}{2} = 1,$$

$$a_3 = \dfrac{a_2}{2} = \dfrac{1}{2}, \, a_4 = \dfrac{a_3}{2} = \dfrac{\frac{1}{2}}{2} = \dfrac{1}{4},$$

$$a_5 = \dfrac{a_4}{2} = \dfrac{\frac{1}{4}}{2} = \dfrac{1}{8}.$$

Therefore, the first five terms are $2, 1, \dfrac{1}{2}, \dfrac{1}{4}, \dfrac{1}{8}.$

A series can be abbreviated by using the Greek letter sigma, Σ, to represent the summation of several terms.

EXAMPLE 4: (A) Express the series $2 + 4 + 6 + \cdots + 20$ in sigma notation. (B) Evaluate $\displaystyle\sum_{k=0}^{5} k^2$.

(A) The series $2 + 4 + 6 + \cdots + 20 = \displaystyle\sum_{i=1}^{10} 2i = 110.$

(B) $\displaystyle\sum_{k=0}^{5} k^2 = 0^2 + 1^2 + 2^2 + 3^2 + 4^2 + 5^2 = 0 + 1 + 4 + 9 + 16 + 25 = 55.$

One of the most common sequences studied at this level is an *arithmetic sequence* (or *arithmetic progression*). Each term differs from the preceding term by a common difference. In general, an arithmetic sequence is denoted by

$$t_1, t_1 + d, t_1 + 2d, t_1 + 3d, \ldots, t_1 + (n-1)d$$

where d is the common difference and $t_n = t_1 + (n-1)d$. The sum of n terms of the series constructed from an arithmetic sequence is given by the formula

$$S_n = \dfrac{n}{2}(t_1 + t_n) \quad \text{or} \quad S_n = \dfrac{n}{2}[2t_1 + (n-1)d]$$

EXAMPLE 5: (A) Find the 28th term of the arithmetic sequence $2, 5, 8, \ldots$. (B) Express the 28 terms of the series of this sequence using sigma notation. (C) Find the sum of the first 28 terms of the series.

(A) $t_n = t_1 + (n-1)d$

$t_{28} = 2 + 27 \cdot 3 = 83$

(B) $\displaystyle\sum_{k=0}^{27}(3k+2)$ or $\displaystyle\sum_{j=1}^{28}(3j-1)$

(C) $S_n = \dfrac{n}{2}(t_1 + t_n)$

$S_{28} = \dfrac{28}{2}(2+83) = 14 \cdot 85 = 1190$

EXAMPLE 6: If $t_8 = 4$ and $t_{12} = -2$, find the first three terms of the arithmetic sequence.

$$t_n = t_1 + (n-1)d$$
$$t_8 = 4 = t_1 + 7d$$
$$t_{12} = -2 = t_1 + 11d$$

To solve these two equations for d, subtract the equations.

$$-6 = 4d$$
$$d = -\frac{3}{2}$$

Substituting in the first equation gives $4 = t_1 + 7\left(-\dfrac{3}{2}\right)$.

Thus,

$$t_1 = 4 + \frac{21}{2} = \frac{29}{2}$$

$$t_2 = \frac{29}{2} + \left(-\frac{3}{2}\right) = \frac{26}{2} = 13$$

$$t_3 = \frac{29}{2} + 2\left(-\frac{3}{2}\right) = \frac{23}{2}$$

The first three terms are $\dfrac{29}{2}, 13, \dfrac{23}{2}$.

EXAMPLE 7: In an arithmetic series, if $S_n = 3n^2 + 2n$, find the first three terms.

When $n = 1$, $S_1 = t_1$. Therefore, $t_1 = 3(1)^2 + 2 \cdot 1 = 5$.

$$S_2 = t_1 + t_2 = 3(2)^2 + 2 \cdot 2 = 16$$
$$5 + t_2 = 16$$
$$t_2 = 11$$

Therefore, $d = 6$, which leads to a third term of 17. Thus, the first three terms are 5, 11, 17.

The terms falling between two given terms of an arithmetic sequence are called *arithmetic means*. If there is only one arithmetic mean between two given terms, it is called the *average* of the two terms.

EXAMPLE 8: Insert three arithmetic means between 1 and 9.

$$1, m_1, m_2, m_3, 9$$
$$t_1, t_2, t_3, t_4, t_5$$

Since $t_1 = 1$ and $t_5 = t_1 + (5-1)d = 9$,

$$1 + 4d = 9$$
$$4d = 8$$
$$d = 2$$

Therefore, the three arithmetic means are 3, 5, and 7.

Another very common sequence studied at this level is a *geometric sequence* (or *geometric progression*). The ratio of any two successive terms is equal to a constant, r, called the *common ratio*. In general, a geometric sequence is denoted by

$$t_1, t_1 r, t_1 r^2, \ldots, t_1 r^{n-1}, \quad \text{where } t_n = t_1 r^{n-1}.$$

The sum of n terms of a series constructed from a geometric sequence is given by

$$S_n = \frac{t_1(1-r^n)}{1-r}.$$

EXAMPLE 9: (A) Find the seventh term of the geometric sequence $1, 2, 4, \ldots$, and (B) the sum of the first seven terms.

(A) $r = \dfrac{t_2}{t_1} = \dfrac{2}{1} = 2; t_7 = t_1 r^{7-1}; t_7 = 1 \cdot 2^6 = 64$

(B) $S_7 = \dfrac{1(1-2^7)}{1-2} = \dfrac{1-128}{-1} = 127$

EXAMPLE 10: The first term of a geometric sequence is 64, and the common ratio is $\dfrac{1}{4}$. For what value of n is $t_n = \dfrac{1}{4}$?

$$t_n = t_1 r^{n-1}; \frac{1}{4} = 64\left(\frac{1}{4}\right)^{n-1}$$

$$\frac{1}{4} = 4^3\left(\frac{1}{4}\right)^{n-1}; \frac{1}{4} = \left(\frac{1}{4}\right)^{-3} \cdot \left(\frac{1}{4}\right)^{n-1}$$

$$\frac{1}{4} = \left(\frac{1}{4}\right)^{-3+n-1}; \left(\frac{1}{4}\right)^{1} = \left(\frac{1}{4}\right)^{n-4}$$

Therefore, $1 = n - 4$ and $n = 5$.

In a geometric sequence, if $|r| < 1$, the sum of the series approaches a limit as n approaches infinity. In the formula $S_n = \dfrac{t_1(1-r^n)}{1-r}$, if $|r| < 1$, the term $r^n \to 0$ as $n \to \infty$.

Therefore, as long as $|r| < 1$, $\displaystyle\lim_{n \to \infty} S_n = \dfrac{t_1}{1-r}$.

EXAMPLE 11: Evaluate (A) $\displaystyle\lim_{n\to\infty}\sum_{k=1}^{n}\frac{1}{2^k}$ and

(B) $\displaystyle\sum_{j=0}^{\infty}(-3)^{-j}$.

Both problems ask the same question: Find the limit of the sum of the geometric series as $n \to \infty$.

(A) When the first few terms, $\dfrac{1}{2}+\dfrac{1}{4}+\dfrac{1}{8}+\cdots$, are listed,

it can be seen that the common ratio is $r = \dfrac{1}{2}$. Therefore,

$$\lim_{n\to\infty} S_n = \frac{\dfrac{1}{2}}{1-\dfrac{1}{2}} = 1.$$

(B) When the first few terms, $\dfrac{1}{1}-\dfrac{1}{3}+\dfrac{1}{9}-\cdots$, are listed,

it can be seen that the common ratio is $r = -\dfrac{1}{3}$. Therefore,

$$\lim_{n\to\infty} S_n = \frac{1}{1-\left(-\dfrac{1}{3}\right)} = \frac{1}{\dfrac{4}{3}} = \frac{3}{4}.$$

EXAMPLE 12: Find the exact value of the repeating decimal 0.4545 . . . .

This can be represented by a geometric series, $0.45 + 0.0045 + 0.000045 + \cdots$, with $t_1 = 0.45$ and $r = 0.01$.
Since $|r| < 1$,

$$\lim_{n\to\infty} S_n = \frac{0.45}{1-0.01} = \frac{0.45}{0.99} = \frac{45}{99} = \frac{5}{11}.$$

The terms falling between two given terms of a geometric sequence are called *geometric means*. If there is only one geometric mean between two given terms, it is called the *mean proportional* of the two terms.

EXAMPLE 13: Insert five geometric means between $\dfrac{1}{8}$ and 8.

$$\frac{1}{8}, m_1, m_2, m_3, m_4, m_5, 8$$
$$t_1, t_2, t_3, t_4, t_5, t_6, t_7$$

Since $t_1 = \dfrac{1}{8}$ and $t_7 = t_1 r^{n-1} = 8$,

$$\frac{1}{8}r^6 = 8$$
$$r^6 = 64$$
$$r = \pm(64)^{1/6} = \pm(2^6)^{1/6} = \pm 2.$$

Thus, there are two sets of geometric means.

	m_1	m_2	m_3	m_4	m_5
If $r = +2$,	$\dfrac{1}{4}$	$\dfrac{1}{2}$	1	2	4
If $r = -2$,	$-\dfrac{1}{4}$	$\dfrac{1}{2}$	-1	2	-4

EXAMPLE 14: Given the sequence 2, x, y, 9, if the first three terms form an arithmetic sequence and the last three terms form a geometric sequence, find x and y.

From the arithmetic sequence $\begin{cases} x = 2+d \\ y = 2+2d \end{cases}$, substitute to eliminate d.

$$y = 2 + 2(x-2)$$
$$y = 2 + 2x - 4$$
$$*y = 2x - 2$$

From the geometric sequence $\begin{cases} 9 = yr \\ y = xr \end{cases}$, substitute to eliminate r.

$$9 = y \cdot \frac{y}{x}$$
$$*9x = y^2$$

Use the two equations with the * to eliminate y:
$$9x = (2x-2)^2$$
$$9x = 4x^2 - 8x + 4$$
$$4x^2 - 17x + 4 = 0$$
$$(4x-1)(x-4) = 0$$
$$4x - 1 = 0 \quad\text{or}\quad x - 4 = 0$$

Thus, $x = \dfrac{1}{4}$ or 4.

Substitute in $y = 2x - 2$:

$$\text{if } x = \frac{1}{4}, \; y = -\frac{3}{2}$$
$$\text{if } x = 4, \; y = 6.$$

EXERCISES

1. If $a_1 = 3$ and $a_n = n + a_{n-1}$, the sum of the first five terms is

   (A) 30
   (B) 17
   (C) 42
   (D) 45
   (E) 68

2. If x, y, and z are three consecutive terms in an arithmetic sequence, which one of the following is true?

   (A) $y = \dfrac{1}{2}(x+z)$
   (B) $y = \sqrt{xz}$
   (C) $y = \sqrt{x+z}$
   (D) $y = z + x$
   (E) $z = \dfrac{1}{2}(x+y)$

3. If the repeating decimal $0.23\overline{737}\ldots$ is written as a fraction in lowest terms, the sum of the numerator and denominator is

(A) 245
(B) 1237
(C) 16
(D) 47
(E) 334

4. The first three terms of a geometric sequence are $\sqrt[4]{3}, \sqrt[8]{3}, 1$. The fourth term is

(A) $\sqrt[16]{3}$

(B) $\dfrac{1}{\sqrt[16]{3}}$

(C) $\dfrac{1}{\sqrt[8]{3}}$

(D) $\dfrac{1}{\sqrt[4]{3}}$

(E) $\sqrt[32]{3}$

5. By how much does the arithmetic mean between 1 and 25 exceed the positive geometric mean between 1 and 25?

(A) 5
(B) about 7.1
(C) 8
(D) 12.9
(E) 18

6. Let S_n equal the sum of all the prime numbers less than or equal to n. What is the largest value for n for which $S_n \le 100$?

(A) 23
(B) 14
(C) 29
(D) 10
(E) 28

7. The units digit of $\displaystyle\sum_{i=1}^{10} i!$ is

(A) 0
(B) 3
(C) 5
(D) 7
(E) 9

8. $\displaystyle\sum_{k=1}^{100}(-1)^k \cdot k =$

(A) 4950
(B) 100
(C) 5050
(D) 50
(E) 0

9. In a geometric series $S_\infty = \dfrac{2}{3}$ and $t_1 = \dfrac{2}{7}$. What is r?

(A) $\dfrac{2}{3}$

(B) $-\dfrac{4}{7}$

(C) $\dfrac{2}{7}$

(D) $\dfrac{4}{7}$

(E) $-\dfrac{2}{7}$

5.5   GEOMETRY AND VECTORS

This section contains miscellaneous topics from geometry not already covered that are likely to appear on the Math Level IIC examination.

Transformations: In a plane, a transformation slides, rotates, or stretches a geometric figure from one position to another. If point (x,y) lies on a graph and is transformed into point (x',y'), the formula for sliding (called a *translation*) is of the form

$$\begin{cases} x' = x + r \\ y' = y + s \end{cases}.$$

The formula for stretching is of the form

$$\begin{cases} x' = hx \\ y' = kx \end{cases}.$$

The formula for rotating is beyond the scope of this book.

EXAMPLE 1: Given the equation $y = x^2 + 4x + 6$, find the equation obtained by performing the translation

$$\begin{cases} x' = x + 2 \\ y' = y - 2 \end{cases}.$$

Solving the translation equations for x and y and substituting gives

$$y' + 2 = (x' - 2)^2 + 4(x' - 2) + 6$$

$$y' + 2 = (x')^2 - 4x' + 4 + 4x' - 8 + 6$$

$$y' = (x')^2$$

EXAMPLE 2: If the graph of $f(x)$ is given in the figure, what does the graph of $f(2x - 3)$ look like?

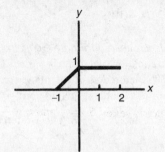

To avoid confusion, let x in $f(2x - 3)$ be called x'. When $2x' - 3$ replaces the original x, the transformation relating the original graph to the desired graph is $x = 2x' - 3$. Therefore, $x' = \dfrac{x + 3}{2}$. When the graph is used to find y and this equation is used to find x', it is possible to obtain points on the desired graph.

When $x = -1$, $y = 0$ and $x' = 1$.

When $x = 0$, $y = 1$ and $x' = \dfrac{3}{2}$.

When $x = 1$, $y = 1$ and $x' = 2$.

When $x = 2$, $y = 1$ and $x' = \dfrac{5}{2}$.

When points (x', y) are graphed, the following figure is obtained:

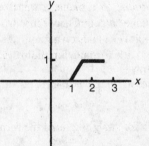

EXAMPLE 3: The shaded region in the figure is acted on by the transformation T, which transforms any point (x, y) into point $(x, x+y)$. What is this figure transformed into?

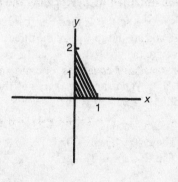

Taking points on the border of the shaded region and transforming them by T gives an idea of the shape of the new region.

$(0,0) \xrightarrow{T} (0,0)$ $\left(\dfrac{1}{4},\dfrac{3}{2}\right) \xrightarrow{T} \left(\dfrac{1}{4},\dfrac{7}{4}\right)$ $\left(\dfrac{3}{4},0\right) \xrightarrow{T} \left(\dfrac{3}{4},\dfrac{3}{4}\right)$

$(0,1) \xrightarrow{T} (0,1)$ $\left(\dfrac{1}{2},0\right) \xrightarrow{T} \left(\dfrac{1}{2},\dfrac{1}{2}\right)$ $\left(\dfrac{3}{4},\dfrac{1}{2}\right) \xrightarrow{T} \left(\dfrac{3}{4},\dfrac{5}{4}\right)$

$(0,2) \xrightarrow{T} (0,2)$ $\left(\dfrac{1}{4},0\right) \xrightarrow{T} \left(\dfrac{1}{4},\dfrac{1}{4}\right)$ $\left(\dfrac{1}{2},1\right) \xrightarrow{T} \left(\dfrac{1}{2},\dfrac{3}{2}\right)$

$(1,0) \xrightarrow{T} (1,1)$

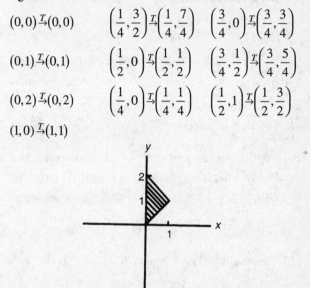

Vectors: A vector in a plane is defined to be an ordered pair of real numbers. A vector in space is defined as an ordered triple of real numbers. On a coordinate system, a vector is usually represented by an arrow whose initial point is the original and whose terminal point is at the ordered pair (or triple) that named the vector. Vector quantities always have a magnitude or *norm* (the length of the arrow) and direction (the angle the arrow makes with the positive x-axis).

All properties of two-dimensional vectors can be extended to three-dimensional vectors. We will express the properties in terms of two-dimensional vectors for convenience. If vector $\vec{V}$ is designated by (v_1, v_2) and vector $\vec{U}$ is designated by (u_1, u_2), vector $\overrightarrow{U + V}$ is designated by $(u_1 + v_1, u_2 + v_2)$ and called the *resultant* of $\vec{U}$ and $\vec{V}$. Vector $-\vec{V}$ has the same magnitude as $\vec{V}$ but has a direction opposite that of $\vec{V}$.

Every vector $\vec{V}$ can be expressed in terms of any other two (three) nonparallel vectors. In many instances unit vectors parallel to the x- and y-axes are used. If vector $\vec{i} = (1,0)$ and vector $\vec{j} = (0,1)$ [and, in three dimensions, $\vec{k} = (0,0,1)$], any vector $\vec{V} = ai + bj$, where a and b are real numbers. A unit vector parallel to $\vec{V}$ can be determined by dividing $\vec{V}$ by its norm, denoted by $\|\vec{V}\|$ and equal to $\sqrt{(v_1)^2 + (v_2)^2}$.

It is possible to determine algebraically whether two vectors are perpendicular by defining the *dot product* or *inner product* of two vectors, $\vec{V}(v_1, v_2)$ and $\vec{U}(u_1, u_2)$.

$$\vec{V} \cdot \vec{U} = v_1 u_1 + v_2 u_2$$

Notice that the dot product of two vectors is a *real number*, not a vector. Two vectors, $\vec{V}$ and $\vec{U}$, are perpendicular if and only if $\vec{V} \cdot \vec{U} = 0$.

EXAMPLE 4: Let vector $\vec{V}$ = (2,3) and vector $\vec{U}$ = (6,–4). (A) What is the resultant of $\vec{U}$ and $\vec{V}$? (B) What is the norm of $\vec{U}$? (C) Express $\vec{V}$ in terms of $\vec{i}$ and $\vec{j}$. (D) Are $\vec{U}$ and $\vec{V}$ perpendicular?

(A) The resultant, $\overrightarrow{U+V}$, equals $(6+2, -4+3) = (8, -1)$.

(B) The norm of $\vec{U}$, $\|\vec{U}\| = \sqrt{36+16} = \sqrt{52} = 2\sqrt{13}$.

(C) $\vec{V} = 2\vec{i} + 3\vec{j}$. To verify this, use the definitions of $\vec{i}$ and $\vec{j}$. $\vec{V} = 2(1,0) + 3(0,1) = (2,0) + (0,3) = (2,3) = \vec{V}$.

(D) $\vec{U} \cdot \vec{V} = 6 \cdot 2 + (-4) \cdot 3 = 12 - 12 = 0$. Therefore, $\vec{U}$ and $\vec{V}$ are perpendicular because the dot product is equal to zero.

EXAMPLE 5: If $\vec{U}$ = (–1,4) and the resultant of $\vec{U}$ and $\vec{V}$ is (4,5), find $\vec{V}$.

Let $\vec{V} = (v_1, v_2)$. The resultant $\overrightarrow{U+V} = (-1,4) + (v_1, v_2) = (4,5)$. Therefore, $(-1 + v_1, 4 + v_2) = (4,5)$, which implies that $-1 + v_1 = 4$ and $4 + v_2 = 5$. Thus, $v_1 = 5$ and $v_2 = 1$. $\vec{V} = (5,1)$.

EXAMPLE 6: Express vector $\vec{V}$ = (3,–7) as a linear combination of (i.e., in terms of) vectors $\vec{U}$ = (–6,8) and $\vec{W}$ = (9,–13).

This question requires that two real numbers, a and b, be found such that $\vec{V} = a\vec{U} + b\vec{W}$.

$$\vec{V} = (3,-7) = a(-6,8) + b(9,-13)$$
$$(3,-7) = (-6a, 8a) + (9b, -13b)$$
$$(3,-7) = (-6a + 9b, 8a - 13b)$$

Therefore, $3 = 6a + 9b$ and $-7 = 8a - 13b$.

Solving these two equations simultaneously gives $a = 4$ and $b = 3$.

Thus, $\vec{V} = 4\vec{U} + 3\vec{W}$.

Three-Dimensional Coordinate Geometry: In three dimensions, the equation of a plane is of the form $Ax + By + Cz + D = 0$. The intercepts of the plane can be found by setting two of the variables equal to zero.

Let $y = z = 0$, and $x = -\dfrac{D}{A}$. The x-intercept occurs at point $\left(-\dfrac{D}{A}, 0, 0\right)$.

Let $x = z = 0$, and $y = -\dfrac{D}{B}$. The y-intercept occurs at point $\left(0, -\dfrac{D}{B}, 0\right)$.

Let $x = y = 0$, and $z = -\dfrac{D}{C}$. The z-intercept occurs at point $\left(0, 0, -\dfrac{D}{C}\right)$.

The line that is the intersection of the plane and one of the coordinate planes is called a *trace*. The equation of the xy-trace is obtained by setting $z = 0$ in the equation of the plane. The xy-trace is the line whose equation is $Ax + By + D = 0$.

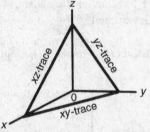

In three dimensions the equation of a line must be expressed as a set of three parametric equations:

$$x = x_0 + c_1 d$$
$$y = y_0 + c_2 d$$
$$z = z_0 + c_3 d$$

where d is the parameter, (x_0, y_0, z_0) are the coordinates of any point on the line, and c_1, c_2, and c_3 are called *direction numbers*. If (x_1, y_1, z_1) are the coordinates of any other point on the line, one set of values for the direction numbers is $c_1 = x_1 - x_0$, $c_2 = y_1 - y_0$, and $c_3 = z_1 - z_0$.

If $(c_1)^2 + (c_2)^2 + (c_3)^2 = 1$, then c_1, c_2, c_3 are called *direction cosines* and are the cosines of the three angles that the line makes with the positive x-, y-, and z-axes.

EXAMPLE 7: What are the traces of the plane with the equation $3x + 2y - 4z = 12$?

Set $z = 0$. The xy-trace is $3x + 2y = 12$.
Set $y = 0$. The xz-trace is $3x - 4z = 12$.
Set $x = 0$. The yz-trace is $2y - 4z = 12$.

EXAMPLE 8: (A) Find an equation of the line passing through the points whose coordinates are (1,2,3) and (4,–2,6). (B) Find the direction cosines of the line.

(A) Let $(x_0, y_0, z_0) = (1,2,3)$. $c_1 = 4 - 1 = 3$, $c_2 = -2 - 2 = -4$, and $c_3 = 6 - 3 = 3$. One set of equations for the line is

$$x = 1 + 3d$$
$$y = 2 - 4d$$
$$z = 3 + 3d$$

(B) To find direction cosines, it is necessary to find the distance between the two points used to determine the direction numbers.

$$\text{Distance} = \sqrt{(4-1)^2 + (-2-2)^2 + (6-3)^2}$$
$$= \sqrt{3^2 + (-4)^2 + 3^2} = \sqrt{9 + 16 + 9} = \sqrt{34}.$$

With direction numbers 3, –4, and 3, the direction cosines become $\dfrac{3}{\sqrt{34}}$, $\dfrac{-4}{\sqrt{34}}$, and $\dfrac{3}{\sqrt{34}}$.

EXAMPLE 9: How far is the plane whose equation is 3x + 4y – 5z + 12 = 0 from the origin?

Extending the formula for the distance between a point and a line to a formula for the distance between a point and a plane gives

$$\text{Distance} = \frac{|Ax_1 + By_1 + Cz_1 + D|}{\sqrt{A^2 + B^2 + C^2}}$$

In this problem $(x_1, y_1, z_1) = (0,0,0)$ and $A = 3$, $B = 4$, $C = -5$.

$$\text{Distance} = \frac{|3 \cdot 0 + 4 \cdot 0 - 5 \cdot 0 + 12|}{\sqrt{3^2 + 4^2 + (-5)^2}} = \frac{|12|}{\sqrt{9 + 16 + 25}}$$

$$= \frac{12}{\sqrt{50}} = \frac{12}{5\sqrt{2}} = \frac{6\sqrt{2}}{5} \approx 1.697$$

Solid Figures: A solid of revolution is obtained by taking a plane figure and rotating it about some line in the plane that does not intersect the figure to form a solid figure.

EXAMPLE 10: If the segment of line y = –2x + 2, which lies in quadrant I (Fig. a) is rotated about the y-axis, a cone is formed (Fig. b). What is the volume of the cone?

$$V = \frac{1}{3}\pi r^2 h$$

$$V = \frac{1}{3}\pi \cdot 1^2 \cdot 2 = \frac{2\pi}{3}$$

Many three-dimensional problems involve situations where it is necessary to picture the solid figures and their relationships to one another.

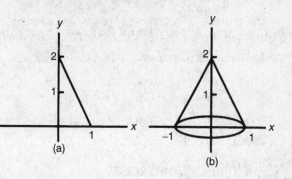
(a)   (b)

EXAMPLE 11: What figure is formed by the set of points at a fixed distance, d, from a line and at the same time 2d units from a point, P, on the line?

The set of points d units from the line is a cylinder of radius d, with the line as the central axis of the cylinder. The set of points $2d$ units from P is a sphere with center P and radius $2d$. Since the radius of the sphere is greater than that of the cylinder, the set of points satisfying the conditions is the intersection of the cylinder piercing the sphere. This intersection consists of two parallel circles with centers on the line, perpendicular to the line, and equidistant from the fixed point P.

EXAMPLE 12: In Example 11, how far apart are the two parallel circles?

A partial picture of the situation is shown below.

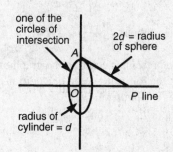

one of the circles of intersection
A
$2d$ = radius of sphere
O
P line
radius of cylinder = d

Distance OP is one-half the distance between the circles, and it is a side of a right triangle whose other two sides are known. Using the Pythagorean theorem gives

$$\overline{OP}^2 + d^2 = (2d)^2$$
$$\overline{OP}^2 = 4d^2 - d^2$$

Therefore, $OP = \sqrt{3}d$.

The distance between the circles is $2\sqrt{3}d$.

EXAMPLE 13: A cone is inscribed in a cylinder. What is the ratio of the volume of the cone to the volume of the cylinder?

The radius of the base and the heights of the cone and the cylinder are equal. The formula for the volume of the cone is $V_1 = \frac{1}{3}\pi r^2 h$. The formula for the volume of the cylinder is $V_2 = \pi r^2 h$. Therefore,

$$\frac{V_1}{V_2} = \frac{\frac{1}{3}\pi r^2 h}{\pi r^2 h} = \frac{\frac{1}{3}}{1} = \frac{1}{3}.$$

EXERCISES

1. In the figure, the graph of $f(x)$ has two transformations performed on it. First it is rotated 180° about the origin, and then it is reflected about the x-axis. Which of the choices below is the function expression of the resulting curve?

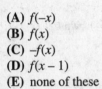

(A) $f(-x)$
(B) $f(x)$
(C) $-f(x)$
(D) $f(x-1)$
(E) none of these

2. If (x,y) represents a point on the graph of $y = x + 2$, which of the following could be a portion of the graph of the set of points $\left(x, \sqrt{y}\right)$?

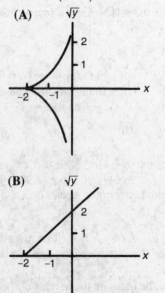

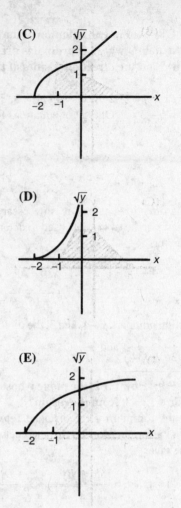

3. In the figure, S is the set of points in the shaded region. Which of the choices below represents the set of all points $(x + y, y)$, where (x,y) is a point in S?

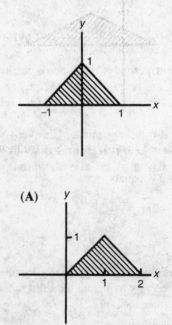

(B)

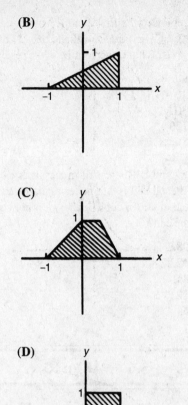

(C)

(D)

(E)

4. If $\vec{V} = 2\vec{i} + 3\vec{j}$ and $\vec{U} = \vec{i} - 5\vec{j}$, the resultant vector

of $2\vec{U} + 3\vec{V}$ equals

(A) $3\vec{i} - 2\vec{j}$

(B) $5\vec{i} + \vec{j}$

(C) $7\vec{i} - 9\vec{j}$

(D) $8\vec{i} - \vec{j}$

(E) $2\vec{i} + 3\vec{j}$

5. A unit vector perpendicular to vector $\vec{V} = (3, -4)$ is

(A) $(4, 3)$

(B) $\left(\dfrac{3}{5}, \dfrac{4}{5}\right)$

(C) $\left(-\dfrac{3}{5}, -\dfrac{4}{5}\right)$

(D) $\left(-\dfrac{4}{5}, -\dfrac{3}{5}\right)$

(E) $\left(-\dfrac{4}{5}, \dfrac{3}{5}\right)$

6. The plane $2x + 3y - 4z = 5$ intersects the x-axis at the point where the x-coordinate is

(A) 2

(B) 2.5

(C) $1\dfrac{2}{3}$

(D) -1.25

(E) 5

7. The distance between two points in space, $P_1(x, -1, -1)$ and $P_2(3, -3, 1)$, is 3. Find the possible values of x.

(A) only 1 and 2

(B) only 2 and 3

(C) only -2 and -3

(D) only 2 and 4

(E) only -2 and -4

8. Which of the following is a trace of the plane $3x + 5y - 7z + 8 = 0$?

(A) $3x + 5y = 8$

(B) $3x + 7z + 8 = 0$

(C) $5y - 7z = 8$

(D) $7z + 3x = 8$

(E) $7z - 5y = 8$

9. Which of the following is the set of parametric equations for the line in space that passes through points $(2, 4, 1)$ and $(-1, 5, 2)$?

(A) $x = -1 + 2d,\ y = 5 + 4d,\ z = 2 + d$

(B) $x = -1 + d,\ y = 5 - d,\ z = 2 - d$

(C) $x = 2 - d,\ y = 4 + 5d,\ z = 1 + 2d$

(D) $x = 2 - 3d,\ y = 4 + d,\ z = 1 + d$

(E) $x = 2 + d,\ y = 4 + 9d,\ z = 1 + 3d$

10. If the portion of the graph of $x^2 + y^2 = 4$ that lies in quadrant I is revolved around the x-axis, the volume of the resulting figure is

(A) 3.1
(B) 6.3
(C) 12.6
(D) 16.8
(E) 33.5

11. A right circular cone whose base radius is 12 is inscribed in a sphere of radius 13. What is the volume of the cone?

(A) 720
(B) $\dfrac{2197\pi}{3}$
(C) 864
(D) 1440
(E) 2592

12. Given a right circular cylinder such that the volume has the same numerical value as the total surface area. The smallest integral value for the radius of the cylinder is

(A) 1
(B) 2
(C) 3
(D) 4
(E) This value cannot be determined.

13. A square pyramid is inscribed in a cylinder whose base radius is 4 and height is 9. What is the volume of the pyramid?

(A) 48
(B) 144
(C) 288
(D) 96
(E) 108

14. The set of points in space equidistant from two fixed points is

(A) one point
(B) one line
(C) a plane
(D) a hyperbola
(E) none of the above

15. What is the length of the xz-trace, between the positive x-axis and the positive z-axis, of the plane whose equation is $2x + 3y + 4z = 7$?

(A) 2.3
(B) 3.9
(C) 4.2
(D) 2.9
(E) 4.6

16. The region in the first quadrant bounded by the line $3x + 2y = 7$ and the coordinate axes is rotated about the x-axis. What is the volume of the resulting figure?

(A) 90
(B) 20
(C) 30
(D) 120
(E) 8

17. Three tennis balls just fit in a cylindrical tennis ball can. If each ball is 2.5 inches in diameter, what is the volume of air (in cubic inches) left between the balls and the can?

(A) 98.2
(B) 122.7
(C) 8.2
(D) 12.3
(E) 65.4

5.6 VARIATION

In a *direct variation* the ratio of the variables is a constant. The first variable mentioned is in the numerator, and all others form a product in the denominator. Thus, as the first variable increases (or decreases), the second variable (or product of variables) must also increase (or decrease) in order to maintain a constant value.

EXAMPLE 1: Give a formula for the statement "x varies directly as y, the square of z, and the square root of a."

$$\frac{x}{yz^2\sqrt{a}} = K, \quad \text{where } K \text{ is a constant.}$$

In an *inverse variation* the product of all variables is equal to a constant. Thus, as the first variable increases (or decreases), the second variable (or product of variables) must decrease (or increase) in order to maintain a constant value.

EXAMPLE 2: Give a formula for the statement "x varies inversely as the square of y and the cube of z."

$$xy^2z^3 = K, \quad \text{where } K \text{ is a constant.}$$

In any variation that combines both direct and inverse variation, the formula is put together using the procedure of a direct variation (a ratio) and then an inverse variation (a product). The result is equal to a constant.

EXAMPLE 3: Give a formula for the statement "z varies directly as x and inversely as y."

$$\frac{z \cdot y}{x} = K, \quad \text{where } K \text{ is a constant.}$$

EXAMPLE 4: Give a formula for the statement "*a* varies as *b* and the square of *c* and inversely as *x* and the square root of *y*."

$$\frac{ax\sqrt{y}}{bc^2} = K, \quad \text{where } K \text{ is a constant.}$$

In general, the formula for a *combined variation* is set up as a ratio (direct variation) until there is some indication that what follows is an inverse variation (usually just the word *inversely*). From that point on a product is formed.

EXAMPLE 5: Give a formula for the statement "*T* varies jointly as *L*, *P*, and the square of *D* and inversely as *R* and *Q*."

$$\frac{T \cdot RQ}{LPD^2} = K, \quad \text{where } K \text{ is a constant.}$$

In all these examples *K* is called the *constant of variation* or the *constant of proportionality*.

EXAMPLE 6: If *x* varies as *y* and inversely as *z*, and *x* = 2 when *y* = 4 and *z* = 3, find (A) the constant of variation, and (B) the value of *z* when *x* = 8 and *y* = 2.

The formula is $\frac{xz}{y} = K$.

(A) $\frac{2 \cdot 3}{4} = K = \frac{3}{2} = $ the constant of variation.

(B) $\frac{8z}{2} = K = \frac{3}{2}$

$8z = 3$

Therefore, $z = \frac{3}{8}$.

EXAMPLE 7: If *x* varies as *y* and *z*², and inversely as the square root of *w*, what is the effect on *y* when *x* is doubled, *z* is halved, and *w* is multiplied by 4?

The formula is $\frac{x\sqrt{w}}{yz^2} = K$.

For all practical purposes, the quickest way to solve a problem of this type is to arbitrarily choose a value for each variable and compute *K*. For example, let *x* = 1, *y* = 3, *z* = 2, and *w* = 4 (a perfect square since $\sqrt{w}$ is needed).

$$\frac{1\sqrt{4}}{3 \cdot 2^2} = K = \frac{1 \cdot 2}{3 \cdot 4} = \frac{1}{6}$$

The new values of the variables are *x* = 2, *z* = 1, and *w* = 16.
Let *y* = y_2 just to emphasize a new value of *y*.

$$\frac{2\sqrt{16}}{y_2 \cdot 1} = \frac{1}{6}$$

$$\frac{2 \cdot 4}{y^2} = \frac{1}{6}$$

$$y_2 = 48$$

The original value of *y* was 3. The new value of *y* is 48.

Therefore, for any values of *x*, *y*, *z*, and *w*, when *x* is doubled, *z* is halved, and *w* is multiplied by 4, *y* is multiplied by 16.

EXERCISES

1. If *x* varies directly as *y*, which of the following statements must be true?

   I. Their product is a constant.
   II. Their sum is a constant.
   III. Their quotient is a constant.

   (A) only I
   (B) only II
   (C) only III
   (D) only I and II
   (E) only II and III

2. If $\{(x,y) : (4,3),(-2,a)\}$ consists of pairs of numbers that belong to the relationship "*y* varies inversely as x^2," what does *a* equal?

   (A) $-\frac{3}{2}$
   (B) -6
   (C) $-\frac{2}{3}$
   (D) 12
   (E) $\frac{3}{4}$

3. *D* varies as *R* and inversely as the square of *M*. If *R* is divided by 3 and *M* is doubled, what is the effect on *D*?

   (A) *D* is unchanged.
   (B) *D* is divided by 12.
   (C) *D* is multiplied by 12.
   (D) *D* is divided by 6.
   (E) *D* is multiplied by 6.

5.7 LOGIC

There are three major statements in logic:
   I. A *conjunction* is a statement of the form "*A* and *B*" $(A \wedge B)$. It is considered to be true if *A* and *B* are both true. The *negation* of a conjunction, denoted by $(A \wedge B)'$, or $\sim(A \wedge B)$, is equivalent to the statement "the negation of *A* or the negation of *B*," which is denoted by $A' \vee B'$ or $(\sim A) \vee (\sim B)$.

   II. A *disjunction* is a statement of the form "*A* or *B*" $(A \vee B)$. It is considered to be true if either *A* or *B* or both are true. The *negation* of a disjunction, denoted by $(A \vee B)'$ or $\sim(A \vee B)$, is equivalent to the statement "the negation of A *and* the negation of B," which is denoted by $A' \wedge B'$ or $(\sim A) \wedge (\sim B)$.

III. An *implication* is a statement of the form "if A, then B" $(A \rightarrow B)$. It is considered to be true *except* when A is true and B is false. The negation of an implication, denoted by $(A \rightarrow B)'$ or $\sim(A \rightarrow B)$, is equivalent to the statement "A and the negation of B," which is denoted by $A \wedge B'$ or $A \wedge (\sim B)$. There are many forms of an implication that are all equivalent:

if A, then B
A implies B
A only if B
B, if A
A is sufficient for B
B is necessary for A

The double implication "A if and only if B," denoted by "A iff B" or $A \leftrightarrow B$, is equivalent to the statement "if A, then B and if B, then A."

Associated with every implication, $A \rightarrow B$, are three other statements:

1. The converse is $B \rightarrow A$.
2. The inverse is $A' \rightarrow B'$.
3. The contrapositive $B' \rightarrow A'$ or $(\sim B) \rightarrow (\sim A)$, which is equivalent to the original implication $A \rightarrow B$. (If $A \rightarrow B$ is true, then $B' \rightarrow A'$ is also true. If $A \rightarrow B$ is false, then $B' \rightarrow A'$ is also false.)

The negation of the statement "for all A" is "for some not-A." The negation of the statement "for some A" is "for all not-A."

A statement that is always true is called a *tautology*.

EXAMPLE 1: State the negation of each of the following:
(A) **All men are mortal.**
(B) **For some x, x = 2.**
(C) **Some boys play tennis.**
(D) **All cats are not dogs.**

(A) Some men are not mortal.
(B) For all x, $x \neq 2$.
(C) All boys do not play tennis.
(D) Some cats are dogs.

EXAMPLE 2: State the negation of $(p \wedge q) \rightarrow p$.
The negation is $(p \wedge q) \wedge p'$, which is always false because there is no way that p and p' can both be true.

EXAMPLE 3: Is the statement "I will drive only if I am wrong," true or false if it is known that I am always wrong?
Let $p = I$ *will drive* and $q = I$ *am wrong*. The statement is equivalent to $p \rightarrow q$, which is true for all cases except when p is true and q is false. Since it is known that q is true, the statement itself must always be true.

EXAMPLE 4: What conclusion can be drawn from the following statements?

The boy is handsome, or he is short and fat.
If he is short, then he is blond.
He is not blond.

Let $H = he$ *is handsome*, let $S = he$ *is short*, let $F = he$ *is fat*, and let $B = he$ *is blond*. The three statements become:

$$H \vee (S \wedge F)$$
$$S \rightarrow B$$
$$B'$$

Since B' is true, B is false. The only way that $S \rightarrow B$ can be true when B is false is when S is also false. Since S is false, $S \wedge F$ is false regardless of whether F is true or false. Since $S \wedge F$ is false, the only way $H \vee (S \wedge F)$ can be true is if H is true. Therefore, the conclusion is "The boy is handsome."

EXERCISES

1. The statement $(p \vee q) \rightarrow p$ is *false* if
    (A) p is true and q is true
    (B) p is true and q is false
    (C) p is false and q is true
    (D) p is false and q is false
    (E) the statement is a tautology

2. Which of the following is equivalent to the statement "Having equal radii is necessary for two circles to have equal areas"?
    I. Having equal areas is sufficient for two circles to have equal radii.
    II. Two circles have equal areas only if they have equal radii.
    III. Having equal radii implies that two circles have equal areas.

    (A) only I
    (B) only III
    (C) only I and II
    (D) only II and III
    (E) I, II, and III

3. Given the statement "If $x = 2$, then $x^2 = 4$." The negation of this statement is
    (A) $x \neq 2$, and $x^2 \neq 4$
    (B) $x = 2$, and $x^2 \neq 4$
    (C) $x \neq 2$ or $x^2 = 4$
    (D) $x \neq 2$, and $x^2 = 4$
    (E) $x \neq 2$ or $x^2 \neq 4$

4. Given these statements:

   I. Some numbers are not prime.
   II. No primes are squares.

   If *some* means "at least one," it can be concluded from I and II that

   (A) some numbers are squares
   (B) some squares are not numbers
   (C) some numbers are not squares
   (D) no numbers are squares
   (E) none of the above is a conclusion of I and II

5. In a particular town the following facts are true:

   I. Some smarties do not smoke cigars.
   II. All men smoke cigars.

   A necessary conclusion is

   (A) some smarties are men
   (B) some smarties are not men
   (C) no smartie is a man
   (D) some men are not smarties
   (E) no man is a smartie

6. Given the statement "The student will not pass the course only if he does not come to class," which one of the following can be concluded?

   (A) If the student does not pass the course, then he probably missed too many classes.
   (B) If a student comes to class, then he may pass the course.
   (C) If a student comes to class, then he will pass the course.
   (D) If a student passes the course, then he came to class.
   (E) If a student does not come to class, then he will not pass the course.

7. The contrapositive of $p \to q'$ is

   (A) $p' \to q$
   (B) $p' \to q'$
   (C) $q' \to p$
   (D) $q \to p'$
   (E) $q' \to p'$

5.8 STATISTICS

One final subject that may appear on the Math Level IIC test is statistics, but only the most elementary topics considered. Among these are measures of central tendency (averages), frequency distributions, and a simple measure of dispersion (range).

Descriptive statistics consists of methods for describing numerical information in an organized fashion. Probably the most common measure of central tendency, the arithmetic mean, can be determined regardless of whether or not the data are organized. The arithmetic mean is obtained by adding up all the items of data and dividing by the number of items. For example, the mean of 18 test scores in Table 1 is

$$\frac{\Sigma \text{ scores}}{18} = \frac{1123}{18} \approx 62.39$$

In order to draw other conclusions from the test scores in Table 1, the scores would have to be ordered into a frequency distribution (Table 2). A frequency distribution

TABLE 1		
67	62	65
62	57	59
67	60	65
66	62	65
63	55	64
63	59	62

TABLE 2

Data, N	Frequency, f
55	1
57	1
59	2
60	1
62	4
63	2
64	1
65	3
66	1
67	2

consists of the data organized in a table (usually from smallest value to largest value) with the number of times each value occurs.

Many things about the data can be easily determined from a frequency distribution. Three of them are as follows:

1. The *range* of the data is the difference between the largest and smallest values. The range of the data in Table 2 is 12. (67 − 55 = 12).
2. The *mode* of the data is the value that occurs most often. The mode of these data is 62 (there are four 62s).
3. The *median* of the data is the middle score after the data have been ordered. (If there is an even number of scores, the mean of the two middle scores is the median.) The median of these data is

   62.5. $\left(\dfrac{62 + 63}{2} = 62.5 \right).$

Although any one of the mean, median, and mode could correctly be called the average of the data, *average* usually refers to the mean.

EXAMPLE 1: Consider this frequency distribution:

Data	Frequency
0	2
1	3
2	5
3	8
4	2

Find the values of the mode, median, and mean.

Mode $= 3$

Median $= \dfrac{2+3}{2} = 2.5$

Mean $= \dfrac{0+3+10+24+8}{20} = \dfrac{45}{20} = 2.25$

EXAMPLE 2: If the set of data 1, 2, 3, 1, 5, 7, x is to have a mode and x is equal to one of the other elements of the set, what value must x have?
Since there are two 1s and one of every other value, x must equal 1. If it were to equal any of the other numbers, there would be two values with two elements. Thus, the set of data would not have a mode.

EXAMPLE 3: Given this set of data: 15, 24, 28, 32, 35, x, where x is the largest element. If the range is 35, what is the value of x?
Since the range is the difference between the largest element and the smallest element, $x - 15 = 35$. Therefore $x = 50$.

EXAMPLE 4: Given this set of integers: 1, 2, 3, 3, 4, 1, x. If the median is 2, what is the value of x?
Since this set of data has seven elements, the median is the middle number after the data have been ordered. Since the median is 2, the value of x must be any integer ≤ 2.

EXAMPLE 5: Given this set of integers: 1, 2, 3, 3, 4, 1, x. If mean = median = 2, what is the value of x?

Mean $= \dfrac{1+2+3+3+4+1+x}{7} = \dfrac{14+x}{7} = 2$

$$14 + x = 14$$

Therefore, $x = 0$.

EXERCISES

1. In statistics, when the word *average* is mentioned, it refers to

   (A) mean
   (B) median
   (C) mode
   (D) mean or median
   (E) mean, median, or mode

2. If the range of a set of integers is 2 and the mean is 50, which of the following statements must be true?

   I. The mode is 50.
   II. The median is 50.
   III. There are exactly three items of data

   (A) only I
   (B) only II
   (C) only III
   (D) I and II
   (E) I, II, and III

3. If the range of the set of data 1, 1, 2, 2, 3, 3, 3, x is equal to the mean and x is an integer, then x must be

   (A) -1
   (B) -2
   (C) 0
   (D) 1
   (E) There are no values of x that satisfy the stated conditions.

4. In the following frequency distribution, if the mean is to be equal to one of the elements of data, how many 4s must there be?

Data	Frequency
0	1
2	3
3	2
4	?
5	1

   (A) 0
   (B) 2
   (C) 3
   (D) 4
   (E) 5

5. If the mean, median, and mode are calculated from the data in this frequency distribution, which of the following statements is true?

Data	Frequency
2	4
4	3
6	2
8	2

   (A) mode < median < mean
   (B) mean = median
   (C) mode < mean < median
   (D) mean < median < mode
   (E) median < mode < mean

5.9   ODDS AND ENDS

Each year one or two problems on the Math Level IIC examination involve topics that do not fall into any of the categories discussed above. Also, occasionally, definitions are made within problems, and then a question is asked using these definitions. Usually these questions are quite easy if you take the time to read them carefully.

EXAMPLE 1: If a 2 by 2 determinant $\begin{vmatrix} a & b \\ c & d \end{vmatrix}$ is defined to equal $ad - bc$, what does x equal if the determinant $\begin{vmatrix} 2 & 3 \\ x & 5 \end{vmatrix} = 0$?

$$2 \cdot 5 - 3x = 0$$
$$3x = 10$$
$$x = \frac{10}{3}$$

EXAMPLE 2: If the operation $*$ is defined as follows, $a * b = ab - b$, what does $3 * 5$ equal?

$$3 * 5 = 3 \cdot 5 - 5 = 15 - 5 = 10$$

EXAMPLE 3: If the operations $*$ and # are defined so that $a * b = 2a - b$ and $a \# b = \dfrac{a}{b} + b$, does $(3 * 2) \# 4 = 3 * (2 \# 4)$? Explain.

$$(3 * 2) \# 4 = (2 \cdot 3 - 2) \# 4 = 4 \# 4 = \frac{4}{4} + 4 = 5$$

$$3 * (2 \# 4) = 3 * \left(\frac{2}{4} + 4 \right) = 3 * 4 \frac{1}{2} = 2 \cdot 3 - 4 \frac{1}{2} = 6 - 4 \frac{1}{2}$$

$$= 1 \frac{1}{2}$$

Therefore, the expressions are not equal.

EXAMPLE 4: Following the sequence of instructions, tell what number(s) will be printed.

1. Let $x = 2$.
2. Let $y = x + 2$.
3. If $x \cdot y < 10$, then print the value of x, replace x by $x + 3$, and return to step 2.
    If $x \cdot y > 10$, then stop.

Starting with $x = 2$ and $y = 4$, then $x \cdot y = 8 < 10$. Print 2, replace x by 5 and $y = 7$. Then $x \cdot y = 35 \not< 10$. Stop.

Thus, the only number that is printed is 2.

EXAMPLE 5: $P(A|B)$ is the probability of event A happening after event B has already happened. Event A is "a red card is drawn from a deck of 52 cards." Event B is "a black card is drawn from a deck of 52 cards." What does $P(A|B)$ equal?

Since event B is given as having been accomplished, a black card has been drawn from the deck and only 51 cards are left in the deck: 26 red cards and 25 black cards.

Therefore, $P(A|B) = \dfrac{26}{51}$.

EXAMPLE 6: If $f^*(x)$ is defined to be $\dfrac{[f(x)]^2 - f(x)}{x}$ and $f(x) = x^x$, what does $f^* \, 2^2$ equal?

$$f^*(2) = \frac{(2^2)^2 - 2^2}{2} = \frac{4^2 - 4}{2} = \frac{16 - 4}{2} = 6.$$

EXAMPLE 7: If $\begin{Bmatrix} ax + by = c \\ dx + ey = f \end{Bmatrix}$ are solved simultaneously, and $x = 5, y = 2$, and $\begin{vmatrix} a & b \\ d & e \end{vmatrix} = 2$, what does $\begin{vmatrix} c & b \\ f & e \end{vmatrix}$ equal?

Using determinants to solve the pair of equations gives

$$x = \frac{\begin{vmatrix} c & b \\ f & e \end{vmatrix}}{\begin{vmatrix} a & b \\ d & e \end{vmatrix}} \quad \text{and} \quad y = \frac{\begin{vmatrix} a & c \\ d & f \end{vmatrix}}{\begin{vmatrix} a & b \\ d & e \end{vmatrix}}.$$

In this case, $x = 5 = \dfrac{\begin{vmatrix} c & b \\ f & e \end{vmatrix}}{2}.$

Therefore, $\begin{vmatrix} c & b \\ f & e \end{vmatrix} = 10.$

EXAMPLE 8: At the dog pound there are 47 dogs: 16 are large, 18 are brown, and 20 are neither. How many large, brown dogs are there?

Let L = the set of large dogs, let B = the set of brown dogs, and let N = the set of dogs that are neither large nor brown. A Venn diagram is helpful.

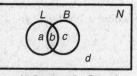

All Dogs at the Pound

Label the number of dogs in each set with a variable, a, b, c, or d. Thus,

$$a + b + c + d = 47$$
$$a + b = 16$$
$$b + c = 18$$
$$d = 20$$

Multiplying the first equation by -1 and adding all four equations results in $b = 7$.

Therefore, there are 7 large, brown dogs.

EXERCISES

1. A *harmonic sequence* is a set of numbers such that their reciprocals, taken in order, form an arithmetic sequence. If S_n = the sum of the first n terms of the harmonic sequence and the first three terms are 2, 3, and 6, then S_4 equals

(A) 11
(B) 12
(C) 1
(D) 22
(E) The value of S_4 cannot be determined.

2. If the pair of equations $\begin{cases} 2x + 3y = 4 \\ 5x + 2y = 8 \end{cases}$ is solved using determinants, the determinant appearing in the denominator of the solution has a value of

(A) −11
(B) −4
(C) −32
(D) 4
(E) 16

3. In the statement "$a = b \pmod n$," b is the remainder when a is divided by n. What is the smallest value of n such that $125 = 7 \pmod n$?

(A) 47
(B) 118
(C) 19
(D) 8
(E) 59

4. If $*$ is a binary operation defined by $a * b = \dfrac{a+b}{2}$ and $\#$ is a binary operation defined by $a \# b = \sqrt{ab}$, for what values of a and b does $a * b = a \# b$?

(A) all real numbers
(B) no real numbers
(C) 0 only
(D) 1 only
(E) only when $a = b$

5. A binary operation $*$ is defined on the set of ordered pairs of real numbers as follows: $(a,b) * (x,y) = (a + x, b - y)$. If $(1,2) * (1,1) = (p,q) * (1,2)$, then (p,q) equals

(A) (1,1)
(B) (1,2)
(C) (1,3)
(D) (0,2)
(E) $\left(2, \dfrac{1}{2}\right)$

6. Carrying out the following instructions in order, find what number is printed first.

1. Let $a = x = 2$.
2. Go to step 4.
3. Replace x by the value of $a + 2$.
4. If $a < 6$, replace a by $x + 2$ and go to step 3; otherwise print the value of x.

(A) 10
(B) 8
(C) 6
(D) 12
(E) 2

7. If darts are thrown at a target like the one in the figure (and every dart hits the figure), what is the probability that the first dart will hit outside the circle?

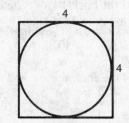

(A) 0.2
(B) 0.8
(C) 0.3
(D) 0.5
(E) 0.6

8. In a group of 100 birds, 85 had a long beak, 45 had gray feathers, and 38 had both a long beak and gray feathers. How many did not have a long beak and did not have gray feathers?

(A) 15
(B) 55
(C) 62
(D) 8
(E) 30

9. If the operation $\#$ is defined by $a \# b = \sqrt{\sin a + \sec b}$, $3 \# 5 =$

(A) 0.7
(B) 1.0
(C) 1.9
(D) 2.3
(E) 3.7

10. If $a * b$ is a binary operation defined by

$$a * b = \frac{\sqrt[3]{a} + \sqrt[3]{b}}{\sqrt{ab}}, 3 * \pi =$$

(A) 1.14
(B) 1.93
(C) 1.60
(D) 0.95
(E) 2.31

ANSWERS AND EXPLANATIONS

In these solutions the following notation is used:

a: active—Calculator use is necessary or, at a minimum, extremely helpful.

n: neutral—Answers can be found without a calculator, but a calculator may help.

i: inactive—Calculator use is not helpful and may even be a hindrance.

Part 5.1 Permutations and Combinations

1. n **C** $\dfrac{(5+3)!}{5!+3!} = \dfrac{8!}{5!+3!}$

$= \dfrac{8 \cdot 7 \cdot 6 \cdot 5 \cdot 4 \cdot 3 \cdot 2 \cdot 1}{(5 \cdot 4 \cdot 3 \cdot 2 \cdot 1) + (3 \cdot 2 \cdot 1)}$

$= \dfrac{40320}{120 + 6} = 320.$

2. n **B** Number available = $5 \cdot 8 \cdot 3 = 120.$

3. n **B** There are $\binom{6}{4} = 15$ different groups of 4 shells out of the 6 available. In each of these groups the circular permutation can be turned over. Therefore, number of bracelets $= 15 \cdot \dfrac{3!}{2} = 45.$

4. n **C** Permutation with repetitions: $\dfrac{5!}{2!2!} = 30.$

5. n **C** A combination: $\binom{52}{2} = \dfrac{52 \cdot 51}{2 \cdot 1} = 1326.$

6. i **D** The value of x must be less than 4. The only value that works is 0.

7. i **D** Only 7 can go in the units place. Any one of 4 numbers can go in the other two places. Number of numbers = $4 \cdot 4 \cdot 1 = 16.$

8. n **C** In order to draw lines through each of 2 points, there are 8 points to be chosen 2 at a time. Number of lines $= \binom{8}{2} = \dfrac{8 \cdot 7}{2 \cdot 1} = 28.$

Part 5.2 Binomial Theorem

1. a **C** The middle term is the sixth term, which is $\binom{10}{5}x^5\left(-\dfrac{1}{x}\right)^5 = \dfrac{10 \cdot 9 \cdot 8 \cdot 7 \cdot 6}{5 \cdot 4 \cdot 3 \cdot 2 \cdot 1} = -252.$

2. a **C** The seventh term is $\binom{8}{6}(a^3)^2\left(-\dfrac{1}{a}\right)^6 = \dfrac{8 \cdot 7}{2 \cdot 1} \cdot a^6 \cdot \dfrac{1}{a^6} = 28.$

3. a **B** Since x^5 is a factor of each term, the term in the binomial expansion containing x^{12} is $-\binom{12}{6} \cdot 1^6 \cdot (x^2)^6.$

Therefore, the coefficient is $-\dfrac{12!}{6!6!} = -924.$

4. a **B** The third term is

$\binom{\frac{1}{3}}{2} \cdot (8x)^{-5/3}(-y)^2 = \dfrac{\frac{1}{3} \cdot \frac{-2}{3}}{2 \cdot 1} \cdot \dfrac{1}{32}x^{-5/3}y^2$

$= -\dfrac{1}{9} \cdot \dfrac{1}{32}x^{-5/3}y^2.$

The coefficient of the third term is $-\dfrac{1}{288}.$

Part 5.3 Probability

1. i **C** There is 1 way to get a 2, and there are 2 ways to get a 3, 4 ways to get a 5, 6 ways to get a 7, 2 ways to get an 11. Out of 36 elements in the sample space, 15 successes are possible.

$P(\text{prime}) = \dfrac{15}{36} = \dfrac{5}{12}.$

2. i **D** The probability of getting neither a head nor a 4 is $\dfrac{1}{2} \cdot \dfrac{5}{6} = \dfrac{5}{12}.$ Therefore, probability of getting either is $1 - \dfrac{5}{12} = \dfrac{7}{12}.$

3. n **E** The probability that none of the cards was a spade $= \dfrac{39}{52} \cdot \dfrac{39}{52} \cdot \dfrac{39}{52} = \dfrac{3}{4} \cdot \dfrac{3}{4} \cdot \dfrac{3}{4} = \dfrac{27}{64}.$ Probability that 1 was a spade $= 1 - \dfrac{27}{64} = \dfrac{37}{64}.$

4. i **E** The only situation when neither of these sets is satisfied occurs when three tails appear. $P(A \cup B) = \dfrac{7}{8}.$

5. n   E   If all 3 are boys, none are girls.

$$P(3B) = \frac{\binom{12}{3}\binom{4}{0}}{\binom{16}{3}},$$ where $\binom{12}{3}$ is the number of ways 3 of 12 boys can be selected, $\binom{4}{0}$ is the number of ways 0 of 4 girls can be selected, and $\binom{16}{3}$ is the number of ways 3 of 16 students can be selected. Therefore, $P(3B) = \dfrac{11}{28}$.

6. i   B   Probability of both items being nondefective $= \dfrac{5}{8} \cdot \dfrac{3}{5} = \dfrac{3}{8}$.

7. n   B $\binom{6}{3}$ is the number of ways 3 men can be selected. $\binom{3}{2}$ is the number of ways 2 women can be selected. $\binom{9}{5}$ is the total number of ways people can be selected to fill 5 rooms.

$$P(3 \text{ men, } 2 \text{ women}) = \frac{\binom{6}{3}\binom{3}{2}}{\binom{9}{5}} = \frac{10}{21}.$$

8. a   C   There are $\binom{8}{5}$ different selections containing 5 good items. In any one of these selections, 5 are good with a probability of 0.8 each, and 3 are bad with a probability of 0.2 each. Therefore,

total probability $= \binom{8}{5}(0.8)^5(0.2)^3$

$$= \frac{(8)(7)(6)(5)(4)}{(5)(4)(3)(2)(1)}(0.8)^5(0.2)^3$$

$$= 56(0.8)^5(0.2)^3 \approx 1.47$$

Part 5.4 Sequences and Series

1. n   D $a_2 = 5$, $a_3 = 8$, $a_4 = 12$, $a_5 = 17$. Therefore, $S_5 = 45$.

2. i   A $y = x + d$, $z = y + d$. Substituting for d gives $y = x + (z - y)$. Therefore, $y = \dfrac{1}{2}(x + z)$.

3. n   A   The decimal $0.23\overline{37} = 0.2 + (0.037 + 0.00037 + 0.0000037 + \cdots)$, which is $0.2 +$ an infinite geometric series with a common ratio of 0.01.

$$S_n = 0.2 + \frac{0.037}{0.99} = \frac{2}{10} + \frac{37}{990} = \frac{235}{990} = \frac{47}{198}.$$

The sum of the numerator and the denominator is 245.

Alternative Solution: Let $f =$ the fraction equal to $0.23\overline{737}$ .... Since there are two digits in the repeating block, multiply through by 10^2 to get $100f = 23.73\overline{737}$ .... Subtract the equation $f = 0.23\overline{737}$ ... to get $99f = 23.5$. Therefore, $f = \dfrac{23.5}{99} = \dfrac{235}{990} = \dfrac{47}{198}$. The sum of the numerator and denominator is 245.

4. i   C   Terms are $3^{1/4}$, $3^{1/8}$, 1. Common ratio $= 3^{-1/8}$. Therefore, the fourth term is $1 \cdot 3^{-1/8} = 3^{-1/8}$ or $\dfrac{1}{\sqrt[8]{3}}$.

5. i   C   Arithmetic mean $= \dfrac{1+25}{2} = 13$. Geometric mean $= \sqrt{1 \cdot 25} = 5$. The difference is 8.

6. n   A   List the primes and add them until the sum approaches 100: 2,3,5,7,11,13,17,19,23. Sum equals 100.

7. i   B   Adding just units digits of $i!$ gives $1 + 2 + 6 + 4 + 0 + 0 + 0 + \cdots$. Sum of units digits for any $i!$ with $i > 4 = 13$.

8. n   D   The first few terms are $-1 + 2 - 3 + 4 - \cdots + 100$. Split up the even numbers and the odd numbers, getting $(2 + 4 + \cdots + 100) - (1 + 3 + \cdots + 99)$: 2 arithmetic series with 50 terms and a difference of 2.

$$\text{Sum} = \frac{50}{2}(2 + 100) - \frac{50}{2}(1 + 99)$$
$$= 2550 - 2500 = 50$$

Alternative Solution: The pattern of 100 terms is $-1 + 2 - 3 + 4 - \cdots - 99 + 100$. If they are grouped in pairs, $(-1 + 2) + (-3 + 4) + \cdots + (-99 + 100)$, there are fifty 1s. Therefore, the sum is 50.

9. i   D $\dfrac{2}{3} = \dfrac{\frac{2}{7}}{1-r}$. $2 - 2r = \dfrac{6}{7}$. $14 - 14r = 6$. Therefore, $r = \dfrac{4}{7}$.

Part 5.5 Geometry and Vectors

1. i   A   The resulting graph is the same as the given one except that it is situated to the left of the y-axis instead of to the right. Therefore, it is the graph of $f(-x)$.

2. n E Using the equation to determine y, make up a table of values for $\left(x, \sqrt{y}\right)$.

x	0	1	2	3	4	-2	-1
y	2	3	4	5	6	0	1
$\sqrt{y}$	$\sqrt{2}$	$\sqrt{3}$	$\sqrt{4}$	$\sqrt{5}$	$\sqrt{6}$	0	1

This leads to the graph in Choice E.

3. i B Setting up a table of a few representative values leads to Choice B.

x	-1	$-\frac{1}{2}$	$-\frac{1}{2}$	0	0	0	$\frac{1}{2}$	$\frac{1}{2}$	1
y	0	0	$\frac{1}{2}$	0	$\frac{1}{2}$	1	0	$\frac{1}{2}$	0
$x+y$	-1	$-\frac{1}{2}$	0	0	$\frac{1}{2}$	1	$\frac{1}{2}$	1	1

4. i D $2\vec{U} = 2\vec{i} - 10\vec{j}$ and $3\vec{V} = 6\vec{i} + 9\vec{j}$.
$$2\vec{U} + 3\vec{V} = 8\vec{i} - \vec{j}.$$

5. i D The dot product must equal zero. Choices B and D satisfy. However, only Choice D is a unit vector.

6. i B When a plane intersects the x-axis, the y-and z-coordinates are zero. Therefore, $2x = 5$ and $x = \frac{5}{2}$ or 2.5.

7. i D $3 = \sqrt{(x-3)^2 + (-1+3)^2 + (-1-1)^2}$.
Square both sides: $9 = x^2 - 6x + 9 + 4 + 4$. $x^2 - 6x + 8 = 0$. Then $(x-4)(x-2) = 0$, and $x = 2$ or 4.

8. i E To find a trace, set one variable to zero and check to see if the result is an answer.

9. i D Direction numbers must be multiples of $2 - (-1)$, $4 - 5$, $1 - 2$ or $3, -1, -1$.

10. a D If the quarter-circle is revolved about the x-axis, a hemisphere of radius 2 is formed.
$$A = \frac{2}{3}\pi r^3 = \frac{16\pi}{3} \approx 16.8.$$

11. n C The altitude is 18.
$$V = \frac{1}{3}\pi r^2 h = \frac{1}{3}\pi \cdot 144 \cdot 18 = 864\pi.$$

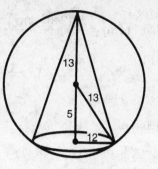

12. i C $V = \pi r^2 h$. Total area $= 2\pi r^2 + 2\pi rh$. $\pi r^2 h = \pi r(2r + 2h)$, which gives $rh = 2r + 2h$. Therefore, $h = \dfrac{2r}{r-2}$. Since h must be positive, the smallest integral value of r is 3.

13. n D If the base radius is 4, the side of the square base is $4\sqrt{2}$. Volume of pyramid $= \frac{1}{3}$ (area of base) $h = \frac{1}{3} \cdot 32 \cdot 9 = 96$.

14. i C In space, the set of points is the perpendicular bisector plane of the segment joining the two fixed points.

15. a B The xz-trace is obtained by letting $y = 0$ in the equation of the plane in order to get $2x + 4z = 7$. The x-intercept $= \frac{7}{2}$ and the z-intercept $= \frac{7}{4}$. The length of the segment between $(0, 0, 1.75)$ and
$$(3.5, 0, 0) = \sqrt{(0-3.5)^2 + (0-0)^2 + (1.75-0)^2}$$
$$= \sqrt{15.3125} \approx 3.9$$

16. a C The line intersects the y-axis at $\frac{7}{2}$ and the x-axis at $\frac{7}{3}$. When the region is rotated about the x-axis, a cone is formed with radius $\frac{7}{2}$ and height $\frac{7}{3}$.
$$V = \frac{1}{3}\pi r^2 h = \frac{1}{3}\pi\left(\frac{7}{2}\right)^2\left(\frac{7}{3}\right) = \frac{7^3\pi}{36} \approx 30.$$

17. a D Volume of one tennis ball $= \frac{4}{3}\pi r^3 = \frac{4}{3}\pi\left(\frac{2.5}{2}\right)^3 = \frac{\pi(2.5)^3}{6}$.

Volume of three tennis balls $= \dfrac{\pi(2.5)^3}{2}$.

Volume of can $= \pi r^2 h = \pi\left(\frac{2.5}{2}\right)^2 (3)(2.5)$
$$= \frac{3\pi(2.5)^3}{4}.$$

Volume of air $= \dfrac{3\pi(2.5)^3}{4} - \dfrac{\pi(2.5)^3}{2}$

$$= \dfrac{\pi(2.5)^3}{4} \approx 12.3.$$

Part 5.6 Variation

1. i   C $\dfrac{x}{y} = K$ by definition of direct variation.

2. i   D $yx^2 = K$. Substitute $(4,3)$ to find $K = 48$. $a \cdot 4 = 48$, and so $a = 12$.

3. i   B $\dfrac{DM^2}{R} = K$. Let $D = 1$, $M = 2$, and $R = 3$ (arbitrary choices), then $K = \dfrac{4}{3}$. Making the indicated changes in the values of the variables results in $R = 1$, $M = 4$, and $K = \dfrac{4}{3}$. Thus, $D = \dfrac{1}{12}$. Therefore, the original value of D (1) was divided by 12.

<u>Alternative Solution</u>: $\dfrac{D_1 \cdot M^2}{R} = K$. $\dfrac{D_2 \cdot (2M)^2}{R} = K = \dfrac{D_1 \cdot M^2}{R}$. Therefore, $\dfrac{3 \cdot 4D_2 M^2}{\cancel{R}} = \dfrac{D_1 \cdot \cancel{M^2}}{\cancel{R}}$ and $D_2 = \dfrac{D_1}{12}$.

Part 5.7 Logic

1. i   C   The statement is false only when $(p \lor q)$ is true and p is false. If p is false, $(p \lor q)$ is true only when q is true.

2. i   C   The statement "q is necessary for p" is equivalent to "p is sufficient for q" (I) and "p only is q" (II), but *not* "q implies p" (III).

3. i   B   The negation of "if p, then q" is "p and not q."

4. i   E   Counterexamples of each answer can be found.

5. i   B   Since "some smarties do not smoke cigars" and "all men smoke cigars," a necessary conclusion is "some smarties are not men."

6. i   C   The statement "p', only if c'" is equivalent to "if p', then c'," which is equivalent to its contrapositive, "if c, then p."

7. i   D   By definition the contrapositive is $q \to p'$.

Part 5.8 Statistics

1. i   E   Although the arithmetic mean is usually what is meant, any one of the three is considered an average.

2. i   B   Since the values are integers, the range is 2, and the mean is 50, the possible numbers in the set are 49, 50, and 51.
  I. The set could consist of n 49s and n 51s, and so 50 would not even be an element of the set. Therefore, the mode is not necessarily 50, and I is false.
  II. Since the mean is 50, there must be the same number of 49s as 51s. Thus, the median is 50, and II is true.
  III. This is obviously false.

3. i   D   Mean $= \dfrac{2+4+9+x}{8} = \dfrac{15+x}{8}$. The range could equal $x - 1$ if $x > 3$; $3 - x$ if $x < 1$; or 2 if $1 < x < 3$. Therefore, $\dfrac{15+x}{8} = x - 1$; $\dfrac{15+x}{8} = 3 - x$; or $\dfrac{15+x}{8} = 2$. Solving for x in each case gives $x = \dfrac{23}{7}$; $x = 1$; or $x = 1$ again. Since x is an integer, the answer can only be 1.

4. i   D   Check the mean against each of the data values. Mean $= \dfrac{6+6+4x+5}{7+x} = 2$; $\dfrac{17+4x}{7+x} = 2$; $17 + 4x = 14 + 2x$; $x = -\dfrac{3}{2}$. Mean $= \dfrac{17+4x}{7+x} = 3$; $17 + 4x = 21 + 3x$; $x = 4$; a positive integer; correct. The mean can be checked against 4 and 5 in similar fashion to show that only four 4s satisfy the condition of the problem.

5. n   A   Mode $= 2$; median $=$ middle number $= 4$; mean $= \dfrac{8+12+12+16}{11} = 4\dfrac{4}{11}$. Therefore, mode < median < mean.

Part 5.9 Odds and Ends

1. i   E   Since 2, 3, 6 are three terms of a harmonic sequence, $\dfrac{1}{2}, \dfrac{1}{3}, \dfrac{1}{6}$ are three terms of an arithmetic sequence. The fourth term of the arithmetic sequence is 0 since the difference is $\dfrac{1}{6}$. Because the reciprocal of zero does not exist, four terms of the harmonic sequence do not exist.

2. i A The denominator determinant is made up of the coefficients of the variables. It is

$$\begin{vmatrix} 2 & -3 \\ 5 & 2 \end{vmatrix} = 4 - (+15) = -11.$$

3. n E The division can be written in the form $n \cdot Q + 7 = 125$, and so $n \cdot Q = 118 = 2 \cdot 59$. To get a remainder of 7, n must be greater than 7. Since 59 is a prime, the answer is E.

4. i E $\dfrac{a+b}{2} = \sqrt{ab}$. Multiplying by 2 and squaring gives

$$a^2 + 2ab + b^2 = 4ab \quad a^2 - 2ab + b^2 = 0$$
$$(a-b)^2 = 0 \quad a = b$$

5. i C $(1,2) * (1,1) = (2,1)$ and $(p,q) * (1,2) = (p+1, q-2)$. Therefore $p + 1 = 2$ and $q - 2 = 1$, and so $p = 1, q = 3$.

6. i A Keeping track of the values of a and x, you have

Step

step	1	4	3	4	3
a	2	4		8	
x	2		6		10

Print 10.

7. a A The probability of hitting target is 1. The probability of hitting inside the circle $= \dfrac{\text{area of circle}}{\text{area of square}} = \dfrac{4\pi}{16}$. Probability of hitting outside the circle $= 1 - \dfrac{4\pi}{16} \approx 0.2$.

8. i D In the Venn diagram below, $a =$ birds that have neither long beaks nor gray feathers.

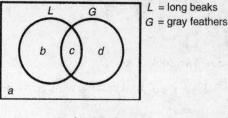

L = long beaks
G = gray feathers

$$\begin{aligned} a + b + c + d &= 100 \\ b + c &= 85 \\ c + d &= 45 \\ c &= 38 \end{aligned}$$

Substituting the value of c, and then the values of b and d, gives $a = 8$.

9. a C Put your calculator in radian mode $3 \# 5 = \sqrt{\sin 3 + \sec 5}$.

$$\sqrt{\sec 5} = \frac{1}{\cos 5} \approx \frac{1}{0.28366} \approx 3.525.$$

Therefore,

$$\sqrt{\sin 3 + \sec 5} \approx \sqrt{0.1411 + 3.525} \approx 1.9.$$

10. a D $3 * \pi = \dfrac{\sqrt[3]{3} + \sqrt[3]{\pi}}{\sqrt{3\pi}}$. Using your calculator to find $\sqrt[3]{3} \approx 1.4422, \sqrt[3]{\pi} \approx 1.4646$, and $\sqrt{3x} \approx 3.0699$ gives an answer of approximately 0.95.

GRAPHING CALCULATORS

PART

3

INTRODUCTION

CHAPTER

6

This section will help you use the power of a graphing calculator to improve your score on the Level IIC Test. If you have never used a graphing calculator, there is enough descriptive detail to give you the basics of its operation and guide you through the specific applications. If you are an experienced user, this section serves as a reference for using a graphing calculator to solve the kinds of problems you will encounter on the Level IIC Test.

Although there are several manufacturers and models of graphing calculators on the market, we are most familiar with the TI-83 (including the TI-83 Plus and TI-83 Silver Edition), so we adopted this model for instructional purposes. Users of this book who have other models made by other manufacturers should have little difficulty finding the keystrokes on their calculators that correspond to those described for the TI-83.[1]

The chapters that follow provide keystroke detail on the TI-83's general features, as well as the methods that are used to solve specific types of problems. Screen images that result from these keystrokes are also displayed to help users who have calculators in hand to check the accuracy of their work. Readers may nonetheless wish to consult their User's Manual to supplement the instruction given in this book.

Basic features and operations of the calculator are covered in Chapter 7. The logic of the arrangement of the keys is outlined first. The main types of screens that you will use to do the work are described next. A brief over-

view of the computational logic of the calculator is given next, followed by tips on editing entries into the calculator. Selections of computational mode are described, with recommendations for the Math Level IIC Test. The menu keys and the variety of functions listed within these menus are outlined next. The final section describes "support functions" that facilitate the use of use of the TI-83.

Chapter 8 provides keystroke detail for using the TI-83 to solve a broad range of problems that might be encountered on the Math Level IIC Test. The chapter is organized around a group of questions that contain descriptive material supported by 20 examples. For convenience, the questions are listed below:

- How do I find a good window for a graph or group of graphs?
- How do I evaluate a function at a specific value?
- How do I find the real zeros of a function?
- How do I find a relative maximum or minimum of a function?
- How do I graph an equation that is not a function?
- How do I use graphs to solve equations?
- How do I use tables to locate the zeros of a polynomial?
- How can I use a graph to solve a polynomial inequality?
- How do I use a graph to check for symmetry?
- How do I use a table to find the limit of a rational function?
- How do I use graphs to solve problems involving absolute value and greatest integer functions?

---

[1]Excluded are newer generation graphing calculators with computer algebra systems, such as the TI-89.

- How do I generate a sequence?
- How can I find the sum, product, or mean?
- How so I use the factorial, permutation, and combination commands?
- How do I use the calculator to iterate a function?
- How do I graph parametric equations on my calculator?

Finally, Chapter 9 includes five short, simple programs for formulas you may need to use on the Math Level IIC Test:

- The distance between two points
- The midpoint of a segment
- The distance between a point and a line
- The angle between two lines in the plane
- The Quadratic Formula

Screens showing the program commands and screens showing the execution and output of these programs are included.

BASIC
OPERATIONS
CHAPTER
7

KEYS

The photo on the next page shows the TI-83 calculator keyboard. The top row (right under the screen) contains the keys that govern graphing and tabulation. The four arrow keys move the cursor around the screen. The STAT, MATH, MATRX, PRGM, and VARS keys lead to menus of commands in those categories. The 2nd key and ALPHA key activate the corresponding functions imprinted above the keys. When referring to 2nd or ALPHA prefixes, the prefix itself will not be used after its first mention. For example, after its first mention, 2nd CALC will just be called CALC.

SCREENS

There are three main "screens" that come into play: the Home Screen, where computations are made; the Y = Screen, where formulas are defined; and the Graphing Screen, where graphs are displayed. Sample screens are shown below.

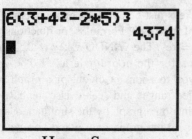

Home Screen

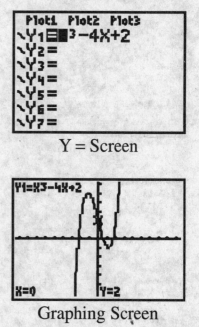

Y = Screen

Graphing Screen

You can return to the Home Screen from either of the other two screens by keying 2nd QUIT.

You may also use the 2nd TABLE and STAT/EDIT/Edit screens to solve problems. The TABLE screen displays functional values in tabular, rather than graph form. This format may be more convenient, for example, when trying to locate a zero between two integers. The STAT/EDIT/Edit screen, on the other hand, provides a spreadsheet capability. You can do arithmetic on a list or you can combine lists arithmetically. Sample TABLE and STAT/EDIT/Edit screens are shown below.

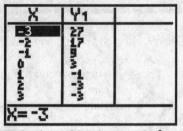

Tabulated Values of $x^2 - 5x + 3$

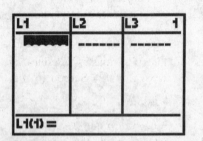

Blank STAT/EDIT/Edit Screen

COMPUTATION

There are two important features that graphing calculators share with the newer scientific calculators. Both follow the conventional algebraic order of operations, including parentheses, and both can be used to evaluate complicated numerical expressions. Both also have im-

plied multiplication (i.e., it is unnecessary to enter the × key) with the use of parentheses. With these features it is rarely necessary to write down an intermediate value in a multistep expression.

EDITING

Editing capabilities are another feature of graphing calculators that is shared by newer scientific calculators. On the TI-83, editing can be done using the DEL (delete) and 2nd INS (insert) key. If you make an error while making an entry into the calculator, simply overwrite or delete the error and insert the correct expression.

The 2nd ENTRY key (above ENTER) pastes previous Home Screen entries back into the Home Screen for editing—either to correct errors or to make small changes to the computation. For example, if you evaluate an expression on the Home Screen and need to perform the same operation again, but with one different value, you need not rekey the entire command; simply key ENTRY, and edit the command as desired. This feature is especially useful if the original command involves multiple keystrokes.

The 2nd ANS key can also be used as an edit tool. Suppose, for example, that you want to compute a quantity, say $4 + 8(3 - 17) = -108$ and you realize that you want to divide 45 by this quantity. Instead of rekeying -108, you can just key $45 \div$ ANS as shown below. This is particularly useful if you don't want to round off the ANS value to rekey it. Another convenient feature changes (decimal) answers to fractions. You can access this feature by keying MATH, followed by ENTER twice. This activates the MATH menu and applies the command to change the answer to a fraction ($\triangleright$ Frac).

```
4+8(3-17)
                  -108
45/Ans
         -.4166666667
Ans▶Frac
                 -5/12
```

Sample Use of ANS and ANS $\triangleright$ FRAC

GRAPHING COMMANDS

The five keys just below the screen govern the graphing features of the TI-83. Formulas for functions are entered via the Y = key. The WINDOW key is used to set ranges and scales for the coordinate axes. The ZOOM key enables you to zoom in or out of a graph or to select prespecified ranges and scales. Use the TRACE key to locate points on a graph by the simultaneous flashing of the cursor on the graph and giving you its x and y coordinates. You can trace a graph using the left and right arrows. If you have more than one graph on a screen, you can move from one to another by using the up and down arrows. The GRAPH key produces a graph or returns to the graphing screen if a graph has already been produced. Examples of specific graphing commands are given in the next chapter.

MODE

The MODE key gives you choices for several of the calculator's important features. The left-most choice on each line is the default choice. The MODE screen is shown below.

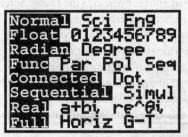

Default Setting on MODE Screen

You alway should use the Normal and Float modes on the Math Level IIC Test. When working a trigonometry problem, you must decide whether to use Radian or Degree measure. You will use Function mode for most problems, but you occasionally will need to change to Polar or Parametric mode for these types of problems.

When graphing, pixels are connected by segments in Connected mode, but they are simply darkened when in Dot mode. Only use Dot mode when you know that there are discontinuities in the graph, as these will be concealed in Connected mode. When graphing two or more functions on the same screen, Sequential mode graphs them in sequence while Simul mode graphs them simultaneously. There is rarely good reason to choose Simul mode over Sequential mode.

Real mode is appropriate for most problems, but it does not display imaginary number results unless imaginary numbers are input. If you get an imaginary number result with real inputs while in Real mode, you will get an error message. On the other hand, $a + bi$ mode displays imaginary number results when they occur, even if the inputs are not imaginary. For example $\sqrt{-36}$ would result in $6i$. You should use $a + bi$ mode on the Math Level IIC Test.

MENU KEYS

As noted above, the menu keys STAT, MATH, MATRX, PRGM, VARS, and their 2nd functions LIST, TEST, ANGLE, DRAW, and DISTR offer a vast array of specialized capabilities in these categories. The submenus

and functions you will need to use on the Math Level IIC Test are summarized below. The use of these functions is described in the next section on specific applications.

MATH/MATH:	>Frac—changes decimal values to fractional form
MATH/NUM:	abs(—absolute value int(—greatest integer
MATH/PRB:	nPr—number of permutations of r out of n objects nCr—number of combinations of r out of n objects !—factorial
VARS/Y-VARS:	Function—pastes Y-defined functions from the Y = Screen to the Home Screen
LIST/OPS:	seq(—generates a user-defined sequence of numbers
LIST/MATH:	mean(—finds the mean of a list of numbers
LIST/MATH:	sum(—finds the sum of the numbers in a list
LIST/MATH:	prod(—finds the product of the numbers in a list

SUPPORT FUNCTIONS

You can adjust the contrast (make the screen darker or lighter) by keying 2nd up arrow or 2nd down arrow, respectively. When you do this, a digit is briefly displayed in the upper right corner of the screen. With fresh batteries, this digit should be about 2 when the screen is at the right contrast. If the digit is 7 or more when the contrast isn't sufficient, you should consider changing the battery soon. (There is a "change battery" warning when you really must change the battery.)

You can also "jump down" or "jump up" a list of menus, programs, or data by keying ALPHA down arrow or ALPHA up arrow, respectively. Programs are listed in alphabetical order. If, for example, you want to execute the QUAD (Quadratic Formula) program from a long list of programs, simply key ALPHA Q and the screen will scroll through all the programs directly to those beginning with the letter Q. Or, if you need to edit the 23rd number in a list, rather than arrow down and scroll through the first 22 numbers, you can get there faster by keying ALPHA down arrow.

The STO and 2nd RCL keys enable you to store values into the variables ALPHA A through Z, or into lists 2nd L_1 through 2nd L_6, and retrieve them. You should not need to store individual values into the ALPHA locations to solve problems on the Math Level IIC Test, but it may at times be necessary to store sequences you generate into lists. More on this can be found in the specific applications that follow.

SPECIFIC APPLICATIONS

CHAPTER

8

This section, written in question-and-answer format, describes specific TI-83 applications that are on the Math Level IIC Test. The decision whether to use a graphing calculator on a particular test item depends on (a) your experience with the graphing calculator, especially in the course or courses you took where the topic was taught; and (b) the nature of the answer choices for the test item.

As a general rule, when answer choices are exact (such as square roots or expressions involving π), you are expected to solve the problem without a calculator (graphing or scientific). You should assess your ability to do this and use a calculator to obtain a decimal approximation only if you decide that you cannot otherwise solve the problem. You can then use your calculator to evaluate answer choices until a match is achieved. If you are able to formulate a graphing calculator solution immediately, this strategy may be best, even if it means evaluating exact answer choices one at a time until the correct one is found.

If the answer choices for a test item are all in decimal form, you are most likely expected to use a calculator or graphing calculator to arrive at the correct answer choice. Again, if most of your classroom experience is with symbolic (noncalculator) strategies, your best approach is to attempt to solve the problem this way until you reach a calculator-ready answer. Then, use your calculator to evaluate the decimal form of your answer to find the correct answer choice. However, if your classroom experi-

ence was based on graphical and tabulation strategies to solve problems, your graphing calculator should give you the correct answer choice more quickly and efficiently than will a symbolic approach. This is the strategy that is addressed primarily in this section.

A few words about graphing calculator syntax are in order before beginning the specific applications. Calculator syntax refers to the structure of calculator commands. These are not always the same as when written algebraically. For example, the mathematical expression $\sin^2 x$ is written as $\sin(x)^2$ in calculator syntax. Also, several of the menu commands, such as absolute value, include a left parenthesis: Abs(, and the user must remember to key the right parenthesis to conclude this command, especially if it is part of a more complex expression. Parentheses that are not necessary in algebra are used in calculator syntax.

For example, an algebraic expression such as $\dfrac{5+6}{3+4}$ does not need parentheses because the division bar serves as a grouping symbol, but when this expression is entered into the calculator, it must be entered as $(5 + 6)/(3 + 4)$.

Graphing calculators are not infallible. Their algorithms for locating points of intersection, zeros, and relative extrema can break down if a function is not well-behaved near the x coordinate of the point in question. For example, the algorithm may not be able to find a zero if it occurs at a point where a function has a vertical or near-vertical tangent line. Fortunately, it is unlikely that such situations would be encountered on the Math Level IIC Test.

How do I find a good window for a graph or group of graphs?

A good window is defined as one that captures the features of a graph or several graphs that enable you to answer a question. The main features of a graph are its turning points, vertical asymptotes, x and y intercepts, and "end behavior" (the behavior of y as x approaches positive or negative infinity). A good window for the graphs of more than one function should include their point or points of intersection.

Unfortunately, there is no single formula for finding a good window. Practice, to gain experience, is the best way to identify good windows. If an equation(s) contain(s) "small" numerical values, ZStandard (the standard window), which gives a –10 to 10 window in both x and y directions, is a good starting point. Depending on what you see, you can Zoom In or Zoom Out to get a clearer picture of the graph's essential features. In other cases, you must evaluate the function or functions at one or more values to set the window manually. Some examples are given below.

EXAMPLE 1: Where does $P(x) = x^3 + 18x - 30$ have a zero?
If you only need one zero, ZStandard will do the job (one zero at $x \approx 0.48$), as shown in the window below.

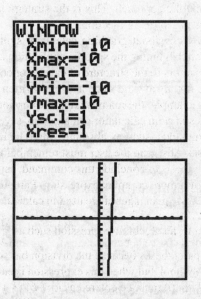

If you need all of the zeros, however, you need to establish whether the graph crosses the x axis at some other point(s). One approach is to take a very "large" window and graph the function in this window, as shown below.

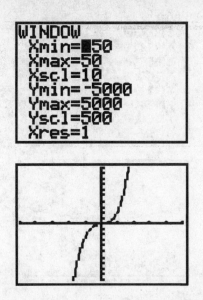

It looks as if there is only one zero, but you should Zoom In several times and adjust the window manually to convince yourself.

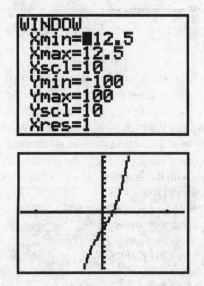

EXAMPLE 2: Find the maximum value of the function $f(x) = \sqrt{1 + \sin^2 x}$ on the interval $\left[-\dfrac{\pi}{4}, \dfrac{\pi}{4} \right]$. This problem is simpler than the one in the previous example because the problem specifies the x values to be considered. Key $-\pi/4$, $\pi/4$, and $\pi/8$ for Xmin, Xmax, and Xscl, respectively. Note that the calculator automatically converts these values to decimal form. If you remember that $\sin x \le 1$, so $1 + \sin^2 x \le 2$, you could deduce $0 \le \sqrt{1 + \sin^2 x} \le 2$, giving you Ymin and Ymax values of 0 and 2, respectively. Otherwise, start Ymin and Ymax at –10 and 10, and adjust as necessary. The window and graph for this example are shown below.

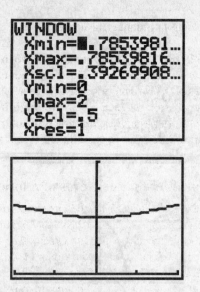

You can see from the graph that the maximum occurs at the endpoints, $-\dfrac{\pi}{4}$ and $\dfrac{\pi}{4}$. Thus, the maximum value of approximately 1.225 can be found by tracing to either of these endpoints.

EXAMPLE 3: What is the maximum value of $6\sin x\cos x$? A good place to start with the window for this function is ZTrig, which automatically sets an $x\varepsilon[-2\pi,2\pi]$ and $y\varepsilon[-4,4]$ window.[2] As can be seen from the graph below, this proves to be sufficient.

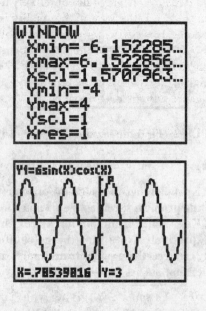

The maximum value of 3 for this function occurs when $x \approx 0.785$.

EXAMPLE 4: Find the area of the region enclosed by the graph of the polar curve $r = \dfrac{1}{\sin\theta + \cos\theta}$ and the x and y axes. First, set the MODE to Pol[ar]. Unless you know what the polar graph looks like, start again with ZStandard, which sets θ min and θ max at 0 and 2π, respectively, with 48 increments of 0.131 (θ step) for polar graphs. The other parameters are the same as when function graphing with ZStandard. The result of this graph is a line, as indicated in the far left figure below. The area described in the problem is too small to see, so adjust the window settings: Xmin = 0, Xmax = 2, Xscl = 0.5, Ymin = 0, Ymax = 2, Yscl = 0.5 to get the graph shown in the far right window.

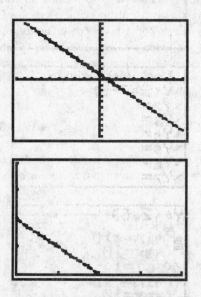

The region is an isosceles right triangle with legs of unit length. (The figure is distorted because the screen isn't square.) The area of this triangle is 0.5.

How do I evaluate a function at a specific value?

You have many options to accomplish function evaluation. The best choice depends on the nature and context of the problem.

- If the formula for the function is fairly simple, and you need only one value, simply key in the formula on the Home Screen, with the numerical value in place of x.
- If the formula for the function is complex, key it into one of the Y = lines. Then key VARS/Y-VARS/Function/Y_k and Y_k will be returned to the Home Screen. Then key (x value) followed by ENTER, and the functional value is returned.

---

[2]You frequently will see the phrase "Plot the graph of . . . in an $x\varepsilon[a,b]$ and $y\varepsilon[c,d]$ window" throughout the book. The variables a, b, c, d stand for Xmin, Xmax, Ymin, and Ymax, respectively.

- Another option is to key the function formula into one of the Y = lines as above, plot the graph of the function in a suitable window, and key 2nd CALC/value, which returns to the Graphing Screen with X = displayed at the bottom. Key the desired value of X; key ENTER, and the cursor is shown on the graph at that value of X, with both X and Y values displayed at the bottom of the screen.

Screens for these options to find $f(2.6)$ for the function $f(x) = x^3 - 6x + 2$ are shown below.

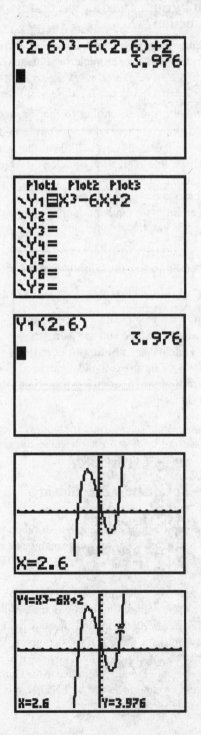

All three options produce the answer 3.976.

EXAMPLE 5: What is the remainder when $3x^3 + 2x^2 - 5x - 8$ is divided by $x + 2$? The Remainder Theorem tells you that the remainder in this case is the polynomial evaluated when $x = -2$. Although it requires more keystrokes, the safest method is to enter the polynomial into Y_1, return to the Home Screen, and enter $Y_1(-2)$.

How do I find the real zeros of a function?

The real zeros of a function are the x intercepts of its graph. First, plot a graph of the function that includes the x intercept(s) of interest. Key CALC/zero. The calculator asks for a left bound, so move the cursor to some point on the graph to the left of the x intercept and key ENTER. The cursor then asks for a right bound, so move the cursor to some point on the graph to the right of the x intercept and key ENTER. Next, the calculator prompts Guess, so put the cursor as close to the x intercept as you can, and key ENTER again. The coordinates of the x intercept (where the y coordinate is either 0 or a number very close to 0) are then displayed.

Of course, functions can have more than one real zero. As illustrated in the example below, the problem must provide enough information to let you know which zero you need to find.

EXAMPLE 6: Find the smallest positive real zero of $y = x^3 - 6x + 2$. Plot the graph of $y = x^3 - 6x + 2$ in the standard window. Although not absolutely necessary, Zoom In to get a better view of the smaller of the two positive zeros. Key CALC/zero and select points to the left and right of this point. Note the two pointers indicating these left and right bounds. Finally, move the cursor as close as you can to the zero and key ENTER. The sequence of screens is shown below.

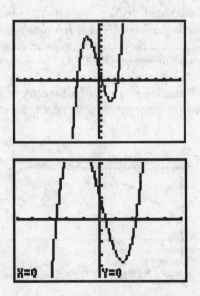

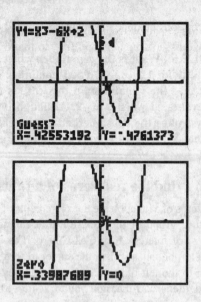

The far right screen shows the smaller zero $x \approx 0.340$.

How do I find a relative maximum or minimum of a function?

Relative minima and maxima of a function are turning points on its graph. The procedure for finding these values is very similar to finding a zero of a function. This time, the relative maximum value must appear in the graph. After keying CALC/maximum, select points to the left and right of the relative maximum. Place the cursor as close to the relative maximum as possible, and key ENTER. The x and y coordinates of the relative maximum will be displayed at the bottom of the screen. Use the same procedure with the MINIMUM key to locate a relative minimum.

EXAMPLE 7: If $0 \leq x \leq 2\pi$, for what value of x does the function $\sin\frac{1}{3}x$ achieve its maximum? Graph $y = \sin((1/3)x)$ in an $x\varepsilon[0,2\pi]$ and $y\varepsilon[-1,1]$ window. Key CALC/maximum and ENTER left and right bounds of the maximum point. Scroll to your guess for the maximum point and key ENTER. The x and y coordinates of the maximum point will be displayed at the bottom of the screen, as illustrated by the screens below. The function achieves its maximum of 1 when $x \approx 4.712$.

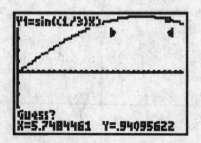

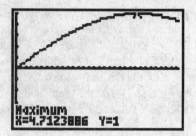

How do I graph an equation that is not a function?

Not all equations are or can be transformed into equations that define functions. For example, the equation of a circle, $x^2 + y^2 = 25$, defines two functions: $y = \sqrt{25 - x^2}$ and $y = -\sqrt{25 - x^2}$, corresponding to the top and bottom halves of the circle. Other conic sections share this characteristic. On the Math Level IIC Test, the main reason you need to be able to graph an equation that is not a function is to answer questions about symmetry. Therefore, you will have to do some algebra to solve for y, typically as $y = \pm f(x)$. Enter the equation for $f(x)$ into y_1; then enter $-y_1$ into y_2 and graph both functions.

EXAMPLE 8: Assess the symmetry of $x^2 - y^2 = 1$. Solving for y, $y^2 = x^2 - 1$, so $y = \pm\sqrt{x^2 - 1}$. First plot the graphs of $y = \sqrt{x^2 - 1}$ and $y = -y_1$ in an $x\varepsilon[-2,2]$ and $y\varepsilon[-2,2]$ window. Since you are using the visual image to draw conclusions about the equation, you need to key ZOOM/Zsquare to avoid distortion. The screens below show that the equation is symmetrical about both axes and the origin.

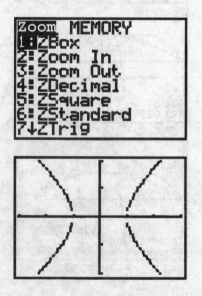

How do I use graphs to solve equations?

First, you need to enter the two sides of the equation as functions, say one into Y_1 and the other into Y_2. Solutions to the equation are points where these two graphs intersect. As was the case with the zeros of a function, a problem must specify which of several possible solutions it's asking you to find. Once you've determined which point of intersection gives the solution, you must first find a window that shows the point of intersection clearly. Then key CALC/intersect, which returns you to the Graphing Screen. The calculator asks you to designate the "first curve." Key ENTER and the cursor jumps to another curve and asks for the "second curve." If there are only two graphs on the screen, key ENTER again, which specifies the other graph as the second curve. (You rarely will be working with more than two graphs, but if there should be more than two on the screen, you must designate which two you are working with.)

The calculator then asks for your "guess," and you move the cursor as close as you can to the point of intersection and key ENTER again. After a moment, the cursor moves, if necessary, to the point of intersection and the x and y coordinates of the point of intersection are displayed at the bottom of the screen.

EXAMPLE 9: Solve $\sin 2x = \sin x$ in the interval $(0,\pi)$. With your calculator in radian mode, plot the graphs of $y = \sin(2x)$ and $y = \sin(x)$ in an $x\varepsilon[0,\pi]$ and $y\varepsilon[-2,2]$ window. The screen below results from keying CALC/ intersect, ENTER, ENTER, and moving the cursor close to the point of intersection. The second screen results when ENTER is keyed once again, with the solution $x \approx 1.047$ displayed at the bottom. Note that the two selected curves have + signs on them.

An alternative approach to using graphs to solve an equation is to transform the equation first so that one side is zero. Then follow the procedures for finding a zero of the function that results on the nonzero side of the equation. In the example given above, you would find the zeros of the function $\sin 2x - \sin x$ on the interval $(0,\pi)$.

How do I use tables to locate the zeros of a polynomial?

The Math Level IIC Test may include questions about the location of the zeros of a polynomial. The TABLE feature of the TI-83 provides a way to locate a zero of a polynomial between two integers or, for that matter, between any two values.

EXAMPLE 10: Between which two consecutive integers is there a zero of $P(x) = 28x^3 - 11x^2 + 15x - 12$? Enter the formula for $P(x)$ into Y_1, and key 2nd TBLSET. As shown in the screen below, set TblStart to –3 and set ΔTbl to 1. Set both Indepnt and Depend to Auto. This sets the calculator to build a table of values, starting at $x = -3$ and increasing x by 1. There is no particular reason for starting at –3, as once you key 2nd TABLE, you may move up or down by 1 in either the positive or negative direction. The TBLSET and corresponding TABLE screens are shown below.

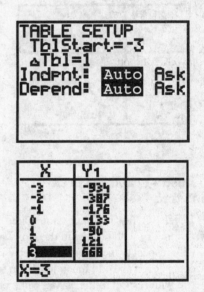

The screen shows that Y_1 $(P(x))$ changes sign between 1 and 2, so you can conclude that the polynomial has a zero between those two integers.

How can I use a graph to solve a polynomial inequality?

EXAMPLE 11: Solve the inequality $6x^2 - 11x > 7$. In this problem you have the functions $Y_1 = 6x^2 - 11x$ and

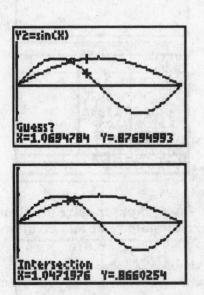

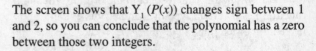

$Y_2 = 7$. You need to determine the values of x for which Y_1 lies above Y_2. It is customary, however, to set one side of the inequality equal to zero, $6x^2 - 11x - 7 > 0$, and find the values of x for which this new function lies above the x axis. Plot the graph of $y = 6x^2 - 11x - 7$ in an $x\varepsilon[-3,5]$ and $y\varepsilon[-15,10]$ window. The graph lies above the x axis to the right of the larger zero and the left of the smaller zero. Using CALC/zero, these values are determined to be $x = \dfrac{7}{3}$ and $x = -\dfrac{1}{2}$, respectively. Therefore, the solution consists of all numbers greater than $\dfrac{7}{3}$ or less than $-\dfrac{1}{2}$. The screens for these two zeros are shown below.

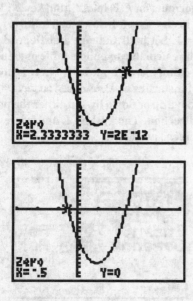

How do I use a graph to check for symmetry?

The Math Level IIC Test may ask about three types of symmetry: symmetry with respect to the x or y axis, and symmetry with respect to the origin. Symmetry is related to the concepts of even and odd functions. Even functions are symmetric with respect to the y axis, while odd functions are symmetric with respect to the origin. Whether a function has a certain type of symmetry is evident upon inspection of its graph.

EXAMPLE 12: What are the symmetries of the function $f(x) = 2x^4 - 3x^2 + 2$? Plot the graph of this function in the standard window and observe that it is symmetric with respect to the y axis. The window showing this graph appears below.

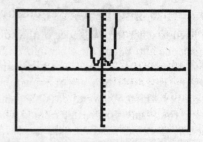

How do I use a table to find the limit of a rational function?

EXAMPLE 13: Find the limit of $\dfrac{3x^2 - 3}{x - 1}$ as x approaches 1. Key $(3x^2 - 3)/(x - 1)$ into Y_1, and key 2n TBLSET. Select Indpnt: Ask and Depend: Auto, and key TABLE. To let x approach 1, enter values progressively closer to 1, from below and above, as shown in the table screen on the right. It should be pretty clear from the table that the desired limit is 6.

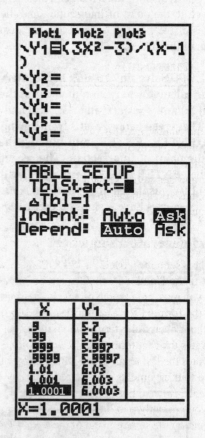

How do I use graphs to solve problems involving absolute value and greatest integer functions?

Equations involving absolute value can be very cumbersome to solve algebraically because they often need to be broken down into cases. A graphical approach makes finding solutions much easier.

EXAMPLE 14: Solve $|2x - 5| = 3x + 4$. Plot the graph of each side of this equation, using MATH/NUM/abs (on the left side), in the standard window. Use CALC/intersect to find the point of intersection at $x = 0.2$, $y = 4.6$. These screens are shown below.

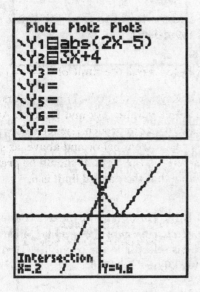

Some of the Math Level IIC Test questions ask you to determine the nth term of a sequence or to find the sum or product of the terms of a sequence. There are formulas for arithmetic and geometric sequences to make these determinations, but you also can use your calculator to generate the sequence and find sums and products.

How do I generate a sequence?

To generate a sequence, use 2nd LIST/OPS/seq, which returns seq(to the Home Screen. The left parenthesis indicates that you must enter some inputs. Four arguments are required: the formula for each term of the sequence; the variable that governs the term; the value of the variable that generates the first term; and the value of the variable that generates the last term. Close with a right parenthesis after these four arguments are entered, and key ENTER. The sequence is listed horizontally in braces.

Since you probably will do something with a sequence (such as finding the sum or mean of the terms), it is best to store the sequence in a list. This can be done either before or after keying ENTER, by keying STO 2nd L_1. You can inspect the sequence in L_1 by keying STAT/EDIT/Edit. (If you don't see L_1, key STAT/EDIT/SetUpEditor, which returns SetUpEditor to the Home Screen. Then key ENTER, followed by STAT/EDIT/Edit again. This time you'll see L_1 on the Home Screen.)

EXAMPLE 15: Generate the first 12 terms of the geometric sequence 2, 6, 18, 54, ... The first term of this sequence is 2, and subsequent terms are the product of 2 and increasing powers of 3. We can think $2 = 2 \times 3^0$; $6 = 2 \times 3^1$; $18 = 2 \times 3^2$, and so on. To generate the desired sequence, we need the powers of 3 to go from 0 to 11 (thus giving 12 terms). The command Seq($2 * 3^X$, $X,0,11$) will generate the desired sequence. (The command Seq($2 * 3^{(X - 1)}$, $X,1,12$) would generate the identical sequence.)

If the problem was to find the seventh term, you would store this sequence in L_1 and key 2nd L_1(7). This sequence is displayed in the screen below.

```
seq(2*3^X,X,0,11
)
{2 6 18 54 162 …
Ans→L1
{2 6 18 54 162 …
L1(7)
            1458
■
```

How can I find the sum, product, or mean?

A question may ask you to find the sum of the terms of an arithmetic or geometric sequence. Or, a question may ask you to demonstrate your knowledge of sigma notation by finding a sum expressed in that notation. You may be asked to find the product of the terms of a sequence. Finally, you may be asked to find the mean or median of a data set that is too large to easily work with manually. These questions and others can be handled by using menu commands in 2nd LIST/MATH, shown on the screen below.

```
NAMES OPS MATH
1:min(
2:max(
3:mean(
4:median(
5:sum(
6:prod(
7↓stdDev(
```

Example 16: Find the sum of the first 20 terms of the arithmetic series 12, 4, –4, ... Each number in the sequence is 8 less than its predecessor, so create the sequence by keying LIST/OPS/seq($12 - 8x,x,0,19$), or its equivalent, seq($12 - 8(x - 1),1,20$). As noted previously, when you key ENTER, the sequence is displayed

horizontally in braces across the Home Screen. Key LIST/MATH/sum(Ans), followed by ENTER, to get the answer –1200. This result can be accomplished in a single step by keying LIST/MATH/sum(seq(12 – 8x,x,0,19)). These results are displayed on the following screen.

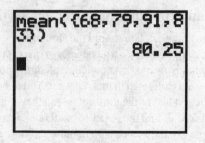

In the example just given, the argument for sum was Ans, in this case a list of numbers in braces { }—2ⁿᵈ( ). Any list presented in this fashion or named (such as L_1, L_2, and so on) can be used as the argument for any of the LIST/MATH commands. For example, if you wanted to find the mean of the 4 numbers 68, 79, 91, 83, simply key LIST/MATH/ mean({68,79,91,83}), then ENTER for the result 80.25. This is shown in the following screen.

```
mean( {68,79,91,8
3} )
              80.25
■
```

How do I use the factorial, permutation, and combination commands?

These commands are used to solve "combinatorics" problems—the number of ways that a set of objects can be selected and/or arranged. Factorial (!) is used to count the number of permutations (distinct arrangements) of objects. If there are n objects, then there are n! distinct ways in which they can be arranged. Work from the Home Screen to do this calculation. For example, to calculate 6!, key 6, followed by MATH/PRB/!, which returns ! following the 6 on the Home Screen. Key ENTER to get the result 720 (6! = 6*5*4*3*2*1).

If a portion r of n objects is selected, the number of distinct ways these r objects can be arranged is nPr (the number of permutations of r objects of n). For example, to find the number of ways of arranging 6 objects of 10, key 10 in the Home Screen, followed by MATH/PRB/nPr, followed by 6. Then key ENTER to get the result 151200. If the order of the r objects is unimportant—you are sim-

ply interested in the number of ways of selecting 6 of 10 objects, regardless of their order—use nCr instead of nPr, to get the result 210. These are shown on the screen below.

```
10 nPr 6
              151200
10 nCr 6
                 210
```

EXAMPLE 17: How many ways can a president, vice president, and secretary be selected in a club of 20 people? Since 3 people are being selected from 20, n = 20 and r = 3. It matters which of the 3 is elected to which office, so use 20 nPr 3 = 6840.

EXAMPLE 18: How many ways can a committee of 3 people be selected in a club of 20 people? Since it doesn't matter who is selected to the committee first, second, or third, use 20 nCr 3 = 1140.

How do I use the calculator to iterate a function?

You may be asked to iterate a function in an exam problem. In this type of problem you are given a starting value and a rule to move from each value to the next. The problem asks you to find a particular value in such a sequence. Both the TI-83 and the newer scientific calculators are well suited to this type of problem, as illustrated in the following example.

EXAMPLE 19: Suppose $a_0 = 3$ and $a_{n+1} = 2a_n – 1$. Find a_4. First, key 3 ENTER. Then key 2 * ANS – 1 ENTER, to get a_1; ENTER again, to get a_2; and ENTER twice more to get the answer a_4. The screen that results from this sequence is shown below.

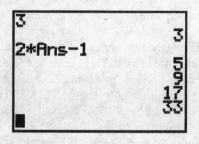

How do I graph parametric equations on my calculator?

Recall that parametric equations are a way of representing points on a plane or in space where the coordinates are represented as functions of a parameter. Points in the plane are typically represented as $x(t)$, $y(t)$, where the parameter t represents time, and x and y are coordinates of a point at time t. When it is possible to eliminate the variable t, y can be expressed as a function of x and the graph of the function can be drawn on your calculator as usual.

If, on the other hand, the parameter cannot be eliminated, you must put your calculator in parametric mode and enter the equations for x and y to draw a graph. The window settings in parametric mode include XMIN, XMAX, XSCL, YMIN, YMAX, YSCL as in function mode, as well as the settings TMIN, TMAX, and TSTEP, that define the smallest, largest, and incremental step values for the parameter. As can be seen from the following example, the parameter need not be denoted by t.

EXAMPLE 20: Sketch the graph of the parametric equations $\begin{cases} x = 2(\theta - \sin \theta) \\ y = 2(1 - \cos \theta) \end{cases}$. **With your calculator in parametric mode, enter the equations for x and y. Since the equations involve trigonometric functions, you need to choose degree or radian mode. If you choose radian mode, a good starting point for TMIN is 0 and for TMAX is 2π. TSTEP determines how smooth the graph will be: the smaller TSTEP, the smoother the graph. A reasonable starting point for TSTEP in this problem is**

$\dfrac{\pi}{24}$, **which will provide 49 data points on the graph. The window settings and resulting graph are shown below.**

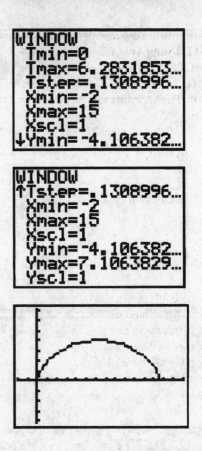

Some experimentation may be necessary to produce a good graphic representation of parametric equations.

PROGRAMS

CHAPTER

9

This chapter describes five programs that you can write for your TI-83 and use as necessary on the Math Level IIC Exam. These programs make it easy for you to use common formulas as part of doing larger problems:

- distance between two points (D2P)
- midpoint of a segment (MIDPT)
- distance between a point and a line (DPL)
- angle between two lines in the plane (ANGLE)
- Quadratic Formula (QUADFORM)

First, you will have to write the programs, which are short and easy to write. Once written, the programs can be executed. The two sections that follow explain how to write and execute the five programs listed above.

Writing the programs

Begin writing each program by keying PRGM/NEW/ ENTER. The cursor appears in ALPHA mode following Name =, and you should type the name of the program (using either the name given above or your own choice of names). Key ENTER after typing the name and a colon (:) will appear for you to enter your first program command.

Key PRGM to display submenus CTL, I/O, and EXEC. These stand for control commands, input/output commands, and execution commands (the latter, to execute one program within another). Key I/O to find the commands "Prompt," and "Display," and key CTL to find the command "Stop." When you key these commands, they are pasted into the program, and you key the remaining characters from the calculator's keyboard, using 2ⁿᵈ and ALPHA as needed.

When a program is executed, the characters that are in quotes (" ") are displayed as text. This makes it easier to follow the correct steps when executing the program. The quote sign (") is obtained by keying ALPHA +.

Once written, programs can only be accessed by keying PRGM/EDIT. The five programs are shown on the screens below. Because the programs are relatively short, there is some overlap between screens.

PROGRAM D2P

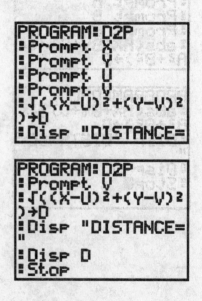

PROGRAM MIDPT

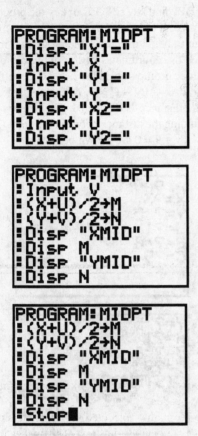

```
PROGRAM:MIDPT
:Disp "X1="
:Input X
:Disp "Y1="
:Input Y
:Disp "X2="
:Input U
:Disp "Y2="
```

```
PROGRAM:MIDPT
:Input V
:(X+U)/2→M
:(Y+V)/2→N
:Disp "XMID"
:Disp M
:Disp "YMID"
:Disp N
```

```
PROGRAM:MIDPT
:(X+U)/2→M
:(Y+V)/2→N
:Disp "XMID"
:Disp M
:Disp "YMID"
:Disp N
:Stop■
```

PROGRAM DPL

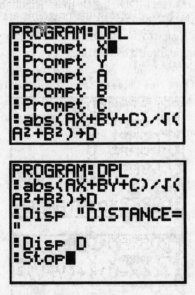

```
PROGRAM:DPL
:Prompt X■
:Prompt Y
:Prompt A
:Prompt B
:Prompt C
:abs(AX+BY+C)/√(
A²+B²)→D
```

```
PROGRAM:DPL
:abs(AX+BY+C)/√(
A²+B²)→D
:Disp "DISTANCE=
"
:Disp D
:Stop■
```

PROGRAM ANGLE

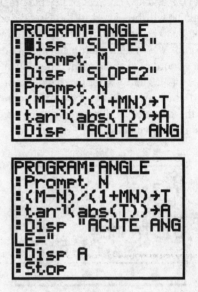

```
PROGRAM:ANGLE
:■isp "SLOPE1"
:Prompt M
:Disp "SLOPE2"
:Prompt N
:(M-N)/(1+MN)→T
:tan⁻¹(abs(T))→A
:Disp "ACUTE ANG
```

```
PROGRAM:ANGLE
:Prompt N
:(M-N)/(1+MN)→T
:tan⁻¹(abs(T))→A
:Disp "ACUTE ANG
LE="
:Disp A
:Stop
```

PROGRAM QUADFORM

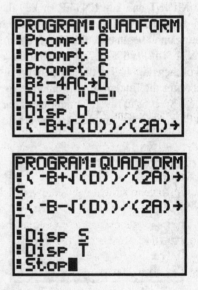

```
PROGRAM:QUADFORM
:Prompt A
:Prompt B
:Prompt C
:B²-4AC→D
:Disp "D="
:Disp D
:(-B+√(D))/(2A)→
```

```
PROGRAM:QUADFORM
:(-B+√(D))/(2A)→
S
:(-B-√(D))/(2A)→
T
:Disp S
:Disp T
:Stop■
```

Executing the programs

To execute a program you must key PRGM/EXEC and either scroll down to the program you want or enter the number of the program you want. The steps you should follow to run each program are described below, and sample values are shown in the screen output for each one.

D2P

Highlight D2P and key ENTER or key the program number. PrgmD2P will be pasted to the Home Screen. Key ENTER again to begin executing the program. When you see the prompt X? key the x coordinate of the point. Prompt Y? for the y coordinate of the point is displayed, followed by the prompts U? and V?, for the x and y coordinates of the second point. After you key the final coordinate and ENTER, the distance is displayed on the screen. This final screen is shown below.

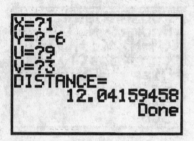

MIDPT

Highlight MIDPT and key ENTER or key the program number. Program MIDPT is pasted to the Home Screen. Key Enter again to begin executing the program. Prompts for X1?, X2?, Y1?, and Y2?, representing the x and y co-ordinates of the endpoints of the segment are displayed in sequence. After the final prompt is entered, the midpoint is displayed. Screens to find the midpoint of the point displayed in the previous example are shown below.

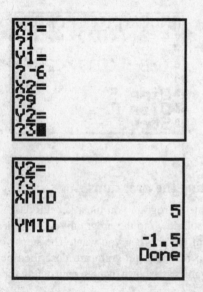

DPL

Highlight DPL and key ENTER or key the program number. Program DPL is pasted to the Home Screen. Key Enter again to begin executing the program. Prompts for X? and Y?, the coordinates of the point, are displayed. The next three prompts are A?, B?, and C?, coefficients of the equation $Ax + By + C = 0$. After the final prompt is entered, the distance is displayed. The screens from the program finding the distance between the point $(3,-5)$ and the line $x - 6y + 9 = 0$ are shown below.

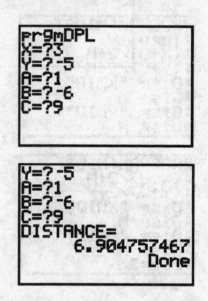

ANGLE

First, be sure to set MODE to "Degree." Highlight ANGLE and key ENTER or key the program number. Program ANGLE is pasted to the Home Screen. This program finds the acute angle formed by two lines in the plane with given slopes. The first prompt M? asks for the slope of one of the lines. Once this value is entered, another prompt N? asks for the slope of the other line. Key ENTER, and the acute angle is given in degrees. A screen showing this sequence appears below.

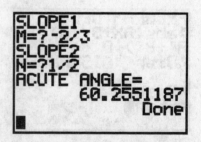

QUADFORM

Highlight QUADFORM and key ENTER or key the program number. Program QUADFORM is pasted to the Home Screen. Prompts for A?, B?, and C?, the coefficients of the equation $Ax^2\ Bx + C = 0$, are displayed. After each is entered, the value of D, the discriminant, and the two solutions are displayed. If the $a + bi$ is selected, imaginary solutions are displayed when they occur. Otherwise, the error message "NONREAL ANS" is displayed. The screens for solving $x^2 + x + 1 = 0$ are shown below.

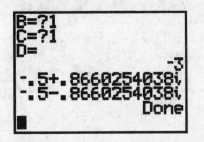

```
prgmQUADFORM
A=?1
B=?1
C=?1
```

```
B=?1
C=?1
D=
                    -3
-.5+.86602540381
-.5-.86602540381
              Done
■
```

MODEL
TESTS
PART
4

ANSWER SHEET FOR MODEL TEST 1

Determine the correct answer for each question. Then, using a no. 2 pencil, blacken completely the oval containing the letter of your choice.

1. Ⓐ Ⓑ Ⓒ Ⓓ Ⓔ
2. Ⓐ Ⓑ Ⓒ Ⓓ Ⓔ
3. Ⓐ Ⓑ Ⓒ Ⓓ Ⓔ
4. Ⓐ Ⓑ Ⓒ Ⓓ Ⓔ
5. Ⓐ Ⓑ Ⓒ Ⓓ Ⓔ
6. Ⓐ Ⓑ Ⓒ Ⓓ Ⓔ
7. Ⓐ Ⓑ Ⓒ Ⓓ Ⓔ
8. Ⓐ Ⓑ Ⓒ Ⓓ Ⓔ
9. Ⓐ Ⓑ Ⓒ Ⓓ Ⓔ
10. Ⓐ Ⓑ Ⓒ Ⓓ Ⓔ
11. Ⓐ Ⓑ Ⓒ Ⓓ Ⓔ
12. Ⓐ Ⓑ Ⓒ Ⓓ Ⓔ
13. Ⓐ Ⓑ Ⓒ Ⓓ Ⓔ
14. Ⓐ Ⓑ Ⓒ Ⓓ Ⓔ
15. Ⓐ Ⓑ Ⓒ Ⓓ Ⓔ
16. Ⓐ Ⓑ Ⓒ Ⓓ Ⓔ
17. Ⓐ Ⓑ Ⓒ Ⓓ Ⓔ

18. Ⓐ Ⓑ Ⓒ Ⓓ Ⓔ
19. Ⓐ Ⓑ Ⓒ Ⓓ Ⓔ
20. Ⓐ Ⓑ Ⓒ Ⓓ Ⓔ
21. Ⓐ Ⓑ Ⓒ Ⓓ Ⓔ
22. Ⓐ Ⓑ Ⓒ Ⓓ Ⓔ
23. Ⓐ Ⓑ Ⓒ Ⓓ Ⓔ
24. Ⓐ Ⓑ Ⓒ Ⓓ Ⓔ
25. Ⓐ Ⓑ Ⓒ Ⓓ Ⓔ
26. Ⓐ Ⓑ Ⓒ Ⓓ Ⓔ
27. Ⓐ Ⓑ Ⓒ Ⓓ Ⓔ
28. Ⓐ Ⓑ Ⓒ Ⓓ Ⓔ
29. Ⓐ Ⓑ Ⓒ Ⓓ Ⓔ
30. Ⓐ Ⓑ Ⓒ Ⓓ Ⓔ
31. Ⓐ Ⓑ Ⓒ Ⓓ Ⓔ
32. Ⓐ Ⓑ Ⓒ Ⓓ Ⓔ
33. Ⓐ Ⓑ Ⓒ Ⓓ Ⓔ
34. Ⓐ Ⓑ Ⓒ Ⓓ Ⓔ

35. Ⓐ Ⓑ Ⓒ Ⓓ Ⓔ
36. Ⓐ Ⓑ Ⓒ Ⓓ Ⓔ
37. Ⓐ Ⓑ Ⓒ Ⓓ Ⓔ
38. Ⓐ Ⓑ Ⓒ Ⓓ Ⓔ
39. Ⓐ Ⓑ Ⓒ Ⓓ Ⓔ
40. Ⓐ Ⓑ Ⓒ Ⓓ Ⓔ
41. Ⓐ Ⓑ Ⓒ Ⓓ Ⓔ
42. Ⓐ Ⓑ Ⓒ Ⓓ Ⓔ
43. Ⓐ Ⓑ Ⓒ Ⓓ Ⓔ
44. Ⓐ Ⓑ Ⓒ Ⓓ Ⓔ
45. Ⓐ Ⓑ Ⓒ Ⓓ Ⓔ
46. Ⓐ Ⓑ Ⓒ Ⓓ Ⓔ
47. Ⓐ Ⓑ Ⓒ Ⓓ Ⓔ
48. Ⓐ Ⓑ Ⓒ Ⓓ Ⓔ
49. Ⓐ Ⓑ Ⓒ Ⓓ Ⓔ
50. Ⓐ Ⓑ Ⓒ Ⓓ Ⓔ

MODEL TEST

1

Tear out the preceding answer sheet. Decide which is the best choice by rounding your answer when appropriate. Blacken the corresponding space on the answer sheet. When finished, check your answers with those at the end of the test. For questions that you got wrong, note the sections containing the material that you must review. Also if you do not fully understand how you arrived at some of the correct answers, you should review the appropriate sections. Finally, fill out the self-evaluation sheet on page 156 in order to pinpoint the topics that give you the most difficulty.

TEST DIRECTIONS

<u>Directions</u>: Decide which answer choice is best. If the exact numerical value is not one of the answer choices, select the closest approximation. Fill in the oval on the answer sheet that corresponds to your choice.

Notes:
(1) You will need to use a scientific or graphing calculator to answer some of the questions.
(2) You will have to decide whether to put your calculator in degree or radian mode for some problems.
(3) All figures that accompany problems are plane figures unless otherwise stated. Figures are drawn as accurately as possible to provide useful information for solving the problem, except when it is stated in a particular problem that the figure is not drawn to scale.
(4) Unless otherwise indicated, the domain of a function is the set of all real numbers for which the functional value is also a real number.

<u>Reference Information.</u> The following formulas are provided for your information.

Volume of a right circular cone with radius r and height h: $V = \frac{1}{3}\pi r^2 h$

Lateral area of a right circular cone if the base has circumference c and the slant height is l: $S = \frac{1}{2}cl$

Volume of a sphere of radius r: $V = \frac{4}{3}\pi r^3$

Surface area of a sphere of radius r: $S = 4\pi r^2$

Volume of a pyramid of base area B and height h: $V = \frac{1}{3}Bh$

---

1. The slope of a line perpendicular to the line whose equation is $\frac{x}{3} - \frac{y}{4} = 1$ is

   (A) $\frac{1}{4}$

   (B) $-\frac{4}{3}$

   (C) $-\frac{3}{4}$

   (D) $\frac{4}{3}$

   (E) -3

2. What is the range of the data set 8, 12, 12, 15, 18?

   (A) 12
   (B) 18
   (C) 13
   (D) 15
   (E) 10

3. What is the set of points in space equidistant from two vertices of an equilateral triangle and 2 inches from the third vertex?

   (A) a circle
   (B) a line segment
   (C) two points
   (D) a parabola
   (E) two parallel lines

4. If $f(x) = \sqrt{2x + 3}$ and $g(x) = x^2 + 1$, then $f(g(2)) =$

   (A) 6.16
   (B) 3.61
   (C) 2.24
   (D) 3.00
   (E) 6.00

5. $\left(-\frac{1}{16}\right)^{2/3} =$

   (A) 0.16
   (B) −0.25
   (C) 6.35
   (D) −0.16
   (E) The value is not a real number.

6. The circumference of circle $x^2 + y^2 - 10y - 36 = 0$ is

   (A) 192
   (B) 38
   (C) 125
   (D) 54
   (E) 49

GO ON TO THE NEXT PAGE

7. What is the value of $\displaystyle\sum_{j=3}^{5} \ln j$?

   (A) 1.6
   (B) 4.8
   (C) 7.8
   (D) 4.1
   (E) 1.9

8. If $f(x) = 2$ for all real numbers x, then $f(x + 2) =$

   (A) 0
   (B) 2
   (C) 4
   (D) x
   (E) The value cannot be determined.

9. The volume of the region between two concentric spheres of radii 2 and 5 is

   (A) 66
   (B) 28
   (C) 368
   (D) 490
   (E) 113

10. The number of terms in the expansion of $(2x^2 - 3y^{1/3})^7$ is

   (A) 6
   (B) 8
   (C) 1
   (D) 7
   (E) 9

11. In right triangle ABC, $AB = 10$, $BC = 8$, $AC = 6$. The sine of $\angle A$ is

   (A) $\dfrac{4}{3}$
   (B) $\dfrac{3}{4}$
   (C) $\dfrac{4}{5}$
   (D) $\dfrac{5}{4}$
   (E) $\dfrac{3}{5}$

12. If $16^x = 4$ and $5^{x+y} = 625$, then $y =$

   (A) 2
   (B) 5
   (C) $\dfrac{25}{2}$
   (D) $\dfrac{7}{2}$
   (E) 1

GO ON TO THE NEXT PAGE

13. If the parameter is eliminated from the equations $x = t^2 + 1$ and $y = 2t$, then the relation between x and y is

(A) $y = x - 1$
(B) $y = 1 - x$
(C) $y^2 = x - 1$
(D) $y^2 = (x - 1)^2$
(E) $y^2 = 4x - 4$

14. Let $f(x)$ be a polynomial function: $f(x) = x^5 + \cdots$. If $f(1) = 0$ and $f(2) = 0$, then $f(x)$ is divisible by

(A) $x - 3$
(B) $x^2 - 2$
(C) $x^2 + 2$
(D) $x^2 - 3x + 2$
(E) $x^2 + 3x + 2$

15. $\mathrm{Sin}\left(\mathrm{Arctan}\,\dfrac{1}{3}\right)$ equals

(A) 0.95
(B) 0.32
(C) 0.33
(D) 0.35
(E) 0.50

16. If $z > 0$, $a = z\cos\theta$, and $b = z\sin\theta$, then $\sqrt{a^2 + b^2} =$

(A) 1
(B) z
(C) $2z$
(D) $z\cos\theta\sin\theta$
(E) $z(\cos\theta + \sin\theta)$

17. Given the statement, "Only if it rains will the trip be cancelled," if it is known that it did not rain, what conclusion can be drawn?

(A) The trip is cancelled.
(B) The trip is not cancelled.
(C) The trip might be cancelled.
(D) No conclusion can be drawn.
(E) None of the above.

18. If $f(x) = \begin{cases} \dfrac{5}{x - 2}, & \text{when } x \neq 2 \\ k, & \text{when } x = 2 \end{cases}$, what must the value of k be in order for $f(x)$ to be a continuous function?

(A) 2
(B) 5
(C) 0
(D) −2
(E) No value of k will make $f(x)$ a continuous function.

GO ON TO THE NEXT PAGE

19. What is the probability that a prime number is less than 7, given that it is less than 13?

(A) $\dfrac{1}{2}$

(B) $\dfrac{3}{4}$

(C) $\dfrac{3}{5}$

(D) $\dfrac{1}{3}$

(E) $\dfrac{2}{5}$

20. The ellipse $4x^2 + 8y^2 = 64$ and the circle $x^2 + y^2 = 9$ intersect at points where the y-coordinate is

(A) ± 1.41
(B) ± 2.24
(C) ± 10.00
(D) ± 2.45
(E) ± 2.65

21. Each term of a sequence, after the first, is inversely proportional to the term preceding it. If the first two terms are 2 and 6, what is the twelfth term?

(A) 2
(B) 6
(C) $2 \cdot 3^{11}$
(D) 46
(E) The twelfth term cannot be determined.

22. A company offers you the use of its computer for a fee. Plan A costs $6 to join and then $9 per hour to use the computer. Plan B costs $25 to join and then $2.25 per hour to use the computer. After how many minutes of use would the cost of plan A be the same as the cost of plan B?

(A) 18,052
(B) 173
(C) 169
(D) 165
(E) 157

23. If the probability that the Giants will win the NFC championship is p and if the probability that the Raiders will win the AFC championship is q, what is the probability that only one of these teams will win its respective championship?

(A) pq
(B) $p + q - 2pq$
(C) $|p - q|$
(D) $1 - pq$
(E) $2pq - p - q$

24. If a geometric sequence begins with the terms $\frac{1}{3}, 1, \ldots$, what is the sum of the first 10 terms?

(A) $9841\frac{1}{3}$
(B) 6561
(C) $3280\frac{1}{3}$
(D) $33\frac{1}{3}$
(E) 6

25. The value of $\dfrac{453!}{450!\,3!}$ is

(A) greater than 10^{100}
(B) between 10^{10} and 10^{100}
(C) between 10^5 and 10^{10}
(D) between 10 and 10^5
(E) less than 10

26. If A is the angle formed by the line $2y = 3x + 7$ and the x-axis, then $\angle A$ equals

(A) $72°$
(B) $56°$
(C) $215°$
(D) $0°$
(E) $-45°$

27. What is the smallest positive x-intercept of the graph of $y = 3\sin 2\left(x + \dfrac{2\pi}{3}\right)$?

(A) 0
(B) 2.09
(C) 1.05
(D) 0.52
(E) 1.31

28. If $(x - 4)^2 + 4(y - 3)^2 = 16$ is graphed, the sum of the distances from any fixed point on the curve to the two foci is

(A) 4
(B) 8
(C) 12
(D) 16
(E) 32

USE THIS SPACE FOR SCRATCH WORK

29. In the equation $x^2 + kx + 54 = 0$, one root is twice the other root. The value(s) of k is (are)

(A) ± 5.2
(B) 15.6
(C) −5.2
(D) ± 15.6
(E) 22.0

30. The remainder obtained when $3x^4 + 7x^3 + 8x^2 - 2x - 3$ is divided by $x + 1$ is

(A) 5
(B) 0
(C) −3
(D) 3
(E) 13

31. If $f(x) = e^x$ and $g(x) = f(x) + f^{-1}(x)$, what does $g(2)$ equal?

(A) 8.1
(B) 7.5
(C) 8.3
(D) 5.1
(E) 7.4

32. $(x + 2)^3 - 3(x + 2)^2 + 3(x + 2) - 1 =$

(A) $(x + 1)^3$
(B) $(x - 1)^3$
(C) $(x + 2)^3$
(D) x^3
(E) $(x + 3)^3$

33. For what values of k does the graph of $\dfrac{(x - 2k)^2}{1} - \dfrac{(y - 3k)^2}{3} = 1$ pass through the origin?

(A) only 0
(B) only 1
(C) ±1
(D) $\pm\sqrt{5}$
(E) no value

GO ON TO THE NEXT PAGE

34. If $\dfrac{1-\cos\theta}{\sin\theta} = \dfrac{\sqrt{3}}{3}$, then $\theta =$

    (A) 15°
    (B) 30°
    (C) 45°
    (D) 60°
    (E) 75°

35. If $x^2 + 3x + 2 < 0$ and $f(x) = x^2 - 3x + 2$, then

    (A) $0 < f(x) < 6$
    (B) $f(x) \geq \dfrac{3}{2}$
    (C) $f(x) > 12$
    (D) $f(x) > 0$
    (E) $6 < f(x) < 12$

36. If $f(x) = |x| + [x]$, the value of $f(-2.5) + f(1.5)$ is

    (A) 3
    (B) 1
    (C) –2
    (D) 1.5
    (E) 2

37. If $(\sec x)(\tan x) < 0$, which of the following must be true?

    I. $\tan x < 0$
    II. $\csc x \cot x < 0$
    III. x is in the third or fourth quadrant

    (A) I only
    (B) II only
    (C) III only
    (D) II and III
    (E) I and II

38. At the end of a meeting all participants shook hands with each other. Twenty-eight handshakes were exchanged. How many people were at the meeting?

    (A) 14
    (B) 7
    (C) 8
    (D) 28
    (E) 56

39. Suppose the graph of $f(x) = 2x^2$ is translated 3 units down and 2 units right. If the resulting graph represents the graph of $g(x)$, what is the value of $g(-1.2)$?

    (A) 2.88
    (B) 17.48
    (C) –0.12
    (D) –1.72
    (E) 37.28

GO ON TO THE NEXT PAGE

40. What is the smallest positive angle that will make $5 - \sin\left(x + \dfrac{\pi}{6}\right)$ a maximum?

(A) 1.05
(B) 2.09
(C) 1.57
(D) 4.19
(E) 5.24

41. If $f(x) = ax + b$, which of the following make(s) $f(x) = f^{-1}(x)$?

  I. $a = -1$, b = any real number
  II. $a = 1$, $b = 0$
  III. a = any real number, $b = 0$

(A) only I
(B) only II
(C) only III
(D) only I and II
(E) only I and III

42. In the figure, $\angle A = 110°$, $a = \sqrt{6}$, and $b = 2$. What is the value of $\angle C$?

(A) 50°
(B) 25°
(C) 20°
(D) 15°
(E) 10°

43. If vector $\vec{v} = \left(1, \sqrt{3}\right)$ and vector $\vec{u} = (3, -2)$, find the value of $\left|3\vec{v} - \vec{u}\right|$.

(A) 52
(B) $2 + 3\sqrt{3}$
(C) 6
(D) $0.2 + 3\sqrt{3}$
(E) 7

44. A sector of a circle, AOB, with a central angle of $\dfrac{2\pi}{5}$ and a radius of 5 is bent to form a cone with vertex at O. What is the volume of the cone that is formed?

(A) 8.17
(B) 6.04
(C) 4.97
(D) 5.13
(E) 12.31

45. In $\triangle ABC$, $a = 2x$, $b = 3x + 2$, $c = \sqrt{12}$, and $\angle C = 60°$. Find x.

(A) 0.50
(B) 0.64
(C) 0.77
(D) 1.64
(E) 1.78

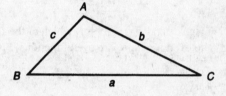

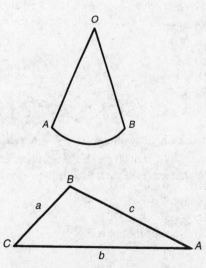

GO ON TO THE NEXT PAGE

46. If point $P(a, a\sqrt{3})$ is 6 units from line $5x + 3y = 12$, then a equals

   (A) only 4.6
   (B) only −2.3
   (C) 2.3 or −4.6
   (D) only 2.3
   (E) −2.3 or 4.6

47. If $f(x) = 3x^2 + 4x + 5$, what must the value of k equal so that the graph of $f(x - k)$ will be symmetric to the y-axis?

   (A) $\dfrac{2}{3}$

   (B) $-\dfrac{2}{3}$

   (C) 0

   (D) −4

   (E) $-\dfrac{4}{3}$

48. If $f(x) = \cos x$ and $g(x) = 2x + 1$, which of the following is an even function (are even functions)?

   I. $f(x) \cdot g(x)$
   II. $f(g(x))$
   III. $g(f(x))$

   (A) only I
   (B) only II
   (C) only III
   (D) only I and II
   (E) only II and III

49. A cylinder whose base radius is 3 is inscribed in a sphere of radius 5. What is the difference between the volume of the sphere and the volume of the cylinder?

   (A) 354
   (B) 297
   (C) 88
   (D) 448
   (E) 1345

50. Under which conditions is $\dfrac{xy}{x - y}$ negative?

   (A) $0 < y < x$
   (B) $x < y < 0$
   (C) $x < 0 < y$
   (D) $y < x < 0$
   (E) None of the above

ANSWER KEY

1. C	6. E	11. C	16. B	21. B	26. B	31. A	36. E	41. D	46. E
2. E	7. D	12. D	17. B	22. C	27. C	32. A	37. C	42. C	47. A
3. A	8. B	13. E	18. E	23. B	28. B	33. C	38. C	43. C	48. C
4. B	9. D	14. D	19. C	24. A	29. D	34. D	39. B	44. D	49. B
5. A	10. B	15. B	20. E	25. C	30. D	35. E	40. D	45. A	50. B

ANSWER EXPLANATIONS

In these solutions the following notation is used:

a: active—Calculator use is necessary or, at a minimum, extremely helpful.

g: Graphing calculator solution is preferred.

i: inactive—Calculator use is not helpful and may even be a hindrance.

1. i **C** Solve for y. $y = \frac{4}{3}x - 4$. Slope $= \frac{4}{3}$. Slope of perpendicular $= -\frac{3}{4}$. [2.2].

2. i **E** Range = largest value – smallest value = $18 - 8 = 10$. [4.7].

3. i **A** The set of points equidistant from two vertices is the perpendicular bisecting plane of the segment joining them. The set of points 2 inches from the third vertex is a sphere with center at the third vertex and radius 2. The plane and sphere intersect in a circle. [5.5].

4. g **B** Enter the function f into Y_1 and the function g into Y_2. Evaluate $Y_1(Y_2(2))$ to get the correct answer choice B.
An alternative solution is to evaluate $g(2) = 5$ and $f(5) = \sqrt{13}$, and either use your calculator to evaluate $\sqrt{13}$ or observe that $3 < \sqrt{13} < 4$, indicating 3.61 as the only feasible answer choice. [1.2].

5. a **A** $\left[\left(-\frac{1}{16}\right)^2\right]^{1/3} = \left(\frac{1}{256}\right)^{1/3} = \sqrt[3]{\frac{1}{256}} \approx 0.16$. [4.2].

6. a **E** Complete the square to get $x^2 + (y-5)^2 = 61$. Radius $= \sqrt{61}$. $C = 2\pi r = 2\pi\sqrt{61} \approx 49$. [4.1].

7. a **D** $\sum_{j=3}^{5} \ln j = \ln 3 + \ln 4 + \ln 5 \approx 4.1$. [5.4].

8. i **B** Regardless of what is substituted for x, $f(x)$ still equals 2. [1.2].

Alternative Solution: $f(x + 2)$ causes the graph of $f(x)$ to be shifted 2 units to the left. Since $f(x) = 2$ for all x, $f(x + 2)$ will also equal 2 for all x.

9. g **D** Enter the formula for the volume of a sphere $(4/3)\pi x^3$ (in the reference list of formulas) into Y_1. Return to the Home Screen, and enter $Y_1(5) - Y_1(2)$ to get the correct answer choice D.

An alternative solution is to evaluate $V = \frac{4}{3}\pi$ $(5^3 - 2^3) = \frac{4}{3}\pi(117)$ directly. [5.5].

10. i **B** The number of terms is always 1 more than the exponent of the binomial if the exponent is a positive integer. Number of terms = 8. [5.2].

11. i **C** $\sin A = \frac{8}{10} = \frac{4}{5}$. [3.1].

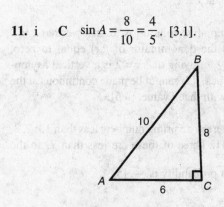

12. i **D** Since the $\frac{1}{2}$ power is the square root, $x = \frac{1}{2}$ because the square root of 16 is 4. Since $5^4 = 625$, $x + y = 4$, so that $y = 4 - \frac{1}{2} = \frac{7}{2}$. [4.2].

13. i **E** $t = \dfrac{y}{2}$. Eliminate the parameter and get

$x = \dfrac{y^2}{4} + 1$ or $y^2 = 4x - 4$. [4.6].

14. i **D** $f(1) = 0$ and $f(2) = 0$ imply that $x - 1$ and $x - 2$ are factors of $f(x)$. Their product, $x^2 - 3x + 2$, is also a factor. [2.4].

15. g **B** Set the calculator in degree mode, and evaluate $\sin(\tan^{-1}(1/3))$ to get the correct answer choice B. An alternative solution is to define

$\theta = \text{Arctan}\left(\dfrac{1}{3}\right)$, so that $\tan\theta = \dfrac{1}{3}$. Using a right triangle, the Pythagorean Theorem, and the

definition, find $\sin\theta = \dfrac{1}{\sqrt{10}} \approx 0.32$. [3.6].

16. i **B** $a^2 = z^2\cos^2\theta$ and $b^2 = z^2\sin^2\theta$, so $a^2 + b^2 = z^2(\cos^2\theta + \sin^2\theta) = z^2$ because $\cos^2\theta + \sin^2\theta = 1$.

Since $\sqrt{z^2} = z$ when $z > 0$, the correct answer choice is B. [3.5].

17. i **B** The statement, "If the trip is cancelled, it rained," is equivalent to its contrapositive, "If it didn't rain, the trip was not cancelled." Since it did not rain, it can be concluded that the trip is not cancelled. [5.7].

18. g **E** Plot the graph of $y = \dfrac{5}{x-2}$ in the standard window, and observe the asymptote at $x = 2$. This says that no value of k that can make $f(x)$ continuous at $x = 2$.

An alternative solution is to observe that $x = 2$ makes the denominator of $f(x)$ equal to zero, thereby implying that $x = 2$ is a vertical asymptote. Thus, $f(x)$ cannot be made continuous at the point with that x value. [4.5].

19. i **C** There are 5 prime numbers less than 13: 2, 3, 5, 7, 11. Three of these are less than 7, so the

correct probability is $\dfrac{3}{5}$. [5.3].

20. a **E** Substituting for x and solving for y gives $4(9 - y^2) + 8y^2 = 64$. $4y^2 = 28$, and so $y^2 = 7$ and

$y = \pm\sqrt{7} \approx \pm 2.65$. [4.1].

21. i **B** $t_n \cdot t_{n+1} = K$. $2 \cdot 6 = K = 12$. Therefore, $6 \cdot t_3 = 12$, and so $t_3 = 2$. Continuing this process gives all odd terms to be 2 and all even terms to be 6. [5.6, 5.4].

22. g **C** Graph the cost of Company A $y = 6 + 9x$ and the cost of Company B $y = 25 + 2.25x$ in a window $x\varepsilon[0,10]$ and $y\varepsilon[0,50]$. Use CALC/intersect to find the x–coordinate of the point of intersection at 2.8148 hours, the "break-even" point. Multiply by 60 to convert this time to the correct answer choice.

An alternative solution is to solve the equation $6 + 9x = 25 + 2.25x$ and multiply the solution by 60 to get the answer of about 169 minutes. [2.1].

23. i **B** The probability that both teams will win is pq. The probability that both will lose is $(1 - p)(1 - q)$. The probability that only one will win is $1 - [pq + (1 - p)(1 - q)] = 1 - (pq + 1 - p - q + pq) = p + q - 2pq$. [5.3].

Alternative Solution: The probability that the Giants will win and the Raiders will lose is $p(1 - q)$. The probability that the Raiders will win and the Giants will lose is $q(1 - p)$. Therefore, the probability that either one of these results will occur is $p(1 - q) + q(1 - p) = p + q - 2pq$.

24. g **A** Calculate the common ratio as $\dfrac{1}{1/3} = 3$. The

first term is $\dfrac{1}{3}$ so the nth term is $t_n = \left(\dfrac{1}{3}\right)3^{n-1}$.

Use the sum and sequence features of your calculator to evaluate the sum of the first 10 terms in the generated sequence:

LIST/MATH/sum(LIST/OPS/

seq((1/3)3^X, X, 0.9)) = 9841.333 $\cdots$

The range is 0 to 9 instead of 1 to 10 because the formula for t_n uses the exponent $n - 1$.

An alternative solution is to use the formula for the sum of a geometric series:

$$S_n = \dfrac{t_1(1 - r^n)}{1 - r} = \dfrac{(1/3)(1 - 3^{10})}{1 - 3} = 9841\dfrac{1}{3}.$$

[5.4].

25. g **C** No calculator currently on the market can compute 453!, so doing this problem requires some knowledge of factorial arithmetic. The easiest solution to the problem is to observe that

$\dfrac{453!}{450!3!}$ is the number of combinations of 453

taken 3 at a time ($_{453}C_3$). Enter 453MATH/PRB/nCr3 into your calculator to find that the correct answer choice is C.

An alternative solution is to simplify $\dfrac{453!}{450!3!}$ to

$$\dfrac{453 \cdot 452 \cdot 451}{3 \cdot 2 \cdot 1} = 1.5 \cdots \times 10^7. \quad [5.1].$$

26. a **B** Solve for y: $y = \dfrac{3}{2}x + \dfrac{7}{2}$. Slope $= \dfrac{\Delta y}{\Delta x} = \dfrac{3}{2}$.

Tan A also equals $\dfrac{\Delta y}{\Delta x}$. Therefore, $\tan A = \dfrac{3}{2}$.

$\text{Tan}^{-1}\left(\dfrac{3}{2}\right) = \angle A \approx 56°. \quad [2.2].$

27. g **C** Put your calculator in radian mode and graph $y = 3\sin(2(x + 2\pi/3))$ using Ztrig. Since the answer choices are pretty far apart, you can determine the correct answer choice by tracing to the first positive x-intercept 1.05.

An alternative solution can be found by solving $3\sin(2(x + 2\pi/3)) = 0$. This means that $x + 2\pi/3 = 0, \pi, \ldots$, The equation $x + 2\pi/3 = \pi$ yields the smallest positive solution $x = \dfrac{\pi}{3} \approx 1.05. \quad [3.5].$

28. i **B** Divide the equation through by 16 to get $\dfrac{(x-4)^2}{16} + \dfrac{(y-3)^2}{4} = 1$. This is the equation of an ellipse with $a^2 = 16$. The sum of the distances to the foci $= 2a = 8. \quad [4.1].$

29. a **D** If the roots are r and $2r$, their sum $= -\dfrac{b}{a} = 3r = -\dfrac{k}{1}$ and their product $= \dfrac{c}{a} = 2r^2 = \dfrac{54}{1}$. Therefore, $r = \pm\sqrt{27}$ and $k = \pm 3\sqrt{27} \approx \pm 15.6. \quad [2.3].$

Alternative Solution: If the roots are r and $2r$, $(x - r)(x - 2r) = 0$. Multiply to obtain $x^2 - 3r + 2r^2 = 0$, which represents $x^2 + kx + 54 = 0$. Thus, $-3r = k$ and $2r^2 = 54$. Since $r = -\dfrac{k}{3}$, then $2\left(-\dfrac{k}{3}\right) = 54$ and $k = \pm 3\sqrt{27} \approx 15.6$.

30. i **D** Substituting -1 for x gives 3. $[2.4].$

Alternative Solution: Using synthetic division gives

-1	3	7	8	-2	-3
		-3	-4	-4	6
	3	4	4	-6	$\underline{3}$ = remainder

31. a **A** The inverse of $f(x) = e^x$ is $f^{-1}(x) = \ln x$. $g(2) = e^2 + \ln 2 \approx 8.1. \quad [4.2].$

32. i **A** This expression is of the form $A^3 - 3A^2 + 3A - 1$, which equals $(A - 1)^3$. Substituting $x + 2$ for A gives Choice A. $[5.2].$

33. i **C** If the graph passes through the origin, $x = 0$ and $y = 0$, then $\dfrac{4k^2}{1} - \dfrac{9k^2}{3} = 1$. $k^2 = 1$, and so $k = \pm 1. \quad [4.1].$

34. g **D** Graph $y = \dfrac{1 - \cos x}{\sin x}$ and $y = \dfrac{\sqrt{3}}{3}$ using Ztrig in degree mode. Find the point of intersection with CALC/intersect to arrive at the correct answer choice D.

An alternative solution uses the identities $\tan\dfrac{\theta}{2} = \dfrac{1 - \cos x}{\sin x}$ and $\tan 30° = \dfrac{\sqrt{3}}{3}$ to deduce $\dfrac{\theta}{2} = 30°$, so $\theta = 60°. \quad [3.5].$

35. g **E** The problem is asking for the range of $f(x)$ values for values of x that satisfy the inequality. First graph the inequality in Y_1, starting with the standard window and zooming in until the x values for the portion of the graph that falls below the x-axis can be identified as the interval $(-2,-1)$. Then enter the formula for $f(x)$ in Y_2. Although it can be done graphically, the simplest way to find the range of values of $f(x)$ that correspond to $x\varepsilon(-2,-1)$ is to use the TABLE function. Deselect Y_1 and enter TBLSET and set TblStart to -2, $\Delta Tbl = 0.1$, and Indpnt and Depend to Auto. Then enter TABLE and observe that the Y_2 values range from 12 to 6 as x ranges from -2 to -1, yielding the correct answer choice E.

An alternative solution is to solve the inequality algebraically by solving the associated equation $x^2 + 3x + 2 = 0$ and testing points. The left side of the equation factors as $(x + 2)(x + 1)$, and the Zero Product Property implies that $x = -2$ or $x = -1$. Points inside the interval $(-2,-1)$ satisfy the inequality, while those outside it do not. Since the graph of $f(x)$ is a parabola and $f(-2) = 12$ and $f(-1) = 6$, $f(x)$ takes the range of values between 12 and 6. $[4.3, 2.3].$

36. g **E** Recall that the notation [x] means the greatest integer less than or equal to x. Enter abs(x) + int(x) into Y_1. Return to the Home Screen, and enter $Y_1(-2.5) + Y_1(1.5)$ to get the correct answer choice E.

An alternative solution evaluates |−2.5| + [−2.5] + |1.5| + [1.5] without the aid of a calculator. Of these 4 values, only [−2.5] is tricky since [−2.5] = −3, not −2. Thus, |−2.5| + [−2.5] + |1.5| + [1.5] = 2.5 − 3 + 1.5 + 1 = 2. [4.3, 4.4].

37. i **C** Set up the following table.

	Q1	Q2	Q3	Q4
sec x	+	−	−	+
tan x	+	−	+	−
cot x	+	−	+	−
csc x	+	+	−	−

The product sec x tan x is negative only when its factors have different signs, so III is the only true statement. [3.1].

38. i **C** $\binom{x \text{ people}}{2} = 28$. $\frac{x(x-1)}{2 \cdot 1} = 28$. $x^2 - x = 56$. $x = 8$. [5.1].

39. a **B** Since the function g is f translated 3 down and 2 right, $g(x) = f(x-2) - 3$. Therefore, $g(-1.2) = f(-3.2) - 3 = 2(-3.2)^2 - 3 = 18.48$. [5.5].

40. g **D** Set your calculator to radian mode and graph the function in an $x\varepsilon[0,2\pi]$ and $y\varepsilon[3,7]$ window. Use CALC/maximum to find 4.19 as the smallest positive x where a maximum occurs.

An alternative solution is to observe that the function will be a maximum when $\sin\left(x+\frac{\pi}{6}\right) = -1$.

This will happen when $x + \frac{\pi}{6} = \frac{3\pi}{2}$, or when

$x = \frac{4\pi}{3} \approx 4.19$. [3.4].

41. a **D** The graph of f must be symmetric about the line y = x. In I, f becomes y = −1x + b, which is symmetric about y = x. In II, f becomes y = x, which is symmetric about y = x since it is y = x. In III, f becomes y = ax, which is not necessarily symmetric about y = x. [1.3].

42. a **C** Law of sines:
$$\frac{\sin 110°}{\sqrt{6}} = \frac{\sin B}{2}; \sin B = \frac{2\sin 110°}{\sqrt{6}} \approx 0.7673.$$
$\text{Sin}^{-1}(0.7673) = \angle B = 50°$. Therefore, $\angle C = 180° - 110° - 50° = 20°$. [3.7].

43. a **C** $\left|3\vec{v} - \vec{u}\right| = \left|(3\sqrt{3}) - (3, -2)\right| = \left|(0, 3\sqrt{3} + 2)\right| = \sqrt{0^2 + (3\sqrt{3} + 2)^2} = 3\sqrt{3} + 2 \approx 7$. [5.5].

44. a **D** Circumference of base of cone = length of arc $AB = 5$. $\frac{2\pi}{5} = 2\pi$. Circumference = $2\pi r = 2\pi$. Therefore, radius of base = 1. Height of cone = $\sqrt{24}$. Therefore, volume = $\frac{1}{3}\pi r^2 h = \frac{1}{3}\pi \cdot 1 \cdot \sqrt{24} \approx 5.13$. [5.5, 3.2].

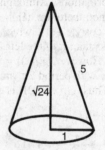

45. a **A** Law of cosines:

$$12 = (2x)^2 + (3x+2)^2 - 2 \cdot 2x \cdot (3x+2)\cos 60°.$$
$$12 = 4x^2 + 9x^2 + 12x + 4 - (12x^2 + 8x) \cdot \frac{1}{2}.$$

$7x^2 + 8x - 8 = 0$. Solving gives
$$x = \frac{-8 \pm \sqrt{64 + 224}}{14} = \frac{-8 \pm \sqrt{288}}{14}$$
Since a side of a triangle must be positive, x can equal only $\frac{-8 + \sqrt{288}}{14} \approx 0.64$. [3.7].

46. a **E** $\frac{|5a + 3(3\sqrt{3})a - 12|}{\sqrt{5^2 + 3^2}} = 6$. $|5a + 3\sqrt{3}a - 12| = 6\sqrt{34}$. Thus, $5a + 3\sqrt{3}a - 12 = 6\sqrt{34}$ or $5a + 3\sqrt{3}a - 12 = -6\sqrt{34}$, and $(5 + 3\sqrt{3})a = 12 + 6\sqrt{34}$ or $(5 + 3\sqrt{3})a = 12 - 6\sqrt{34}$. Therefore,

$$a = \frac{12 + 6\sqrt{34}}{5 + 3\sqrt{3}} \approx 4.6 \text{ or } a = \frac{12 - 6\sqrt{34}}{5 + 3\sqrt{3}} \approx -2.3.$$

[1.1].

47. g **A** Graph $y = 3x^2 + 4x + 5$ in the standard window, and observe that the graph must be moved slightly to the right to be symmetric to the y-axis. Therefore, k must be positive, and A is the only possible answer choice. If there had been other positive answer choices, use CALC/minimum to find the vertex of the parabola and observe that its x coordinate is $-0.66666...$. If the function entered into Y_1, set $Y_2 = Y_1\left(x - \dfrac{2}{3}\right)$ and graph Y_2 to verify this answer. [5.5].

48. g **C** Use ZTrig to plot the graphs of $y = (\cos x) \cdot (2x + 1)$, $y = \cos(2x + 1)$, and $y = 2(\cos x) + 1$ to see that only the third graph is symmetric about the y-axis and thus represents an even function.

An alternative solution is to use your knowledge of transformations. Although f is an even function, g is not; therefore, (I) $f \cdot g$ is not even. Also, $f(g(x)) = \cos(2x + 1)$, which is a cosine curve shifted less than π to the left. Thus, $f(g(x))$ (II) is not even. However, $g(f(x)) = 2\cos x + 1$ is a cosine curve with period 2π, amplitude 2, shifted 1 unit up. Thus, $g(f(x))$ (III) is even. [1.4].

49. a **B** Height of cylinder is 8.

Volume of sphere $= \dfrac{4}{3}\pi r^3 = \dfrac{4}{3}\pi(125) = \dfrac{500\pi}{3}$.

Volume of cylinder $= \pi r^2 h = \pi(9)8$.

Difference $= \dfrac{500}{3}\pi - 72\pi \approx 523.6 - 226.7$

≈ 297.

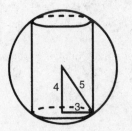

50. i **B** In answer choice B, x and y have the same sign, and x is less than y. Therefore xy is positive, $x - y$ is negative, and the quotient is negative. The numerators and denominators in answer choices A, C, and D both have the same sign, so the quotients are positive.

SELF-EVALUATION CHART FOR MODEL TEST 1

SUBJECT AREA	QUESTIONS	NUMBER OF RIGHT WRONG OMITTED

Mark correct answers with C, wrong answers with X, and omitted answers with O.

Algebra
(9 questions)
Review section

5	7	12	22	24	29	30	32	35
4.2	5.4	4.2	5.9	5.4	2.3	2.4	5.2	4.3

_____ _____ _____

Solid Geometry
(4 questions)
Review section

3	9	44	49
5.5	5.5	5.5	5.5

_____ _____ _____

Coordinate Geometry
(6 questions)
Review section

1	6	20	28	33	46
2.2	4.1	4.1	4.1	4.1	2.2

_____ _____ _____

Trigonometry
(11 questions)
Review section

11	15	16	17	26	27	34	37	40	42	45
3.1	3.6	3.5	5.7	2.2	3.5	3.5	3.1	3.4	3.7	3.7

_____ _____ _____

Functions
(9 questions)
Review section

4	8	14	18	31	36	41	47	48
1.2	1.2	2.4	4.5	4.2	4.3	1.3	2.3	1.4

_____ _____ _____

Miscellaneous
(11 questions)
Review section

2	10	13	19	21	23	25	38	39	43	50
4.7	5.2	4.6	5.3	5.6	5.3	5.1	5.1	5.5	5.5	5.9

_____ _____ _____

TOTALS _____ _____ _____

Raw score = (number right) − $\frac{1}{4}$ (number wrong) = _____

Round your raw score to the nearest whole number = _____

Evaluate Your Performance
Model Test 1

Rating	Number Right
Excellent	41–50
Very Good	33–40
Above Average	25–32
Average	15–24
Below Average	Below 15

ANSWER SHEET FOR MODEL TEST 2

Determine the correct answer for each question. Then, using a no. 2 pencil, blacken completely the oval containing the letter of your choice.

1. Ⓐ Ⓑ Ⓒ Ⓓ Ⓔ
2. Ⓐ Ⓑ Ⓒ Ⓓ Ⓔ
3. Ⓐ Ⓑ Ⓒ Ⓓ Ⓔ
4. Ⓐ Ⓑ Ⓒ Ⓓ Ⓔ
5. Ⓐ Ⓑ Ⓒ Ⓓ Ⓔ
6. Ⓐ Ⓑ Ⓒ Ⓓ Ⓔ
7. Ⓐ Ⓑ Ⓒ Ⓓ Ⓔ
8. Ⓐ Ⓑ Ⓒ Ⓓ Ⓔ
9. Ⓐ Ⓑ Ⓒ Ⓓ Ⓔ
10. Ⓐ Ⓑ Ⓒ Ⓓ Ⓔ
11. Ⓐ Ⓑ Ⓒ Ⓓ Ⓔ
12. Ⓐ Ⓑ Ⓒ Ⓓ Ⓔ
13. Ⓐ Ⓑ Ⓒ Ⓓ Ⓔ
14. Ⓐ Ⓑ Ⓒ Ⓓ Ⓔ
15. Ⓐ Ⓑ Ⓒ Ⓓ Ⓔ
16. Ⓐ Ⓑ Ⓒ Ⓓ Ⓔ
17. Ⓐ Ⓑ Ⓒ Ⓓ Ⓔ

18. Ⓐ Ⓑ Ⓒ Ⓓ Ⓔ
19. Ⓐ Ⓑ Ⓒ Ⓓ Ⓔ
20. Ⓐ Ⓑ Ⓒ Ⓓ Ⓔ
21. Ⓐ Ⓑ Ⓒ Ⓓ Ⓔ
22. Ⓐ Ⓑ Ⓒ Ⓓ Ⓔ
23. Ⓐ Ⓑ Ⓒ Ⓓ Ⓔ
24. Ⓐ Ⓑ Ⓒ Ⓓ Ⓔ
25. Ⓐ Ⓑ Ⓒ Ⓓ Ⓔ
26. Ⓐ Ⓑ Ⓒ Ⓓ Ⓔ
27. Ⓐ Ⓑ Ⓒ Ⓓ Ⓔ
28. Ⓐ Ⓑ Ⓒ Ⓓ Ⓔ
29. Ⓐ Ⓑ Ⓒ Ⓓ Ⓔ
30. Ⓐ Ⓑ Ⓒ Ⓓ Ⓔ
31. Ⓐ Ⓑ Ⓒ Ⓓ Ⓔ
32. Ⓐ Ⓑ Ⓒ Ⓓ Ⓔ
33. Ⓐ Ⓑ Ⓒ Ⓓ Ⓔ
34. Ⓐ Ⓑ Ⓒ Ⓓ Ⓔ

35. Ⓐ Ⓑ Ⓒ Ⓓ Ⓔ
36. Ⓐ Ⓑ Ⓒ Ⓓ Ⓔ
37. Ⓐ Ⓑ Ⓒ Ⓓ Ⓔ
38. Ⓐ Ⓑ Ⓒ Ⓓ Ⓔ
39. Ⓐ Ⓑ Ⓒ Ⓓ Ⓔ
40. Ⓐ Ⓑ Ⓒ Ⓓ Ⓔ
41. Ⓐ Ⓑ Ⓒ Ⓓ Ⓔ
42. Ⓐ Ⓑ Ⓒ Ⓓ Ⓔ
43. Ⓐ Ⓑ Ⓒ Ⓓ Ⓔ
44. Ⓐ Ⓑ Ⓒ Ⓓ Ⓔ
45. Ⓐ Ⓑ Ⓒ Ⓓ Ⓔ
46. Ⓐ Ⓑ Ⓒ Ⓓ Ⓔ
47. Ⓐ Ⓑ Ⓒ Ⓓ Ⓔ
48. Ⓐ Ⓑ Ⓒ Ⓓ Ⓔ
49. Ⓐ Ⓑ Ⓒ Ⓓ Ⓔ
50. Ⓐ Ⓑ Ⓒ Ⓓ Ⓔ

MODEL TEST

50 questions   1 hour

Tear out the preceding answer sheet. Decide which is the best choice by rounding your answer when appropriate. Blacken the corresponding space on the answer sheet. When finished, check your answers with those at the end of the test. For questions that you got wrong, note the sections containing the material that you must review. Also if you do not fully understand how you arrived at some of the correct answers, you should review the appropriate sections. Finally, fill out the self-evaluation sheet on page 177 in order to pinpoint the topics that give you the most difficulty.

TEST DIRECTIONS

<u>Directions</u>: Decide which answer choice is best. If the exact numerical value is not one of the answer choices, select the closest approximation. Fill in the oval on the answer sheet that corresponds to your choice.

Notes:
(1) You will need to use a scientific or graphing calculator to answer some of the questions.
(2) You will have to decide whether to put your calculator in degree or radian mode for some problems.
(3) All figures that accompany problems are plane figures unless otherwise stated. Figures are drawn as accurately as possible to provide useful information for solving the problem, except when it is stated in a particular problem that the figure is not drawn to scale.
(4) Unless otherwise indicated, the domain of a function is the set of all real numbers for which the functional value is also a real number.

<u>Reference Information.</u> The following formulas are provided for your information.

Volume of a right circular cone with radius r and height h: $V = \frac{1}{3}\pi r^2 h$

Lateral area of a right circular cone if the base has circumference c and the slant height is l: $S = \frac{1}{2}cl$

Volume of a sphere of radius r: $V = \frac{4}{3}\pi r^3$

Surface area of a sphere of radius r: $S = 4\pi r^2$

Volume of a pyramid of base area B and height h: $V = \frac{1}{3}Bh$

---

1. If $f(x) = \dfrac{x-2}{x^2-4}$, for what value(s) of x does the graph of $f(x)$ have a vertical asymptote?

   (A) -2, 0, and 2
   (B) -2 and 2
   (C) 2
   (D) 0
   (E) -2

2. If a regular square pyramid has x pairs of parallel edges, then x equals

   (A) 1
   (B) 2
   (C) 4
   (D) 8
   (E) 12

3. Log $(a^2 - b^2) =$
   (A) $\log a^2 - \log b^2$
   (B) $\log \dfrac{a^2}{b^2}$
   (C) $\log \dfrac{a+b}{a-b}$
   (D) $2 \cdot \log a - 2 \cdot \log b$
   (E) $\log(a+b) + \log(a-b)$

4. The sum of the roots of the equation $\left(x - \sqrt{2}\right)^2 \left(x + \sqrt{3}\right)\left(x - \sqrt{5}\right) = 0$ is

   (A) 3.3
   (B) 1.9
   (C) 2.2
   (D) 6.8
   (E) 2.5

5. If the graph of $x + 2y + 3 = 0$ is perpendicular to the graph of $ax + 3y + 2 = 0$, then a equals

   (A) $\dfrac{3}{2}$
   (B) $-\dfrac{3}{2}$
   (C) 6
   (D) -6
   (E) $\dfrac{2}{3}$

6. The maximum value of $6 \cdot \sin x \cdot \cos x$ is

   (A) $\dfrac{1}{3}$
   (B) 1
   (C) 3
   (D) 6
   (E) $\dfrac{3\sqrt{3}}{2}$

GO ON TO THE NEXT PAGE

7. If $f(r,\theta) = r\cos\theta$, then $f(2,3) =$

   (A) -3.00
   (B) -1.98
   (C) 0.10
   (D) 2.00
   (E) 1.25

8. If 5 and -1 are both zeros of the polynomial $P(x)$, then a factor of $P(x)$ is

   (A) $x^2 - 5$
   (B) $x^2 - 4x + 5$
   (C) $x^2 + 4x - 5$
   (D) $x^2 + 5$
   (E) $x^2 - 4x - 5$

9. $i^{14} + i^{15} + i^{16} + i^{17} =$

   (A) 1
   (B) $2i$
   (C) $1 - i$
   (D) 0
   (E) $2 + 2i$

10. When the graph of $y = \sin 2x$ is drawn for all values of x between $10°$ and $350°$, it crosses the x-axis

    (A) zero times
    (B) one time
    (C) two times
    (D) three times
    (E) six times

11. When $\left(1 - \dfrac{1}{x}\right)^{-6}$ is expanded, the sum of the last three coefficients is

    (A) 10
    (B) 11
    (C) 16
    (D) -11
    (E) The sum cannot be determined.

12. A particular sphere has the property that its surface area has the same numerical value as its volume. What is the length of the radius of this sphere?

    (A) 1
    (B) 2
    (C) 3
    (D) 4
    (E) 6

USE THIS SPACE FOR SCRATCH WORK

GO ON TO THE NEXT PAGE

13. If $f(x,y) = x^4 + x^2y - y^4$, which of the following is (are) true?

   I. $f(x,y) = f(x,-y)$
   II. $f(x,y) = f(-x, y)$
   III. $f(x,y) = f(-x,-y)$

   (A) I only
   (B) II only
   (C) III only
   (D) I and III only
   (E) I, II, and III

14. The pendulum on a clock swings through an angle of 1 radian, and the tip sweeps out an arc of 12 inches. How long is the pendulum?

   (A) 6 inches
   (B) 12 inches
   (C) 24 inches
   (D) $\dfrac{12}{\pi}$ inches
   (E) $\dfrac{24}{\pi}$ inches

15. What is the domain of the function $f(x) = 4 - \sqrt{3x^3 - 7}$?
   (A) $x \le -1.33$ or $x \ge 1.33$
   (B) $x \ge 2.33$
   (C) $x \ge 1.33$
   (D) $x \ge 1.53$
   (E) $x \le -2.33$ or $x \ge 2.33$

16. If $x + y = 90°$, which of the following must be true?

   (A) $\cos x = \cos y$
   (B) $\sin x = -\sin y$
   (C) $\tan x = \cot y$
   (D) $\sin x + \cos y = 1$
   (E) $\tan x + \cot y = 1$

17. The graph of the equation $y = x^3 + 5x + 1$

   (A) does not intersect the x-axis
   (B) intersects the x-axis at one and only one point
   (C) intersects the x-axis at exactly three points
   (D) intersects the x-axis at more than three points
   (E) intersects the x-axis at exactly two points

18. The length of the radius of the sphere $x^2 + y^2 + z^2 + 2x - 4y = 10$ is

   (A) 3.16
   (B) 3.38
   (C) 3.87
   (D) 3.74
   (E) 3.46

GO ON TO THE NEXT PAGE

19. If the roots of the equation $x^2 + bx + c = 0$ are r and s, the value of $\dfrac{(r+s)^2}{rs}$ in terms of b and c is

    (A) $\dfrac{b^2}{c^2}$

    (B) $\dfrac{c^2}{b}$

    (C) $\dfrac{b^2}{c}$

    (D) $-\dfrac{b}{c}$

    (E) $-\dfrac{b^2}{c}$

USE THIS SPACE FOR SCRATCH WORK

20. Which of the following is the solution set for $x(x-3)(x+2) > 0$?

    (A) $x < -2$
    (B) $-2 < x < 3$
    (C) $-2 < x < 3$ or $x > 3$
    (D) $x < -2$ or $0 < x < 3$
    (E) $-2 < x < 0$ or $x > 3$

21. Which of the following is the equation of the circle that has its center at the origin and is tangent to the line with equation $3x - 4y = 10$?

    (A) $x^2 + y^2 = 2$
    (B) $x^2 + y^2 = 4$
    (C) $x^2 + y^2 = 3$
    (D) $x^2 + y^2 = 5$
    (E) $x^2 + y^2 = 10$

22. If $f(x) = 3 - 2x + x^2$, then $\dfrac{f(x+t) - f(x)}{t} =$

    (A) $t^2 + 2xt - 2t$
    (B) $x^2 t^2 - 2xt + 3$
    (C) $t + 2x - 2$
    (D) $2x - 2$
    (E) none of the above

23. If $f(x) = x^3$ and $g(x) = x^2 + 1$, which of the following is an odd function (are odd functions)?

    I. $f(x) \cdot g(x)$
    II. $f(g(x))$
    III. $g(f(x))$

    (A) only I
    (B) only II
    (C) only III
    (D) only II and III
    (E) I, II, and III

GO ON TO THE NEXT PAGE

24. In how many ways can a committee of four be selected from nine men so as to always include a particular man?

(A) 84
(B) 70
(C) 48
(D) 56
(E) 126

25. When $f(x) = \sin x$ and $g(x) = \cos x$ over the interval $(0, 2\pi)$, and $M(f,g)$ is defined to be the maximum of f and g, the graph of $M(f,g)$ looks like which one of the following?

(A)

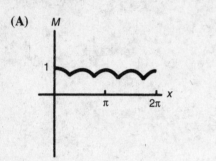

(B)

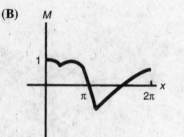

(C)

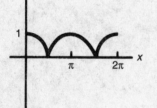

(D)

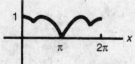

(E)

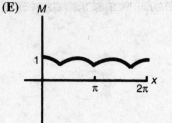

26. If the mean of the set of data 1, 2, 3, 1, 2, 5, x is $3.\overline{27}$, what is the value of x?

    (A) 8.9
    (B) −10.7
    (C) 5.6
    (D) 2.5
    (E) 7.4

27. In $\triangle JKL$, $\sin L = \frac{1}{3}$, $\sin J = \frac{3}{5}$, and $JK = \sqrt{5}$ inches. The length of KL, in inches, is

    (A) 3.0
    (B) 3.9
    (C) 3.5
    (D) 1.7
    (E) 4.0

28. If L varies inversely as the square of D, what is the effect on D when L is multiplied by 4?

    (A) It is multiplied by $\frac{3}{2}$.
    (B) It is multiplied by 4.
    (C) It is multiplied by 2.
    (D) It is divided by 2.
    (E) None of the above effects occurs.

29. Which of the following statements is logically equivalent to: "If he studies, he will pass the course."

    (A) He passed the course; therefore, he studied.
    (B) He did not study; therefore, he will not pass the course.
    (C) He did not pass the course; therefore he did not study.
    (D) He will pass the course only if he studies.
    (E) None of the above.

30. If $f(x, y, z) = \sqrt{2}x + \sqrt{3}y - z$ and $f(a, b, 0) = f(0, a, b)$, then $\frac{b}{a} =$

    (A) 0.32
    (B) 2.7
    (C) 8.6
    (D) 0.12
    (E) 1.18

GO ON TO THE NEXT PAGE

31. In $\triangle ABC$, $a = 1$, $b = 4$, and $\angle C = 30°$. The length of c is

(A) 4.6
(B) 3.6
(C) 3.2
(D) 2.9
(E) 2.3

32. The solution set of $3x + 4y < 0$ lies in which quadrants?

(A) I only
(B) I and II
(C) I, II, and III
(D) II, III, and IV
(E) I, II, III, and IV

33. Which of the following could represent the inverse of the function graphed on the right?

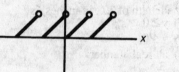

(A)

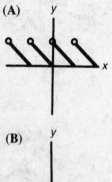

(B)

(C)

(D)

(E)

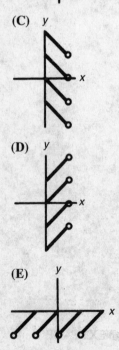

34. If f is a linear function and $f(-2) = 11$, $f(5) = -2$, and $f(x) = 4.3$, what is the value of x?

(A) 3.2
(B) −1.9
(C) 2.9
(D) 1.6
(E) −3.1

35. Which of the following could be a term in the expansion of $(p - q)^{18}$?

(A) $816p^{15}q^3$
(B) $-816p^{15}q^3$
(C) $816p^{16}q^2$
(D) $-816p^{16}q^2$
(E) $-816p^4q^{14}$

36. The range of the function $y = x^{-2/3}$ is

(A) $y < 0$
(B) $y > 0$
(C) $y \geq 0$
(D) $y \leq 0$
(E) all real numbers

37. If $\cos x = \dfrac{4}{5}$ and $\dfrac{3\pi}{2} \leq x \leq 2\pi$, then $\tan 2x =$

(A) $-\dfrac{7}{24}$

(B) $-\dfrac{24}{25}$

(C) $-\dfrac{3}{4}$

(D) $-\dfrac{24}{7}$

(E) $\dfrac{7}{25}$

38. A coin is tossed three times. Given that at least one head appears, what is the probability that exactly two heads will appear?

(A) $\dfrac{3}{8}$

(B) $\dfrac{3}{7}$

(C) $\dfrac{3}{4}$

(D) $\dfrac{5}{8}$

(E) $\dfrac{7}{8}$

GO ON TO THE NEXT PAGE

39. A unit vector parallel to vector $\vec{V} = (2, -3, 6)$ is vector

(A) $(-2, 3, -6)$
(B) $(6, -3, 2)$
(C) $(-0.29, 0.43, -0.86)$
(D) $(0.29, 0.43, -0.86)$
(E) $(-0.36, -0.54, 1.08)$

40. If the values of the function $g(x)$ represent the slope of the line tangent to the graph of the function $f(x)$, shown on the right, at each point (x, y), which of the following could be the graph of $g(x)$?

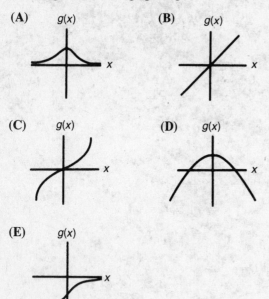

(A) $g(x)$

(B) $g(x)$

(C) $g(x)$

(D) $g(x)$

(E) $g(x)$

41. If $f(x) = \dfrac{x}{x-1}$ and $f^2(x) = f(f(x))$, $f^3(x) = f(f^2(x))$, $\ldots, f^n(x) = f(f^{n-1}(x))$, where n is a positive integer greater than 1, what is the smallest value of n such that $f^n(x) = f(x)$?

(A) 2
(B) 3
(C) 4
(D) 6
(E) No value of n works.

42. A committee of 5 people is to be selected from 6 men and 9 women. If the selection is made randomly, what is the probability that the committee consists of 3 men and 2 women?

(A) $\dfrac{1}{3}$

(B) $\dfrac{240}{1001}$

(C) $\dfrac{1260}{3003}$

(D) $\dfrac{13}{18}$

(E) $\dfrac{1}{9}$

43. Three consecutive terms, in order, of an arithmetic sequence are $x+\sqrt{2}, 2x+\sqrt{3}$, and $5x-\sqrt{5}$. Then x equals

(A) 2.46
(B) 3.56
(C) 2.14
(D) 2.45
(E) 3.24

44. The graph of $xy-4x-2y-4=0$ can be expressed as a set of parametric equations. If $y=\dfrac{4t}{t-3}$ and $x=f(t)$, then $f(t)=$

(A) $t+1$
(B) $t-1$
(C) $3t-3$
(D) $\dfrac{t-3}{4t}$
(E) $\dfrac{t-3}{2}$

45. If $f(x)=ax^2+bx+c$, how must a and b be related so that the graph of $f(x-3)$ will be symmetric about the y-axis?

(A) $a=b$
(B) $b=0$, a is any real number
(C) $b=3a$
(D) $b=6a$
(E) $a=\dfrac{1}{9}b$

46. The graph of $y=\log_5 x$ and $y=\ln 0.5x$ intersect at a point where x equals

(A) 6.24
(B) 1.14
(C) 1.69
(D) 1.05
(E) 5.44

GO ON TO THE NEXT PAGE

47. Which of the following is equivalent to $\sin(A + 30°) + \cos(A + 60°)$ for all values of A?

(A) $\sin A$

(B) $\cos A$

(C) $\sqrt{3} \cdot \sin A + \cos A$

(D) $\sqrt{3} \cdot \sin A$

(E) $\sqrt{3} \cdot \cos A$

48. The area of the region enclosed by the graph of the polar curve $r = \dfrac{1}{\sin\theta + \cos\theta}$ and the x- and y-axes is

(A) 0.48

(B) 0.50

(C) 0.52

(D) 0.98

(E) 1.00

49. A rectangular box has dimensions of length = 6, width = 4, and height = 5. The angle formed by a diagonal of the box with the base of the box contains

(A) 27°

(B) 35°

(C) 40°

(D) 44°

(E) 55°

50. If (x,y) represents a point on the graph of $y = 2x + 1$, which of the following could be a portion of the graph of the set of points (x, y^2)?

(A)

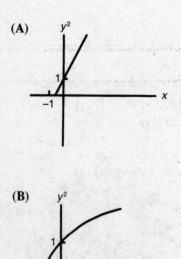

(B)

GO ON TO THE NEXT PAGE

(C)

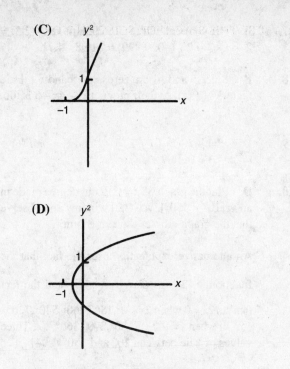

(D)

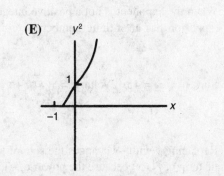

(E)

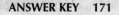

ANSWER KEY

1. E	6. C	11. E	16. C	21. B	26. A	31. C	36. B	41. B	46. A
2. B	7. B	12. C	17. B	22. C	27. E	32. D	37. D	42. B	47. B
3. E	8. B	13. B	18. C	23. A	28. D	33. D	38. B	43. C	48. B
4. A	9. D	14. B	19. C	24. D	29. C	34. D	39. C	44. B	49. B
5. D	10. D	15. C	20. E	25. B	30. D	35. B	40. A	45. D	50. C

ANSWER EXPLANATIONS

In these solutions the following notation is used:

a: active—Calculator use is necessary or, at a minimum, extremely helpful.

g: Graphing calculator is preferred.

i: inactive—Calculator use is not helpful and may even be a hindrance.

1. g E Graph the function f in the standard window and observe the vertical asymptote at $x = -2$.

An alternative solution is to factor the denominator of f as $(x + 2)(x - 2)$; cancel the factor $x - 2$ in the numerator and denominator so that $f(x) = \dfrac{1}{x+2}$; and recall that a function has a vertical asymptote when the denominator is zero and the numerator isn't. There is a hole in the graph of f at $x = 2$. [4.5].

2. i B The only pairs of parallel edges are the opposite sides of the square base. [5.5].

3. i E $\text{Log}(a^2 - b^2) = \log(a + b)(a - b) = \log(a + b) + \log(a - b)$. [4.2].

4. a A The sum of the roots is $\sqrt{2} + \sqrt{2} + (-\sqrt{3}) + \sqrt{5} \approx 1.414 + 1.414 - 1.732 + 2.236 \approx 3.3$. [2.4].

5. i D The slope of the first line is $-\dfrac{1}{2}$, and the slope of the second is $-\dfrac{a}{3}$. To be perpendicular, $-\dfrac{1}{2} = \dfrac{3}{a}$, or $a = -6$. [2.2].

6. g C Plot the graph of $6\sin x\cos x$ using ZTrig and observe that the max of this function is 3. (None of the other answer choices is close. If one were, you could use CALC/max to find the maximum value of the function.)

An alternative solution is to recall that $2x = 2\sin x\cos x$, so that $6\sin x\cos x = 3\sin 2x$. Since the amplitude of $\sin 2x$ is 1, the amplitude of $3\sin 2x$ is 3. [3.5].

7. a B Put your calculator in radian mode. $f(2,3) = 2 \cdot \cos 3 \approx 2 \cdot (-0.98999) \approx -1.98$. [4.7].

8. i B Since 5 and −1 are zeros, $x - 5$ and $x + 1$ are factors of $P(x)$, so their product $x^2 - 4x - 5$ is too. [2.4].

9. i D $i^{14}(1 + i + i^2 + i^3)$. $i^2 = -1$ and $i^3 = -i$. $i^{14}(1 + i - 1 - i) = 0$. [4.7].

10. g D Plot the graph of $y = \sin 2x$ in degree mode in an $x\varepsilon[10° \times 350°]$, $y\varepsilon[-2,2]$ window and observe that the graph crosses the axis 3 times.

An alternative explanation uses the fact that the function $\sin 2x$ has period $\dfrac{2\pi}{2} = \pi$ and the fact that $\sin 2x = 0$ when $2x = 0°,180°,360°,540°,720°, \dots$, or when $x = 0°,90°,180°,270°,360°, \dots$ Three values of x lie between 10° and 350°. [3.4]

11. i E When the exponent is not a positive integer, the expansion has an infinite number of terms. [5.2].

12. i C Surface area $= 4\pi r^2$. Volume $= \dfrac{4}{3}\pi r^3$. $4\pi r^2 = \dfrac{4}{3}\pi r^3$. $r = 3$. [5.5].

13. i B Replacing y with $-y$ changes the sign of the middle term of $f(x,y)$ because the power of y is 1 in that term, while replacing x with $-x$ has no effect since the power of x is even in both terms that contain x. Therefore, only II is true. [1.5].

14. i B $s = r\theta$. $12 = r$. [3.2].

15. a C The domain consists of all numbers that make $3x^3 - 7 \geq 0$. Therefore, $x^3 \geq \dfrac{7}{3}$ and $x \geq 1.33$. [1.1, 2.5].

16. i C Cofunctions of complementary angles are equal. Since x and y are complementary, tan and cot are cofunctions [3.1].

An alternative solution is to choose any values of x and y such that their sum is 90°. (For example, $x = 40°$ and $y = 50°$.) Test the answer choices with your calculator in degree mode to see that only Choice C is true.

17. g **B** Plot the graph of $y = x^3 + 5x + 1$ in the standard window and zoom in a couple of times to see that it crosses the x-axis only once. To make sure you are not missing anything, you should also plot the equation in a $x\varepsilon[-1,1]$, $y\varepsilon[-1,1]$ and an $x\varepsilon[-100,100]$, $y\varepsilon[-100,100]$ window.

An alternative solution is to use Descartes Rule of Signs, which indicates that the graph does not intersect the positive x-axis, but it does intersect the negative x-axis once. [2.4].

18. a **C** Complete the square: $(x^2 + 2x + 1) + (y^2 - 4y + 4) + z^2 = 10 + 1 + 4 = 15$. Therefore, $r = \sqrt{15} \approx 3.87$. [5.5].

19. i **C** In any quadratic equation the sum of the roots $= -\dfrac{b}{a}$ and the product of the roots $= \dfrac{c}{a}$. Also, $r + s = -b$, $(r + s)^2 = b^2$, $rs = c$. Therefore, $\dfrac{(r+s)^2}{rs} = \dfrac{b^2}{c}$. [2.3].

An alternative solution is to observe that if the roots are r and s, $(x - r)(x - s) = 0$. Multiply to obtain $x^2 + (-r - s)r + rs = 0$, which in the form of $x^2 + bx + c = 0$. Thus, $(-r - s) = b$ and $rs = c$. Thus, $(r + s)^2 = b^2$ and $\dfrac{(r+s)^2}{rs} = \dfrac{b^2}{c}$.

20. g **E** Plot the graph of $y = x(x - 3)(x + 2)$ in the standard window, and observe that the graph is above the x-axis when $-2 < x < 0$ or when $x > 3$.

An alternative solution is to find the zeros of the function $x(x - 3)(x + 2)$ as $x = 0, 3, -2$ and test points in the intervals established by these zeros. Points between -2 and 0 and greater than 3 satisfy the inequality. [2.4].

21. g **B** Use the program DPL to calculate the distance between the point $(0,0)$ and the line $3x - 4y = 10$ as 2. Therefore, the radius of the circle is 2, and its equation is $x^2 + y^2 = 4$.

An alternative solution is to use the formula for the distance between a point and a line:

$\text{distance} = \dfrac{3(0) - 4(0) - 10}{\sqrt{3^2 + 4^2}} = \dfrac{10}{5} = 2$ to get the radius. [4.1].

22. i **C**

$$\frac{f(x+t) - f(x)}{t}$$

$$= \frac{3 - 2(x+t) + (x+t)^2 - (3 - 2x + x^2)}{t}$$

$$= \frac{3 - 2x - 2t + x^2 + 2xt + t^2 - 3 + 2x - x^2}{t}$$

$$= 2x - 2 + t. \ [1.2].$$

CAUTION: For calculus students only: This difference quotient looks like the definition of the derivative. However, no limit is taken, so don't jump at $f'(x)$, which is Choice D.

23. g **A** Enter x^3 into Y_1 and $x^2 + 1$ into Y_2. Then enter Y_1Y_2 into Y_3; $Y_1(Y_2)$ into Y_4; and $Y_2(Y_1)$ into Y_5. De-select Y_1 and Y_2. Inspection of Y_3, Y_4, and Y_5 shows that only Y_3 is symmetric about the origin.

An alternative solution is to define each of the three functions as $h(x)$ and check each against the definition of an odd function, $h(-x) = -h(x)$:

$h(-x) = (-x)^3((-x)^2 + 1) = -x^5 - x^3 = -(x^5 + x^3)$
$\quad = -h(x)$

$h(-x) = f((-x)^2 + 1) = (x^2 + 1)^3 \neq -h(x)$

$h(-x) = g((-x)^3) = ((-x)^3)^2 + 1 = x^6 + 1 \neq -h(x)$

[1.4].

24. a **D** Since one particular man must be on the committee, the problem becomes: "Form a committee of 3 from 8 men." Calculate $\dbinom{8}{3} = \dfrac{8 \times 7 \times 6}{3 \times 2 \times 1} = 56$.

[5.1].

25. g **B** Graph MATH/NUM/max $(\sin(x), \cos(x))$ in the window $x\varepsilon[0, 2\pi]$ and $y\varepsilon[-2, 2]$. This graphs the exact function given, and the correct answer B will be evident.

An alternative solution is to sketch graphs of $\sin x$ and $\cos x$ on the same set of axes and highlight the graph of the maximum. [3.4].

26. a **A** Mean $= \dfrac{1 + 2 + 3 + 1 + 2 + 5 + x}{7} = \dfrac{14 + x}{7} =$ $3.\overline{27}$. Therefore, $x = 8.9$. [5.8].

27. a E Law of sines: $\dfrac{\frac{1}{3}}{\sqrt{5}} = \dfrac{\frac{3}{5}}{KL}$; $\dfrac{1}{3}KL = \dfrac{3\sqrt{5}}{5}$.

Therefore, $KL = \dfrac{9\sqrt{5}}{5} \approx 4.0$. [3.7].

Alternative Solution: Drop a perpendicular, KM, to LJ.

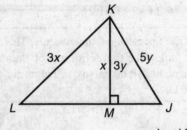

In a right triangle, $\sin x = \dfrac{\text{opposite side}}{\text{hypotenuse}}$, and so

$\sin L = \dfrac{1}{3}$.

Label $KM = x$ and $KL = 3x$. Similarly, $\sin J = \dfrac{3}{5}$,

so label $KM = 3y$ and $JK = 5y$. Therefore, from

the diagram, since $JK = \sqrt{5}$, $y = \dfrac{\sqrt{5}}{5}$. Then

$x = \dfrac{3\sqrt{5}}{5}$, and $KL = \dfrac{9\sqrt{5}}{5} \approx 4.0$.

28. i D $LD^2 = K$. If D is divided by 2 and then squared, the 4 in the denominator will cancel out the 4 times L. [5.6].

29. i C Relative to the statement, answer choice A is the converse, B is the inverse, C is the contrapositive, and D is another form of the inverse. Of these, the contrapositive is the logical equivalent of the original statement. [5.7].

30. a D $f(a,b,0) = \sqrt{2}a + \sqrt{3}b = f(0,a,b) = \sqrt{3}a - b$. $\left(\sqrt{3} + 1\right)b = \left(\sqrt{3} - \sqrt{2}\right)a$. Dividing both sides by $\left(\sqrt{3} + 1\right)a$ gives $\dfrac{b}{a} = \dfrac{\sqrt{3} - \sqrt{2}}{\sqrt{3} + 1} \approx 0.12$. [1.5].

31. a C Law of cosines: $c^2 = 16 + 1 - 8 \cdot \dfrac{\sqrt{3}}{2} = 17 - 4\sqrt{3} \approx 10.07$. Therefore, $c \approx 3.2$. [3.7].

32. g D Use the standard window to graph $y < -\dfrac{3}{4}x$, by moving the cursor all the way left (past Y =) and keying Enter until a "lower triangle" is observed. The shaded portion of the graph will lie in all but Quadrant I.

An alternative solution is to graph the related equation $y = -\dfrac{3}{4}x$ and test points to determine which side of the line contains solutions to the inequality. This will indicate the quadrant that the graph does not enter. [2.2].

33. i D Fold the graph about the line $y = x$, and the resulting graph will be Choice D. [1.3].

34. a D Since these points are on a line, all slopes must be equal. $\text{Slope} = \dfrac{-2-11}{5-(-2)} = \dfrac{-13}{7} = \dfrac{4.3-11}{x-(-2)}$. Cross-multiplying the far right equation gives $-13(x + 2) = 7(-6.7)$, and solving for x yields $x \approx 1.6$. [2.6].

35. i B If the term is even it must be negative, and if it is odd it must be positive. Therefore, the answer must be Choice B or C. The coefficient of B comes from $\binom{18}{3} = 816$. [5.2].

36. g B Plot the graph of $y = x^{-2/3}$ in the standard window and observe that the entire graph lies above the x-axis.

An alternative solution uses the fact that $x^{-2/3} = \dfrac{1}{x^{2/3}}$, and $x^{2/3} = (x^{1/3})^2$, so that all values of x (except zero) are positive. Therefore, the range of $y = x^{-2/3}$ is $y > 0$. [1.1, 4.2].

37. a D Since $\dfrac{3\pi}{2} \le x \le 2\pi$, x is in the fourth quadrant. Taking $\cos^{-1}\dfrac{4}{5}$ on the calculator yields first quadrant angle. Therefore, $x = -\cos^{-1}\dfrac{4}{5}$. To find the correct answer choice of $-\dfrac{24}{7}$, enter $\tan\left(-2\cos^{-1}\dfrac{4}{5}\right)$ and change to fractional form.

An alternative solution can be obtained from the fact that if $\cos x = \dfrac{4}{5}$ in Quadrant IV, then $\tan x = -\dfrac{3}{4}$. Then use the double-angle formula for the tangent, $\tan 2x = \dfrac{2\tan x}{1-\tan^2 x}$, and simplify:

$\tan 2x = \dfrac{-3/2}{1-9/16} = -\dfrac{24}{7}$. [3.1, 3.6].

38. i B There are 8 elements in the sample space of a coin being flipped 3 times. Of these elements, 7 contain 1 head and 3 (HHT, HTH, THH) contain 2 heads. Probability $= \dfrac{3}{7}$. [5.3].

39. a C A unit vector parallel to $\overrightarrow{V} = \dfrac{\overrightarrow{V}}{|\overrightarrow{V}|} \cdot |\overrightarrow{V}| =$

$\sqrt{2^2 + (-3)^2 + 6^2} = 7$. A unit vector is either $\left(\dfrac{2}{7}, -\dfrac{3}{7}, \dfrac{6}{7}\right)$, which is parallel to and in the same direction as $\overrightarrow{V}$, or $\left(-\dfrac{2}{7}, \dfrac{3}{7}, -\dfrac{6}{7}\right)$, which is parallel to and in the direction opposite to $\overrightarrow{V}$. Using your calculator, you find that $-\dfrac{2}{7} \approx -0.29$, which indicates that the correct answer is Choice C. [5.5].

40. n A The slope of a tangent line at any point on the graph of f is a positive number. Therefore, every point on the graph of g must be positive. [5.5, 5.9].

TIP: If you don't know how to do a general problem, think of one or two specific cases. This may lead you to a feasible answer choice.

---

41. i B $f^2(x) = f(f(x)) = f\left(\dfrac{x}{x-1}\right) = \dfrac{\dfrac{x}{x-1}}{\dfrac{x}{x-1}-1} = x$

$f^3(x) = f(f^2(x)) = f(x) = \dfrac{x}{x-1}$. [1.2, 5.9].

42. g B There are $\binom{15}{5}$ ways of selecting a committee of 5 out of 15 people (men and women). There are $\binom{6}{3}$ ways of selecting 3 men of 6, and for each of these, there are $\binom{9}{2}$ ways of selecting 2 women of 9. Therefore, there are $\binom{6}{3}\binom{9}{2}$ ways

of selecting 3 men and 2 women. The probability of selecting 3 men and 2 women is

$\dfrac{\binom{6}{3}\binom{9}{2}}{\binom{15}{5}} = \dfrac{240}{1001}$. Using the calculator command

$nCr = \binom{n}{r}$, enter this expression and change to a fraction. [5.1, 5.3].

43. a C $2x + \sqrt{3} = (x + \sqrt{2}) + d$, and $5x - \sqrt{5} = (2x + \sqrt{3}) + d$. Eliminate d, and $x + \sqrt{3} - \sqrt{2} = 3x - \sqrt{5} - \sqrt{3}$. Thus, $x = \dfrac{2\sqrt{3} - \sqrt{2} + \sqrt{5}}{2} \approx 2.14$. [5.4].

44. i B Substituting for y gives $x\left(\dfrac{4t}{t-3}\right) - 4x - 2\left(\dfrac{4t}{t-3}\right) - 4 = 0$, which simplifies to $12x - 12t + 12 = 0$. Then $x = t - 1$. [4.7].

45. i D Complete the square: $a\left(x^2 + \dfrac{b}{a}x + \dfrac{b^2}{4a^2}\right) + c - \dfrac{b^2}{4a} = a\left(x + \dfrac{b}{2a}\right)^2 - \dfrac{b^2 - 4ac}{4a}$. To have symmetry about the y-axis, $\left(x - 3 + \dfrac{b}{2a}\right)^2 = x^2$. Therefore, $b = 6a$. [2.3, 5.5].

<u>Alternative Solution</u>: The first coordinate of the vertex of a parabola is $\dfrac{-b}{2a}$, and so, in vertex form, $f(x) = a\left(x + \dfrac{b}{2a}\right)^2 + L$ For $f(x-3)$ to be symmetric about the y-axis, $f\left(x - \dfrac{b}{2a}\right) = f(x-3)$. Therefore, $x - \dfrac{b}{2a} = x - 3$, which gives $b = 6a$.

46. g A Use the change of base formula and graph $y = \dfrac{\log x}{\log 5}$ as Y_1. Graph $y = \ln(0.5x)$ as Y_2. The answer choices suggest a window of $x\varepsilon[0,10]$ and $y\varepsilon[0,2]$. Use the CALC/intersect to find the correct answer choice A.

An alternative solution converts the equations to exponential form: $x = 5^y$ and $\dfrac{1}{2}x = e^y$, or $5^y = 2e^y$. Taking the natural log of both sides gives $y\ln 5 = \ln 2 + y$, and solving for y yields

$y = \dfrac{\ln 2}{\ln 5 - 1} \approx 1.14$. Finally substituting back to find x, $5^{1.14} \approx 6.24$. [4.2].

47. i **B** $\sin(A + 30°) = \sin A \cdot \cos 30° + \cos A \cdot \sin 30° = \dfrac{\sqrt{3}}{2} \cdot \sin A + \dfrac{1}{2} \cdot \cos A$. $\cos(A + 60) = \cos A \cdot \cos 60° - \sin A \cdot \sin 60° = \dfrac{1}{2} \cdot \cos A - \dfrac{\sqrt{3}}{2} \cdot \sin A$. Adding the two equations gives $\sin(A + 30°) + \cos(A + 60°) = \cos A$. [3.3].

48. g **B** With your calculator in polar mode, graph $r = \dfrac{1}{\sin\theta + \cos\theta}$ in a $x\varepsilon[-3,3]$ and $y\varepsilon[-3,3]$ window. Let θ run from 0 to 2π in increments of 0.1. The graph of the function and the two axes form an isosceles right triangle of side length 1, so its area is 0.5.

An alternative solution is to multiply both sides of the equation by $\sin\theta + \cos\theta$ to get $r\sin\theta + r\cos\theta = 1$. Since $x = r\cos\theta$ and $y = r\sin\theta$, the rectangular form of the equation is $x + y = 1$. This line makes an isosceles right triangle of unit leg lengths with the axes, the area of which is $\dfrac{1}{2}(1)(1) = 0.5$ [4.7].

49. a **B** The diagonal of the base is $\sqrt{6^2 + 4^2} = \sqrt{52}$. The diagonal of the box is the hypotenuse of a right triangle with one leg $\sqrt{52}$ and the other leg 5. Let θ be the angle formed by the diagonal of the base and the diagonal of the box. $\tan\theta = \dfrac{5}{\sqrt{52}} \approx 0.69337$, and so $\theta = \tan^{-1} 0.69337 \approx 35°$. [3.1]

50. g **C** Plot $y = (2x + 1)^2$ in a window with $x\varepsilon[-2,2]$ and $t\varepsilon[-1,5]$, and observe that choice C is the only possible choice.

An alternative solution is to note that the graph of $y = (2x+1)^2 = 4\left(x + \dfrac{1}{2}\right)^2$ is a parabola with vertex at $\left(-\dfrac{1}{2}, 0\right)$ that opens up. The only answer choice with these properties is C. [2.3].

SELF-EVALUATION CHART FOR MODEL TEST 2

SUBJECT AREA	QUESTIONS	NUMBER OF RIGHT WRONG OMITTED

Mark correct answers with C, wrong answers with X, and omitted answers with O.

Algebra
(9 questions)

3	4	8	9	19	20	28	36	46

Review section

4.2	2.4	2.4	4.7	2.3	2.5	5.6	1.1	4.2

_____ _____ _____

Solid geometry
(4 questions)

2	12	18	39

Review section

5.5	5.5	5.5	5.5

_____ _____ _____

Coordinate geometry
(6 questions)

5	17	21	32	40	50

Review section

2.2	2.4	4.1	3.5	5.5	5.5

_____ _____ _____

Trigonometry
(10 questions)

6	10	14	16	25	27	31	37	47	49

Review section

3.5	3.4	3.2	3.1	3.4	3.7	3.7	3.1	3.3	3.1

_____ _____ _____

Functions
(12 questions)

1	7	13	15	22	23	30	33	34	41	44	45

Review section

4.5	4.7	1.5	2.5	1.2	1.4	1.5	1.3	2.2	1.2	4.7	2.3

_____ _____ _____

Miscellaneous
(9 questions)

11	24	26	29	35	38	42	43	48

Review section

5.2	5.1	5.8	5.7	5.2	5.3	5.3	5.4	4.7

_____ _____ _____

TOTALS   _____ _____ _____

Raw score = (number right) − 1/4 (number wrong) = _____

Round your raw score to the nearest whole number = _____

Evaluate Your Performance
Model Test 2

Rating	Number Right
Excellent	41–50
Very Good	33–40
Above Average	25–32
Average	15–24
Below Average	Below 15

ANSWER SHEET FOR MODEL TEST 3

Determine the correct answer for each question. Then, using a no. 2 pencil, blacken completely the oval containing the letter of your choice.

1. Ⓐ Ⓑ Ⓒ Ⓓ Ⓔ
2. Ⓐ Ⓑ Ⓒ Ⓓ Ⓔ
3. Ⓐ Ⓑ Ⓒ Ⓓ Ⓔ
4. Ⓐ Ⓑ Ⓒ Ⓓ Ⓔ
5. Ⓐ Ⓑ Ⓒ Ⓓ Ⓔ
6. Ⓐ Ⓑ Ⓒ Ⓓ Ⓔ
7. Ⓐ Ⓑ Ⓒ Ⓓ Ⓔ
8. Ⓐ Ⓑ Ⓒ Ⓓ Ⓔ
9. Ⓐ Ⓑ Ⓒ Ⓓ Ⓔ
10. Ⓐ Ⓑ Ⓒ Ⓓ Ⓔ
11. Ⓐ Ⓑ Ⓒ Ⓓ Ⓔ
12. Ⓐ Ⓑ Ⓒ Ⓓ Ⓔ
13. Ⓐ Ⓑ Ⓒ Ⓓ Ⓔ
14. Ⓐ Ⓑ Ⓒ Ⓓ Ⓔ
15. Ⓐ Ⓑ Ⓒ Ⓓ Ⓔ
16. Ⓐ Ⓑ Ⓒ Ⓓ Ⓔ
17. Ⓐ Ⓑ Ⓒ Ⓓ Ⓔ

18. Ⓐ Ⓑ Ⓒ Ⓓ Ⓔ
19. Ⓐ Ⓑ Ⓒ Ⓓ Ⓔ
20. Ⓐ Ⓑ Ⓒ Ⓓ Ⓔ
21. Ⓐ Ⓑ Ⓒ Ⓓ Ⓔ
22. Ⓐ Ⓑ Ⓒ Ⓓ Ⓔ
23. Ⓐ Ⓑ Ⓒ Ⓓ Ⓔ
24. Ⓐ Ⓑ Ⓒ Ⓓ Ⓔ
25. Ⓐ Ⓑ Ⓒ Ⓓ Ⓔ
26. Ⓐ Ⓑ Ⓒ Ⓓ Ⓔ
27. Ⓐ Ⓑ Ⓒ Ⓓ Ⓔ
28. Ⓐ Ⓑ Ⓒ Ⓓ Ⓔ
29. Ⓐ Ⓑ Ⓒ Ⓓ Ⓔ
30. Ⓐ Ⓑ Ⓒ Ⓓ Ⓔ
31. Ⓐ Ⓑ Ⓒ Ⓓ Ⓔ
32. Ⓐ Ⓑ Ⓒ Ⓓ Ⓔ
33. Ⓐ Ⓑ Ⓒ Ⓓ Ⓔ
34. Ⓐ Ⓑ Ⓒ Ⓓ Ⓔ

35. Ⓐ Ⓑ Ⓒ Ⓓ Ⓔ
36. Ⓐ Ⓑ Ⓒ Ⓓ Ⓔ
37. Ⓐ Ⓑ Ⓒ Ⓓ Ⓔ
38. Ⓐ Ⓑ Ⓒ Ⓓ Ⓔ
39. Ⓐ Ⓑ Ⓒ Ⓓ Ⓔ
40. Ⓐ Ⓑ Ⓒ Ⓓ Ⓔ
41. Ⓐ Ⓑ Ⓒ Ⓓ Ⓔ
42. Ⓐ Ⓑ Ⓒ Ⓓ Ⓔ
43. Ⓐ Ⓑ Ⓒ Ⓓ Ⓔ
44. Ⓐ Ⓑ Ⓒ Ⓓ Ⓔ
45. Ⓐ Ⓑ Ⓒ Ⓓ Ⓔ
46. Ⓐ Ⓑ Ⓒ Ⓓ Ⓔ
47. Ⓐ Ⓑ Ⓒ Ⓓ Ⓔ
48. Ⓐ Ⓑ Ⓒ Ⓓ Ⓔ
49. Ⓐ Ⓑ Ⓒ Ⓓ Ⓔ
50. Ⓐ Ⓑ Ⓒ Ⓓ Ⓔ

MODEL TEST

3

Tear out the preceding answer sheet. Decide which is the best choice by rounding your answer when appropriate. Blacken the corresponding space on the answer sheet. When finished, check your answers with those at the end of the test. For questions that you got wrong, note the sections containing the material that you must review. Also, if you do not fully understand how you arrived at some of the correct answers, you should review the appropriate sections. Finally, fill out the self-evaluation sheet on page 198 in order to pinpoint the topics that give you the most difficulty.

TEST DIRECTIONS

Directions: Decide which answer choice is best. If the exact numerical value is not one of the answer choices, select the closest approximation. Fill in the oval on the answer sheet that corresponds to your choice.

Notes:
(1) You will need to use a scientific or graphing calculator to answer some of the questions.
(2) You will have to decide whether to put your calculator in degree or radian mode for some problems.
(3) All figures that accompany problems are plane figures unless otherwise stated. Figures are drawn as accurately as possible to provide useful information for solving the problem, except when it is stated in a particular problem that the figure is not drawn to scale.
(4) Unless otherwise indicated, the domain of a function is the set of all real numbers for which the functional value is also a real number.

Reference Information. The following formulas are provided for your information.

Volume of a right circular cone with radius r and height h: $V = \dfrac{1}{3}\pi r^2 h$

Lateral area of a right circular cone if the base has circumference c and the slant height is l: $S = \dfrac{1}{2}cl$

Volume of a sphere of radius r: $V = \dfrac{4}{3}\pi r^3$

Surface area of a sphere of radius r: $S = 4\pi r^2$

Volume of a pyramid of base area B and height h: $V = \dfrac{1}{3}Bh$

---

1. If $f(x) = 3x - 5$ and $g(y) = y^2 - 1$, then $f(g(z)) =$

    (A) $(3z - 5)^2 - 1$
    (B) $3z^2 - 8$
    (C) $3z^3 - 5z^2 - 3z + 5$
    (D) $z^2 + 3z - 6$
    (E) $9z^2 - 30z - 26$

2. A sphere is tangent to two distinct parallel planes. The set of all points equidistant from the two planes that intersect the sphere is

    (A) a circle
    (B) two points
    (C) a plane
    (D) a straight line segment
    (E) the empty set

3. If x and y are real numbers and $y = \sqrt[3]{4 - x^2}$, what is the maximum value of y?

    (A) 2
    (B) 1.59
    (C) ∞
    (D) 1.41
    (E) 0

4. A cone is inscribed in a hemisphere of radius r so that the base of the cone coincides with the base of the hemisphere. What is the ratio of the volume of the cone to the volume of the hemisphere?

    (A) $\dfrac{r}{2}$
    (B) $\dfrac{r}{3}$
    (C) $\dfrac{2r}{3}$
    (D) $\dfrac{1}{2}$
    (E) $\dfrac{1}{3}$

5. If $\log(\cos \theta) = p$, then $\log(\sec \theta) =$

    (A) $-p$
    (B) $1 - p$
    (C) $\dfrac{1}{p}$
    (D) $-\dfrac{1}{p}$
    (E) p

6. If the parabola $ay^2 + by + c = x$ passes through points $(-4,17)$, $(5,11)$, and $(8,1)$, the value of $a + b + c$ equals

    (A) $\dfrac{17}{43}$

    (B) $\dfrac{5}{2}$

    (C) $\dfrac{14}{27}$

    (D) 8

    (E) -11

7. If, in the figure on the right, θ is the angle between segment PQ and the x-axis, then θ equals

    (A) $23°$
    (B) $54°$
    (C) $25°$
    (D) $71°$
    (E) $67°$

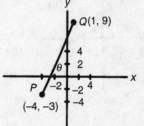

8. If the graph on the right represents the function f defined on interval $[-2,2]$, which of the following could represent the graph of $y = f(2x)$?

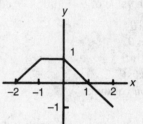

(A)

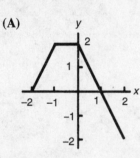

(B)

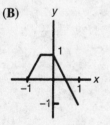

(C)

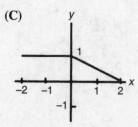

GO ON TO THE NEXT PAGE

(D)

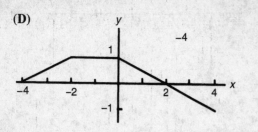

(E)

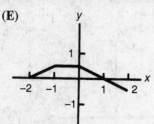

9. For each value of θ, $\sin(90° + \theta)$ equals

    (A) $\sin \theta$
    (B) $\cos \theta$
    (C) $-\sin \theta$
    (D) $-\cos \theta$
    (E) none of the above

10. A purse contains five different coins (penny, nickel, dime, quarter, half-dollar). How many different sums of money can be made using one or more coins?

    (A) 5
    (B) 10
    (C) 32
    (D) 120
    (E) none of the above

11. If $3x^3 - x^2 + 12x - 4 = (x - 2i)(3x - 1)Q(x)$ for all numbers x, then $Q(x) =$

    (A) $x - 2$
    (B) $x + 2$
    (C) $x - 2i$
    (D) $x + 2i$
    (E) $x + i$

GO ON TO THE NEXT PAGE

12. A point in space has coordinates $(7,-3,8)$. How far is the point from the x-axis?

(A) 8.5
(B) 7
(C) 8.1
(D) 11
(E) 7.6

13. If $P(x)$ is a sixth-degree polynomial, $P(1) = 3$, and $P(-1) = -2$, what is the value of the remainder when $P(x)$ is divided by $x - 1$?

(A) -2
(B) 2
(C) -3
(D) 3
(E) The value cannot be determined.

14. $\operatorname{Sec}\left(\sin^{-1}\dfrac{\sqrt{5}}{5}\right)$ equals

(A) 2.15
(B) 1.12
(C) 0.89
(D) 0.98
(E) 1.10

15. An equation in polar form equivalent to $x^2 + y^2 - 4x + 2 = 0$ is

(A) $r = 4\cos\theta + 2$
(B) $r^2 = 4\cos\theta + 2$
(C) $4r = \cos\theta$
(D) $r^2 - 4r\cos\theta + 2 = 0$
(E) $r^2 = 4r\cos\theta$

16. If $\log_5(x - 3) = \log_{25}(x - 1)$, which of the following could be the value of x?

(A) 12
(B) 2
(C) 8
(D) 5
(E) 10

17. Which of the following could be a possible root of $4x^3 - px^2 + qx - 6 = 0$?

(A) $\dfrac{1}{6}$

(B) 4

(C) $\dfrac{3}{2}$

(D) $\dfrac{2}{3}$

(E) $\dfrac{4}{3}$

GO ON TO THE NEXT PAGE

18. Where defined, $\dfrac{\csc x - 1}{\sin x - 1} =$

    (A) $\sin x$
    (B) $-\sin x$
    (C) $\csc x$
    (D) $-\csc x$
    (E) $-\cos x$

19. If $\log_b A = 0.2222$ and $\log_b B = 0.3333$, then the value of $\log_b\left(\sqrt{A} \cdot B^2\right)$ is

    (A) 0.0741
    (B) 0.1111
    (C) 0.5555
    (D) 0.7777
    (E) 0.9999

20. If $f(x) = x^2 + bx + c$ for all x, and if $f(-3) = 0$ and $f(1) = 0$, then $b + c =$

    (A) -5
    (B) -1
    (C) 1
    (D) 5
    (E) 0

21. If $f(x) = \sqrt[3]{4x + 2}$ for all x, then $f^{-1}\left(\dfrac{1}{2}\right) =$

    (A) 0.53
    (B) -0.30
    (C) 0.29
    (D) -0.47
    (E) -0.38

22. The sides of a triangle are 2 inches, 3 inches, and 4 inches. The value of the angle opposite the 3-inch side is

    (A) $43°$
    (B) $47°$
    (C) $64°$
    (D) $137°$
    (E) $133°$

23. The set of points (x,y) that satisfy $(x - 3)(y + 2) > 0$ lies in quadrant(s)

    (A) only I and IV
    (B) only II
    (C) only III and IV
    (D) only I, III, and IV
    (E) I, II, III, and IV

GO ON TO THE NEXT PAGE

24. If $ax^3 + bx^2 + cx + 3 = 0$ when $x = -1$, what is the value of $ax^3 - bx^2 + cx + 3$ when $x = 1$?

    (A) –6
    (B) –3
    (C) 0
    (D) 3
    (E) 6

25. If the operation @ is defined so that $a \mathbin{@} b = \dfrac{a}{1+\dfrac{\pi}{b}}$,

    what is the value of $(1\mathbin{@}2) \mathbin{@} 3$?

    (A) 1.43
    (B) 5.26
    (C) 0.19
    (D) 0.70
    (E) 1.26

26. If h varies as V and inversely as the square of r, which of the following is true?

    (A) If r is increased by 2, V is increased by 4.
    (B) If both r and h are doubled, then V is doubled.
    (C) If r is doubled and h is divided by 4, then V remains unchanged.
    (D) If r is doubled and h is divided by 2, then V remains unchanged.
    (E) None of the above is true.

27. If $\begin{cases} x = t^3 + 9 \\ y = \dfrac{3}{4}t^3 + 7 \end{cases}$ represents a line, what is the y-intercept?

    (A) $\dfrac{3}{4}$
    (B) 9
    (C) 16
    (D) $-\dfrac{27}{4}$
    (E) $\dfrac{1}{4}$

28. Arcsec 1.8 + Arccsc 1.8 equals

    (A) 74°
    (B) 16°
    (C) 90°
    (D) 0°
    (E) 39°

GO ON TO THE NEXT PAGE

29. What is the mean of this set of data: $\sqrt{1}, \sqrt{2}, \sqrt{3}, \sqrt{4},$ $\sqrt{5}, \sqrt{6}, \sqrt{7}$?

(A) 2
(B) 1.65
(C) 2.14
(D) 0.76
(E) 1.93

30. Which of the following functions has (have) an inverse that is also a function?

I. $y = x^2 - 2x + 4$

II. $y = |x + 1|$

III. $y = \sqrt{40 - 9x^2}$

(A) only I
(B) only II
(C) only III
(D) I, II, and III
(E) none of the above

31. The sum of all the numerical coefficients of $(x - y)^{17}$ is

(A) $\binom{17}{8}$

(B) $2 \cdot \binom{17}{9}$

(C) 1
(D) 0
(E) 17

32. The absolute value of the difference between the roots of the equation $1.4x^2 + 2.3x - 5 = 0$ is

(A) 4.12
(B) 1.64
(C) 3.57
(D) 2.17
(E) 3.40

33. If $f(x) = 5x + 2$ and $f(x - 3) + f(x) = 2x$, then $x =$

(A) $\dfrac{7}{3}$

(B) $\dfrac{11}{8}$

(C) $-\dfrac{1}{3}$

(D) $-\dfrac{1}{8}$

(E) $\dfrac{1}{13}$

34. If $2x + 3y - 4 - |2x + 3y - 4| = 0$ for all values of x and y belonging to the set $\{(x,y):p \text{ is true}\}$, then p must be the statement

(A) $2x + 3y > 4$
(B) $2x + 3y = 4$
(C) $2x + 3y < 4$
(D) $2x + 3y \geq 4$
(E) $x = 0$ and $y = 0$

35. If $f(x) = 2^x$, then $f(\log_5 2) =$

(A) 5
(B) 2.32
(C) 0.43
(D) 0.19
(E) 1.35

36. If $f(x) = \sqrt{x - 1}$ and $g(x) = \sin x$, then $g^{-1}\left(f\left(\sqrt{3}\right)\right) =$

(A) 0.75
(B) 1.41
(C) 0.99
(D) 1.12
(E) 1.03

37. A basket contains 10 apples, of which 5 are rotten. What is the probability that a person who buys 4 apples will get none that are rotten?

(A) $\dfrac{1}{2}$
(B) $\dfrac{2}{5}$
(C) $\dfrac{2}{25}$
(D) $\dfrac{1}{4032}$
(E) $\dfrac{1}{42}$

38. The smallest positive value of x satisfying the equation $\tan 5x = -2$ is

(A) 13°
(B) 31°
(C) 23°
(D) 49°
(E) 63°

USE THIS SPACE FOR SCRATCH WORK

GO ON TO THE NEXT PAGE

39. If f represents an even function, which of the following is also an even function (are also even functions)?

 I. $g(x) = f(x + 1)$
 II. $h(x) = f(x) + 1$
 III. $k(x) = f^{-1}(x)$

 (A) only I
 (B) only II
 (C) only III
 (D) II and III
 (E) I and III

40. Two cards are drawn from a regular deck of 52 cards. What is the probability that the cards will be an ace and a 10?

 (A) 0.157
 (B) 0.012
 (C) 0.077
 (D) 0.0004
 (E) 0.009

41. If the slope of line l_1 is $x + 1$, the slope of line l_2 is $x - 2$, and l_1 is perpendicular to l_2, x equals

 (A) 1.62 or –0.62
 (B) 0.62 or –1.62
 (C) only –0.62
 (D) only 1.62
 (E) only –1.62

42. A particle travels in a circular path at 50 centimeters per minute. If it traverses an arc of 30° in 30 seconds, what is the radius of the circular path?

 (A) 2865 centimeters
 (B) 48 centimeters
 (C) 24 centimeters
 (D) 50 centimeters
 (E) 95 centimeters

43. If the line passing through point (a,b) forms a right isosceles triangle with the x- and y-axes, the area of the triangle is

 (A) $\dfrac{(a+b)^2}{2}$
 (B) $\dfrac{a^2 - b^2}{2}$
 (C) $\dfrac{a^2 + b^2}{2}$
 (D) $\dfrac{1}{2}ab$
 (E) The area cannot be determined.

44. A cube is inscribed in a sphere, and a smaller sphere is inscribed in the cube. What is the ratio of the volume of the small sphere to the volume of the large sphere?

(A) 0.50 : 1
(B) 0.33 : 1
(C) 0.58 : 1
(D) 0.19 : 1
(E) 0.71 : 1

USE THIS SPACE FOR SCRATCH WORK

45. If $[x]$ is defined to represent the greatest integer less than or equal to x, and $f(x) = \left| x - [x] - \dfrac{1}{2} \right|$, the graph of $f(x)$ is discontinuous at

(A) no values of x
(B) all integer values of x
(C) all even integer values of x
(D) all odd integer values of x
(E) all odd multiples of $\dfrac{1}{2}$

46. If (x,y) represents a point on the graph of $y = x + 2$, which of the following could be a portion of the graph of the set of points $\left(\sqrt{x}, y \right)$?

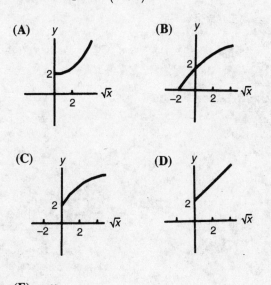

47. $\displaystyle\sum_{i=1}^{1000}\left(\frac{1}{i}-\frac{1}{i+1}\right)=$

    (A) $\dfrac{1}{1000}$

    (B) 0.999

    (C) 0.111

    (D) −0.111

    (E) undefined

48. If $f(ab) = f(a) + f(b)$ for all real numbers in the domain of f, $f(x)$ equals which of the following?

    I. $\dfrac{1}{x}$

    II. e^x

    III. $\log x$

    (A) only I

    (B) only II

    (C) only III

    (D) only I and II

    (E) only II and III

49. If the equation $x^2 + 2(k+2)x + 9k = 0$ has equal roots, $k =$

    (A) 1 or 4

    (B) 0 or 4

    (C) 4 only

    (D) −1 or 4

    (E) 2 or −4

50. What is the amplitude of the graph of $y = a \cos x + b \sin x$?

    (A) $\dfrac{a+b}{2}$

    (B) $a + b$

    (C) $\sqrt{ab}$

    (D) $\sqrt{a^2 + b^2}$

    (E) $(a+b)\sqrt{2}$

ANSWER KEY

1. B	6. D	11. D	16. D	21. D	26. C	31. D	36. E	41. A	46. A
2. A	7. E	12. A	17. C	22. B	27. E	32. A	37. E	42. B	47. B
3. B	8. B	13. D	18. D	23. D	28. C	33. B	38. C	43. A	48. C
4. D	9. B	14. B	19. D	24. E	29. E	34. D	39. B	44. D	49. A
5. A	10. E	15. D	20. B	25. C	30. E	35. E	40. B	45. A	50. D

ANSWER EXPLANATIONS

In these solutions the following notation is used:

   a: active—Calculator use is necessary or, at a minimum, extremely helpful.

   g: Graphing calculator is preferred.

   i: inactive—Calculator use is not helpful and may even be a hindrance.

1. i   **B** $f(g(z)) = f(z^2 - 1) = 3(z^2 - 1) - 5 = 3z^2 - 8$. [1.2].

2. i   **A**   The set of points equidistant from two parallel planes is a plane parallel to both planes and halfway between them. This plane cuts the sphere through its center. The intersection is a circle. [5.5].

3. a   **B**   Plot the graph of $y = \sqrt[3]{4 - x^2}$ and observe that the maximum occurs when $x = 0$. The TRACE function will show that at this point $y \approx 1.59$.

An alternative solution is to observe that since $x^2 \geq 0$, the radicand will never be bigger than 4. Therefore, the maximum value of y is $\sqrt[3]{4} \approx 1.59$. [2.3, 4.1].

4. i   **D**   Volume of cone $= \dfrac{1}{3}\pi r^2 h = \dfrac{1}{3}\pi(r^2)r = \dfrac{1}{3}r^3\pi$. Volume of hemisphere $= \dfrac{2}{3}\pi r^3$. $V_c : V_h = 1 : 2$. [5.5].

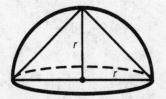

5. i   **A** $\text{Log}(\sec\theta) = \log\left(\dfrac{1}{\cos\theta}\right)$

$= \log 1 - \log(\cos\theta) = 0 - p$.

An alternative solution is to let $\theta =$ any angle, say $60°$. Thus, the problem becomes $\log(\cos\theta) = \log(\cos 60°) = \log 0.5 \approx -0.3010 = p$. Therefore,

$\log(\sec\ 60°) = \log\left(\dfrac{1}{\cos 60°}\right) = \log 2 \approx 0.3010$

$= -p$. [4.2].

6. i   **D**   Substitute point $(8,1)$ into the equation to get $a + b + c = 8$. [4.1].

7. a   **E**   Drop a vertical line from Q and a horizontal line through P to form a right triangle with sides 5, 12, and 13. $\sin\theta = \dfrac{12}{13} \approx 0.9231$. Therefore, $\theta = \text{Sin}^{-1}(0.9231) \approx 67°$. [3.1, 3.7].

8. i   **B**   Substituting $2x$ in place of x leads to the graph in Choice B. [5.5].

9. g   **B**   With your calculator in degree mode, graph $y = \sin(90 + x)$ in the trig window. Observe that the resulting graph is the same as the graph of $y = \cos x$.

An alternative solution uses the formula for the sin of the sum of two angles: $\sin(90° + x) = \sin 90° \cos x + \cos 90° \sin x = \cos x$. [3.1].

10. i   **E**   Each possible sum results from selection of 1, 2, 3, 4, or 5 coins. Each coin is either selected or not, and there are $2^5 - 1 = 31$ ways of doing this if you exclude selecting no coins. [5.1].

11. i   **D**   Since the coefficients are all real numbers, the complex factors must come in conjugate pairs: $x - 2i$ and $x + 2i$. [2.4].

12. a **A** The point is −3 from the x-axis in the y-direction and 8 in the z-direction. Therefore, the distance from the point to the x-axis is the hypotenuse of a right triangle with legs 3 and 8. Distance = $\sqrt{3^2 + 8^2} = \sqrt{73} \approx 8.5$. [5.5].

13. i **D** By the remainder theorem, the remainder when $P(x)$ is divided by $x - 1$ is $P(1) = 3$. [2.4].

14. g **B** With your calculator in radian or degree mode, enter $1 \big/ \left(\cos\left(\sin^{-1} \dfrac{\sqrt{5}}{5} \right) \right)$ to get the correct answer choice.

An alternative solution follows. Let $\sin^{-1} \dfrac{\sqrt{5}}{5} = \theta$,

so $\sin\theta = \dfrac{\sqrt{5}}{5}$. Draw a right triangle with θ as one of the acute angles and use the definition of $\sin\theta$ and the Pythagorean Theorem to label the lengths of the other two sides, as shown below. Again referring to the triangle, $\sec\theta = \dfrac{5}{2\sqrt{5}} \approx 1.12$.

[3.6, 3.5].

15. i **D** $x^2 + y^2 = r^2$ and $x = r \cdot \cos\theta$ give the equation $r^2 - 4r \cos\theta + 2 = 0$. [4.7].

16. g **D** Enter $\log_5(x - 3)$ into Y_1 and $\log_{25}(x - 1)$ into Y_2. The answer choices suggest using a window $x\varepsilon[0,14]$ and $y\varepsilon[-2,2]$. Then use CALC/intersect to find the point of intersection.

An alternative solution is to write both sides of the equation in logs with respect to like bases:

$$\log_5(x - 3) = \frac{\log(x-3)}{\log 5} = \log_{25}(x - 1)$$

$$= \frac{\log(x-1)}{\log 25} = \frac{\log(x-1)}{2\log 5}$$

$$= \frac{1}{2}\log_5(x - 1).$$

Multiplying both sides of this equation by 2 and using a law of logarithms yields $\log_5(x - 3)^2 = \log_5(x - 1)$. Therefore, $x^2 - 6x + 9 = x - 1$, and the solutions of this quadratic equation are 5 and 2. Since 2 is not in the domain of $\log_5(x - 3)$, the correct answer choice is D. [4.2].

17. i **C** $\dfrac{3}{2}$ is the only possibility for a root because 3 is a factor of 6 and 2 is a factor of 4. [2.4].

18. i **D**

$$\frac{\csc x - 1}{\sin x - 1} = \frac{1/\sin x - 1}{\sin x - 1} \cdot \frac{\sin x}{\sin x} = \frac{1 - \sin x}{\sin x(\sin x - 1)}$$

$$= -\frac{1}{\sin x} = -\sec x.$$

An alternative solution can be obtained by substituting a number for x in the given expression (with your calculator in radian mode), and substituting that same x value in the answer choices until you observe a match. [3.5].

19. i **D**

$$\log_b\left(\sqrt{A} \cdot B^2\right) = \frac{1}{2} \cdot \log_b A + 3 \cdot \log B_b$$

$$= \frac{1}{2}(0.2222) + 2(0.3333) = 0.7777$$

[4.2].

Since the base is not given, an alternative solution is to choose any convenient base (e.g., base 10). $A = 10^{0.2222} \approx 1.668$, and $B = 10^{0.3333} \approx 2.154$. Thus, $\sqrt{A} \cdot B^2 \approx 5.994$. Therefore, $\log\left(\sqrt{A} \cdot B^2\right) = \log 5.994 \approx 0.7777$.

20. i **B** −3 and 1 are zeros. Sum of zeros = $-b = -3 + 1 = -2$, and so $b = 2$. Product of zeros = $c = (-3)(1) = -3$. $b + c = -1$. [2.3].

Since −3 and 1 are zeros, $(x + 3)(x - 1) = x^2 + 2x - 3 = 0$, so $b + c = -1$. [2.3]

21. g **D** Enter $\sqrt[3]{4x + 2}$ into Y_1 and $\dfrac{1}{2}$ into Y_2, and plot both in the standard window. Use CALC/intersect to find the point of intersection ≈ -0.47, which is the correct answer choice.

An alternative solution is to solve the equation $\sqrt[3]{4x + 2} = \dfrac{1}{2}$. Cubing both sides yields $4x + 2 = \dfrac{1}{8}$, or $x = -\dfrac{15}{32} \approx -.47$. [1.3].

22. a **B** Put your calculator in degree mode because the answer choices are in degrees: By the law of cosines, $9 = 4 + 16 - 2 \cdot 2 \cdot 4 \cdot \cos x$. Then $\cos x = \dfrac{-11}{-16} = \dfrac{11}{16}$. Therefore, $\cos^{-1}\left(\dfrac{11}{16}\right) \approx 47°$. [3.7].

23. i **D** Sketch the graph and check points in the different regions. [2.5].

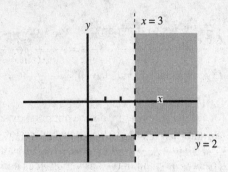

24. i **E** Substituting -1 for x gives $-a + b - c + 3 = 0$. Substituting 1 for x gives $a - b + c + 3 = k$. Adding the two equations gives $k = 6$. [2.4].

25. a **C** $1 @ 2 = \dfrac{1}{1 + \dfrac{\pi}{2}} = \dfrac{1}{2.5708} \approx 0.38898$

$(1 @ 2) @ 3 \approx \dfrac{0.38898}{1 + \dfrac{\pi}{2}} \approx \dfrac{0.38898}{2.0472} \approx 0.19$. [5.9].

26. i **C** The formula for the variation is $\dfrac{hr^2}{V} = K$. Testing the conditions of the selections indicates that the only valid answer is Choice C. [5.6].

27. i **E** The y-intercept occurs when $x = 0$, and the first equation implies that this occurs when $t^3 = -9$. Substituting -9 for t^3 in the second equation yields $y = \dfrac{1}{4}$. [4.6].

28. g **C** $\operatorname{Arc\,sec} 1.8 = x$ implies $\sec x = 1.8$ and $\cos x = \dfrac{1}{1.8}$, or $\operatorname{Arc\,cos}\dfrac{1}{1.8} = x$. Similarly, $\operatorname{Arc\,csc} 1.8 = y$ implies $\operatorname{Arc\,sin}\dfrac{1}{1.8} = y$. With your calculator in degree mode, enter $\cos^{-1}(1/1.8) + \sin^{-1}(1/1.8)$ to get answer choice C.

An alternative solution uses the fact that sine and cosine are cofunctions, which means that the cosine of an angle is the sine of its complement. Since $\sin x = \cos y$ (both equal $\dfrac{1}{1.8}$), x and y must be complements, so their sum is $90°$. [3.1, 3.6].

29. g **E** Enter $\operatorname{mean}\left(\operatorname{seq}\left(\sqrt{x}, x, 1, 7\right)\right)$, and key enter to get 1.93.

An alternative solution is to use your calculator to evaluate the sum of the seven square roots and divide the answer by 7. [5.8].

30. g **E** Plot the graphs of $y = x^2 - 2x + 4$, $y = \operatorname{abs}(x + 1)$, and $y = \sqrt{40 - 9x^2}$ in a standard window. In each case, for some value of y in the range, there are two values of x in the domain that will give that y-value. Therefore, none of the three functions has an inverse that is also a function. [1.3].

31. i **D** Since the signs alternate, the coefficients are symmetric about the middle terms, there is an even number of terms, and the sum of all the coefficients is zero. [5.2].

> **TIP:** If the numbers in a problem are big or awkward, try considering a similar problem with smaller numbers to see whether a pattern shows up. In this case try expanding $(x - y)^3 = x^3 - 3x^2y + 3xy^2 - y^3$, where the sum of the numerical coefficients $= 0$.

32. g **A** Use program QUADFORM to evaluate the two roots. The roots are stored in S and T, so enter $\operatorname{abs}(S - T)$ to find the correct answer choice A.

An alternative solution is to use the Quadratic Formula with a = 1.4, b = 2.3, and c = -5 and your calculator to evaluate the roots 1.239 and -2.882. Then use your calculator to find the absolute value of the difference. [2.3]

33. g **B** Enter $5x + 2$ as Y_1, $Y_1(x - 3) + Y_1(x)$ as Y_2, and $2x$ as Y_3. De-select Y_1 and graph Y_2 and Y_3 in the standard window. Use CALC/intersect to find the point of intersection, return to the Home Screen, enter X, and change to a fraction, to get the correct answer choice B.

An alternative solution evaluates $f(x - 3) = 5(x - 3) + 2 = 5x - 13$, adds this to $5x + 2$, and solves the resulting equation: $10x - 11 = 2x$. [1.2]

34. i **D** $2x + 3y - 4 - |2x + 3y - 4| = 0$ implies that $2x + 3y - 4 = |2x + 3y - 4|$, which is true only when $2x + 3y - 4 \geq 0$. [4.4].

35. a E By the change-of-base theorem, $\log_5 2 = \dfrac{\log_{10} 2}{\log_{10} 5} \approx 0.4307$. Therefore, $2^{0.4307} \approx 1.35$. [4.2].

36. a E $f\left(\sqrt{3}\right) = \sqrt{\sqrt{3} - 1} \approx \sqrt{0.7321} \approx 0.8556$.
$g^{-1}\left(f\left(\sqrt{3}\right)\right) = g^{-1}(0.8556) = \text{Sin}^{-1}(0.8556) \approx$
1.03. [1.3]. Put your calculator in radian mode.

37. a E To get 4 good apples, a person must make 1 of $\binom{5}{4} = 5$ selections of good apples and $\binom{5}{0} = 1$ selection of bad apples. There can be $\binom{10}{4}$ selections of 4 apples from the total of 10. Probability of 4 good apples $= \dfrac{\binom{5}{4}\binom{5}{0}}{\binom{10}{4}} = \dfrac{1}{42}$. [5.2.]

An alternative solution is to observe that since there are 5 good apples in the basket, P(getting a good apple on first pick) $= \dfrac{5}{10}$. On the second pick only 9 apples are left in the basket and 4 of them are good, so P(getting a good apple on second pick) $= \dfrac{4}{9}$. Continuing until 4 good apples are picked gives P(4 good apples) $=$
$\dfrac{5}{10} \cdot \dfrac{4}{9} \cdot \dfrac{3}{8} \cdot \dfrac{2}{7} = \dfrac{1}{42}$.

38. g C Enter $\tan 5x$ into Y_1 and -2 into Y_2. The answer choices are in degrees, so set your calculator to degree mode. Since the problem is looking for the smallest positive angle that make Y_1 and Y_2 equal, set $x \varepsilon [0,100]$ and $y \varepsilon [-4,4]$. Then use CALC/intersect to find the correct answer choice C.

With your calculator in degree mode, find the **reference angle** for $5x$ as $\tan^{-1} 2 \approx 63°$. Since $\tan 5x$ is negative, $5x$ must be in Quadrant II (because you are looking for the smallest positive angle). Thus, $5x = 180° - 63° = 117°$. Therefore,
$x = \dfrac{117}{5} \approx 23°$ [3.5].

39. i B Since f is an even function, $f(-x) = f(x)$. Since the inverse of an even function is not an even function, $h(-x) = f(-x) + 1 = f(x) + 1$. Therefore, $h(x)$ (II) is the only even function. [1.4].

40. a B There are $\binom{4}{1}$ ways to draw 1 of the four aces, $\binom{4}{1}$ ways to draw 1 of the four 10s, and

$\binom{52}{2}$ ways to draw any 2 cards from the deck.
P(getting one ace and one 10) $= \dfrac{\binom{4}{1}\binom{4}{1}}{\binom{52}{2}} = \dfrac{8}{663} \approx$
0.012. [5.3].

An alternative solution is to observe that on the first draw 8 acceptable cards (four aces and four 10s) are available from the 52 cards in the deck. On the second draw only 4 acceptable cards (either the four aces or the four 10s, whichever were not picked on the first draw) are available from the remaining 51 cards. Therefore, P(a 10 and an ace) $= \dfrac{8}{52} \cdot \dfrac{4}{51} = \dfrac{8}{663} \approx 0.012$.

41. g A Two lines are perpendicular if and only if the product of their slopes is -1. Thus, $(x + 1)(x - 2) = -1$. Use program QUADFORM to solve the equivalent equation $x^2 - x - 1 = 0$ to get the correct answer choice A. [2.2].

42. a B The particle travels 25 centimeters in 30 seconds. $s = r\theta$. $\theta = 30° = \dfrac{\pi}{6} \cdot 25 = r \cdot \dfrac{\pi}{6} \cdot r = \dfrac{150}{\pi} \approx 48$. [3.2].

43. i A The line must cut the axes at two points, $(c,0)$ and $(0,c)$. $\dfrac{b-0}{a-c} = \dfrac{b-c}{a-0} =$ slope. Therefore, $c = a + b$. Area $= \dfrac{1}{2}(a+b)^2$. [2.2].

44. a D The diameter of the small sphere equals the side of the cube, say s. The diameter of the large sphere equals the diagonal of the cube, which is $s\sqrt{3}$.

$\dfrac{\text{Small volume}}{\text{Large volume}} = \dfrac{s^3}{\left(s\sqrt{3}\right)^3} = \dfrac{1}{3\sqrt{3}} = \dfrac{\sqrt{3}}{9}$

$\approx \dfrac{1.732}{9} \approx \dfrac{0.19}{1}$

(i.e., volumes of similar figures are to one another as the cubes of linear corresponding parts). [5.5].

45. g A Enter the function into Y_1. The fact that $f(x)$ is an absolute value and the expression inside the absolute value will never exceed 1 suggests the window $x \varepsilon [0,5]$ and $y \varepsilon [0,1]$. An inspection of the graph shows that it is continuous at all values of x.

An alternative solution is to graph the function from $x = 0$ to $x = 1$ by hand. [4.4, 4.6, 4.5].

46. i **A** From the table of values below, the point (x, y) lies on the graph of $y = x + 2$ and the set of points $\left(\sqrt{x}, y\right)$ lies on the graph that you are looking for. A rough sketch indicates that the answer is Choice A. [5.5].

x	0	1	2	3	4
$\sqrt{x}$	0	1	$\sqrt{2}$	$\sqrt{3}$	2
y	2	3	4	5	6

47. a **B** An inspection of the first few terms of the summation indicates a clear pattern of cancellation:

$$\left(\frac{1}{1} - \frac{1}{2}\right) + \left(\frac{1}{2} - \frac{1}{3}\right) + \left(\frac{1}{3} - \frac{1}{4}\right) + \cdots = 1 - \frac{1}{1001}$$

$$= \frac{1000}{1001} \approx 0.999. \text{ [5.4]}.$$

48. i **C** By inspection: If $f(x) = \dfrac{1}{x}$, $f(ab) =$

$\dfrac{1}{ab} = \dfrac{1}{a} \cdot \dfrac{1}{b} = f(a) \cdot f(b)$. False.

If $f(x) = e^x$, $f(ab) = e^{ab} = e^a \cdot e^b = f(a) \cdot f(b)$. False.

If $f(x) = \log x$, $f(ab) = \log ab = \log a + \log b = f(a) + f(b)$. True. [4.2].

49. i **A** In order for its roots to be equal, the left side of the equation must be a perfect square. This can only happen if $(k + 2)^2 = 9k$, or if $k^2 - 5k + 4 = 0$. This factors as $(k - 1)(k - 4) = 0$ to get the correct answer choice A.

50. i **D** Multiply the expression by $\dfrac{\sqrt{a^2 + b^2}}{\sqrt{a^2 + b^2}}$

to get $y = \sqrt{a^2 + b^2}\left(\dfrac{a}{\sqrt{a^2 + b^2}} \cos x\right.$

$\left. + \dfrac{b}{\sqrt{a^2 + b^2}} \sin x\right)$. Since $\left(\dfrac{a}{\sqrt{a^2 + b^2}}\right)^2 +$

$\left(\dfrac{b}{\sqrt{a^2 + b^2}}\right)^2 = 1$, $\dfrac{a}{\sqrt{a^2 + b^2}}$ could represent

$\sin \theta$ and $\dfrac{b}{\sqrt{a^2 + b^2}}$ could represent $\cos \theta$. Thus,

$y = \sqrt{a^2 + b^2} \ (\sin \theta \cos x + \cos \theta \sin x) =$

$\sqrt{a^2 + b^2} \ \sin(\theta + x)$, which has an amplitude of

$\sqrt{a^2 + b^2}$. [3.4, 3.5].

SELF-EVALUATION CHART FOR MODEL TEST 3

SUBJECT AREA	QUESTIONS	NUMBER OF RIGHT WRONG OMITTED

Mark correct answers with C, wrong answers with X, and omitted answers with O.

Algebra
(10 questions)
Review section

3	16	17	19	23	24	26	32	34	49
4.1	4.2	2.4	4.2	2.5	2.4	5.6	2.3	4.4	2.3

___ ___ ___

Solid geometry
(4 questions)
Review section

2	4	12	44
5.5	5.5	5.5	5.5

___ ___ ___

Coordinate geometry
(5 questions)
Review section

6	41	43	45	46
4.1	2.2	2.2	4.5	5.5

___ ___ ___

Trigonometry
(10 questions)
Review section

5	7	9	14	18	22	28	38	42	50
4.2	3.1	3.1	3.6	3.5	3.7	3.6	3.5	3.2	3.4

___ ___ ___

Functions
(12 questions)
Review section

1	8	11	13	20	21	30	33	35	36	39	48
1.2	5.5	2.4	2.4	2.3	1.3	1.3	1.1	4.2	1.3	1.4	4.2

___ ___ ___

Miscellaneous
(9 questions)
Review section

10	15	25	27	29	31	37	40	47
5.1	4.6	5.9	4.6	5.8	5.2	5.2	5.3	5.4

___ ___ ___

TOTALS   ___ ___ ___

Raw score = (number right) − ¼ (number wrong) = _____

Round your raw score to the nearest whole number = _____

Evaluate Your Performance
Model Test 3

Rating	Number Right
Excellent	41–50
Very good	33–40
Above average	25–32
Average	15–24
Below average	Below 15

ANSWER SHEET FOR MODEL TEST 4

Determine the correct answer for each question. Then, using a no. 2 pencil, blacken completely the oval containing the letter of your choice.

1. Ⓐ Ⓑ Ⓒ Ⓓ Ⓔ
2. Ⓐ Ⓑ Ⓒ Ⓓ Ⓔ
3. Ⓐ Ⓑ Ⓒ Ⓓ Ⓔ
4. Ⓐ Ⓑ Ⓒ Ⓓ Ⓔ
5. Ⓐ Ⓑ Ⓒ Ⓓ Ⓔ
6. Ⓐ Ⓑ Ⓒ Ⓓ Ⓔ
7. Ⓐ Ⓑ Ⓒ Ⓓ Ⓔ
8. Ⓐ Ⓑ Ⓒ Ⓓ Ⓔ
9. Ⓐ Ⓑ Ⓒ Ⓓ Ⓔ
10. Ⓐ Ⓑ Ⓒ Ⓓ Ⓔ
11. Ⓐ Ⓑ Ⓒ Ⓓ Ⓔ
12. Ⓐ Ⓑ Ⓒ Ⓓ Ⓔ
13. Ⓐ Ⓑ Ⓒ Ⓓ Ⓔ
14. Ⓐ Ⓑ Ⓒ Ⓓ Ⓔ
15. Ⓐ Ⓑ Ⓒ Ⓓ Ⓔ
16. Ⓐ Ⓑ Ⓒ Ⓓ Ⓔ
17. Ⓐ Ⓑ Ⓒ Ⓓ Ⓔ

18. Ⓐ Ⓑ Ⓒ Ⓓ Ⓔ
19. Ⓐ Ⓑ Ⓒ Ⓓ Ⓔ
20. Ⓐ Ⓑ Ⓒ Ⓓ Ⓔ
21. Ⓐ Ⓑ Ⓒ Ⓓ Ⓔ
22. Ⓐ Ⓑ Ⓒ Ⓓ Ⓔ
23. Ⓐ Ⓑ Ⓒ Ⓓ Ⓔ
24. Ⓐ Ⓑ Ⓒ Ⓓ Ⓔ
25. Ⓐ Ⓑ Ⓒ Ⓓ Ⓔ
26. Ⓐ Ⓑ Ⓒ Ⓓ Ⓔ
27. Ⓐ Ⓑ Ⓒ Ⓓ Ⓔ
28. Ⓐ Ⓑ Ⓒ Ⓓ Ⓔ
29. Ⓐ Ⓑ Ⓒ Ⓓ Ⓔ
30. Ⓐ Ⓑ Ⓒ Ⓓ Ⓔ
31. Ⓐ Ⓑ Ⓒ Ⓓ Ⓔ
32. Ⓐ Ⓑ Ⓒ Ⓓ Ⓔ
33. Ⓐ Ⓑ Ⓒ Ⓓ Ⓔ
34. Ⓐ Ⓑ Ⓒ Ⓓ Ⓔ

35. Ⓐ Ⓑ Ⓒ Ⓓ Ⓔ
36. Ⓐ Ⓑ Ⓒ Ⓓ Ⓔ
37. Ⓐ Ⓑ Ⓒ Ⓓ Ⓔ
38. Ⓐ Ⓑ Ⓒ Ⓓ Ⓔ
39. Ⓐ Ⓑ Ⓒ Ⓓ Ⓔ
40. Ⓐ Ⓑ Ⓒ Ⓓ Ⓔ
41. Ⓐ Ⓑ Ⓒ Ⓓ Ⓔ
42. Ⓐ Ⓑ Ⓒ Ⓓ Ⓔ
43. Ⓐ Ⓑ Ⓒ Ⓓ Ⓔ
44. Ⓐ Ⓑ Ⓒ Ⓓ Ⓔ
45. Ⓐ Ⓑ Ⓒ Ⓓ Ⓔ
46. Ⓐ Ⓑ Ⓒ Ⓓ Ⓔ
47. Ⓐ Ⓑ Ⓒ Ⓓ Ⓔ
48. Ⓐ Ⓑ Ⓒ Ⓓ Ⓔ
49. Ⓐ Ⓑ Ⓒ Ⓓ Ⓔ
50. Ⓐ Ⓑ Ⓒ Ⓓ Ⓔ

MODEL TEST

50 questions   1 hour

Tear out the preceding answer sheet. Decide which is the best choice by rounding your answer when appropriate. Blacken the corresponding space on the answer sheet. When finished, check your answers with those at the end of the test. For questions that you got wrong, note the sections containing the material that you must review. Also, if you do not fully understand how you arrived at some of the correct answers, you should review the appropriate sections. Finally, fill out the self-evaluation sheet on page 217 in order to pinpoint the topics that give you the most difficulty.

TEST DIRECTIONS

Directions: Decide which answer choice is best. If the exact numerical value is not one of the answer choices, select the closest approximation. Fill in the oval on the answer sheet that corresponds to your choice.

Notes:
(1) You will need to use a scientific or graphing calculator to answer some of the questions.
(2) You will have to decide whether to put your calculator in degree or radian mode for some problems.
(3) All figures that accompany problems are plane figures unless otherwise stated. Figures are drawn as accurately as possible to provide useful information for solving the problem, except when it is stated in a particular problem that the figure is not drawn to scale.
(4) Unless otherwise indicated, the domain of a function is the set of all real numbers for which the functional value is also a real number.

Reference Information. The following formulas are provided for your information.

Volume of a right circular cone with radius r and height h: $V = \frac{1}{3}\pi r^2 h$

Lateral area of a right circular cone if the base has circumference c and the slant height is l: $S = \frac{1}{2}cl$

Volume of a sphere of radius r: $V = \frac{4}{3}\pi r^3$

Surface area of a sphere of radius r: $S = 4\pi r^2$

Volume of a pyramid of base area B and height h: $V = \frac{1}{3}Bh$

---

1. The plane whose equation is $3x + 4y - 5z = 60$ intersects the xy-plane in the line whose equation is

    (A) $3x + 4y = 60$
    (B) $x = 20$
    (C) $y = 15$
    (D) $3x - 4y = 0$
    (E) $z = -12$

2. If 5 and $3 + \sqrt{2}$ are zeros of the integral polynomial $P(x) = ax^4 + bx^3 - cx + d$, which of the following must also be a zero?

    I. -5
    II. $3 - \sqrt{2}$
    III. 0

    (A) only I
    (B) only II
    (C) only III
    (D) only I and II
    (E) only II and III

3. If $3x + 4 = 2(y + 2)$, the $y:x$ ratio is

    (A) 2:3
    (B) 3:2
    (C) 1:1
    (D) 7:4
    (E) 4:7

4. The value of $\log_7 \sqrt{3}$ is

    (A) 0.24
    (B) 0.26
    (C) 0.28
    (D) 0.30
    (E) 0.32

5. If $f(x) = \sin x$ and $g(x) = e^x$, then $f(g(\pi)) =$

    (A) 0.39
    (B) -0.91
    (C) -0.73
    (D) 1
    (E) 0.91

6. In the figure, $r \sin \theta$ equals

    (A) a
    (B) b
    (C) $-a$
    (D) $-b$
    (E) $a + b$

GO ON TO THE NEXT PAGE

7. What is the y-intercept of $y = \sqrt{5}\cos\left(x + \frac{\pi}{5}\right)$?

(A) 2.24
(B) 1.81
(C) 0.94
(D) 0.81
(E) 1.63

8. If three noncolinear points determine a plane, how many planes are determined by 10 points, no 3 of which are colinear?

(A) 3
(B) 120
(C) 45
(D) 90
(E) 720

9. If $f(x,y) = \tan x + \tan y$ and $g(x,y) = 1 - \tan x \cdot \tan y$, then $\dfrac{f(1,2)}{g(1,2)} =$

(A) 0
(B) −0.14
(C) 0.58
(D) 0.05
(E) −0.20

10. Compared to the graph of $y = f(x)$, the graph of $y = f(x + 2) - 3$ is

(A) shifted 2 units right and 3 units down.
(B) shifted 2 units left and 3 units down.
(C) shifted 2 units right and 3 units up.
(D) shifted 2 units right and 3 units down.
(E) none of the above.

11. Given the following four equations, where $a \neq b \neq 0$:

I. $ax + by = c$
II. $ax - by = c$
III. $-ax + by = c$
IV. $bx - ay = c$

Which pair represents perpendicular lines?

(A) I and IV
(B) II and IV
(C) II and III
(D) I and II
(E) III and IV

GO ON TO THE NEXT PAGE

12. Which of the following functions is an odd function?

(A) $f(x) = x^3 + 1$

(B) $f(x) = \dfrac{x}{x-1}$

(C) $f(x) = x^3 + x$

(D) $f(x) = 2x^4$

(E) $f(x) = \cos x$

13. If $f(x) = \dfrac{x^2 - 1}{x+1}$, what does $f(i)$ equal, where $i = \sqrt{-1}$?

(A) 0

(B) $\dfrac{2}{i+1}$

(C) $i - 1$

(D) -2

(E) $i + 1$

14. In order for the inverse of $f(x) = \cos(3x)$ to be a function, the domain of f must be limited to

(A) $0° \le x \le 60°$

(B) $0° \le x \le 180°$

(C) $90° \le x \le 270°$

(D) $30° \le x \le 90°$

(E) $-30° \le x \le 30°$

15. $\displaystyle \lim_{x \to a} \dfrac{2x^2 - 3ax + a^2}{x^2 - a^2} =$

(A) $\dfrac{1}{2}$

(B) 0

(C) $\dfrac{3}{2}$

(D) a

(E) The value is undefined.

16. The domain of $f(x) = \dfrac{x^2 - 1}{x^2 - x}$ is

(A) all real numbers

(B) all reals except $x = 1$

(C) all reals except $x = 0$

(D) all reals except $x = -1$

(E) all reals except $x = 0$ or $x = 1$

17. If the ratio of $\sin x$ to $\cos x$ is 1 to 2, then the ratio of $\tan x$ to $\cot x$ is

(A) 1:4

(B) 1:2

(C) 1:1

(D) 2:1

(E) 4:1

GO ON TO THE NEXT PAGE

18. If $\sin A = 0.4321$, what is the tangent of the supplement of $\angle A$?

(A) 0.3965
(B) –0.4791
(C) 2.087
(D) –0.3965
(E) 0.4791

USE THIS SPACE FOR SCRATCH WORK

19. The shaded area in the figure is represented by which of the following?

(A) $(A \cap B) \cap C$
(B) $A \cup (B \cup C)$
(C) $(A \cap B) \cup C$
(D) $A \cap (B \cup C)$
(E) $(A \cup B) \cap C$

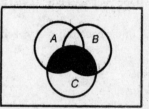

20. If $\dfrac{x+3}{x} < 5$, then the solution set is

(A) $x > \dfrac{3}{4}$

(B) $x \neq \dfrac{3}{4}$

(C) $x > \dfrac{4}{3}$

(D) $x < 0$ or $x > \dfrac{3}{4}$

(E) none of these

21. The function defined by $f(x) = \sqrt{3}\,\cos x + 3 \sin x$ has an amplitude of

(A) 1.27
(B) 1.73
(C) 3.46
(D) 4.73
(E) 5.20

22. What is the equation of the set of points situated in a plane so that the distance between any point and (0,0) is twice the distance between that point and the x-axis?

(A) $3x^2 - y^2 = 0$
(B) $x^2 - 3y^2 = 0$
(C) $x^2 + y^2 - 2y = 0$
(D) $x^2 + y^2 - 2x = 0$
(E) $4x^2 + 3y^2 = 0$

GO ON TO THE NEXT PAGE

23. If p_1 and p_2 are the points of intersection of two circles whose equations are $x^2 + y^2 = 4$ and $(x-2)^2 + (y-2)^2 = 4$, what is the slope of the line perpendicular to the line that passes through p_1 and p_2?

(A) 0
(B) 1
(C) –1
(D) 2
(E) The slope is undefined.

24. The graph of the equation $x^2 - y^2 + 4x - 6y - 12 = 0$ is

(A) a circle
(B) an ellipse
(C) a hyperbola
(D) two intersecting lines
(E) a point

25. Determine q in the equation $4x^2 + qx + 12 = 0$ so that the ratio of the roots is 3.

(A) 0,16,–16
(B) 16,–16
(C) 16
(D) –16
(E) 0

26. $\dfrac{n!}{(n+1)!} + \dfrac{(n+1)!}{(n+2)!} =$

(A) $\dfrac{n(2n+3)}{n+2}$

(B) $\dfrac{2n+3}{(n+1)(n+2)}$

(C) $\dfrac{2n(n+1)!}{(n+2)!}$

(D) $\dfrac{n!}{(n+2)!}$

(E) $\dfrac{1}{(n+2)(n+1)}$

27. In a triangle with sides of 6, 5, and 7, the measure of the largest angle is

(A) 11.5°
(B) 101.5°
(C) 78.5°
(D) 66.5°
(E) 168.5°

28. If $\sin A = 0.8364$ and $\tan A = -1.5258$, $\angle A =$

    (A) 0.99
    (B) 4.13
    (C) 2.56
    (D) 2.15
    (E) 5.29

29. The graph of the curve represented by $\begin{Bmatrix} x = 3 \sin \theta \\ y = 3 \sin \theta \end{Bmatrix}$ is

    (A) a line
    (B) a horizontal line segment
    (C) a circle
    (D) a line segment 3 units long
    (E) a line segment with slope 1

30. If the zeros of the function $f(x)$ are 3, –2, and 1, what are the zeros of $f(x - 3)$?

    (A) 0,–5,–2
    (B) 6,1,4
    (C) 9,–6,3
    (D) –9,6,–3
    (E) $1, -\dfrac{2}{3}, \dfrac{1}{3}$

31. If Δ is defined as $x\Delta y = x - 2y$, which of the following is an expression for $(x\Delta y)\Delta((x\Delta x)\Delta y)$?

    (A) $x - 2y$
    (B) $2x - y$
    (C) $2y$
    (D) x
    (E) $3x + 2y$

32. There are n integers in the solution set of $x(x - 2)(x + 3)(x + 5) < 0$. Therefore, n equals

    (A) 2
    (B) 6
    (C) 4
    (D) 3
    (E) more than 6

33. $(2 \text{ cis } 50°)^3$ written in rectangular form is

    (A) $6.9 + 4i$
    (B) $4 - 6.9i$
    (C) $6.9 - 4i$
    (D) $-6.9 + 4i$
    (E) $-4 + 6.9i$

34. If $A = e^{kt}$, what is the value of k when $A = 10,000$ and $t = 12$?

    (A) 0.06
    (B) 0.58
    (C) 0.33
    (D) 0.77
    (E) 0.82

GO ON TO THE NEXT PAGE

35. If $2\sin 2x = 3\cos 2x$ and $0 \le 2x \le \dfrac{\pi}{2}$, then $x =$

(A) 0.25
(B) 0.52
(C) 0.49
(D) 0.39
(E) 0.63

36. What is the length of the radius of the sphere whose equation is $x^2 + y^2 + z^2 - 4x - 5y + 6z = 0$?

(A) 6.75
(B) 4.39
(C) 2.60
(D) 19.25
(E) 3.46

37. If $f(x) = x^2 + 1$ and $f(g(x)) = 4x - 3$, then $g(x) =$

(A) $2\sqrt{x} - 1$

(B) $2\sqrt{x - 1}$

(C) $\sqrt{x} - 4$

(D) $\sqrt{4x + 4}$

(E) $\dfrac{\sqrt{x - 1}}{4}$

38. What are the coordinates of the point on the line $7x - 3y = 11$ that is closest to the origin?

(A) (1.57,0)
(B) (1.37,0.47)
(C) (1.33,–0.57)
(D) (1.43,–0.47)
(E) (1.27,–0.67)

39. If $f(x) = x + \sqrt{3x + 7}$, on what interval is $f(x) \le 7$?

(A) $(-\infty, 3]$
(B) $(-\infty, 3], [14, \infty)$
(C) $[-2.\overline{3}, 3]$
(D) $[-2.\overline{3}, 3], [14, \infty)$
(E) $[3, 14]$

40. If a square prism is inscribed in a right circular cylinder of radius 4 and height 10, to the nearest whole number, what is the total surface area of the prism?

(A) 192
(B) 88
(C) 226
(D) 290
(E) 320

USE THIS SPACE FOR SCRATCH WORK

41. The lines $4x - 7y - 3 = 0$ and $8x - 14y + 11 = 0$ are parallel. The perpendicular distance between them is

(A) 1.05
(B) 2.11
(C) 0.95
(D) 1.11
(E) 2.05

42. The first three terms of a geometric series are 6, 2, and $\dfrac{2}{3}$. Find the sum of the first 100 terms.

(A) 8.67
(B) 9
(C) 9.33
(D) 9.5
(E) 9.67

43. For all positive angles less than 360°, if $\csc(2x + 30°) = \cos(3y - 15°)$, the sum of x and y is

(A) 185°
(B) 65°
(C) 35°
(D) 215°
(E) 95°

44. The lines $3x - 7y = -15$ and $4x + 2y = 9$ and the x-axis intersect to form a triangle. How many degrees are in the angle of the triangle where the two lines intersect?

(A) 116.6°
(B) 93.4°
(C) 139.8°
(D) 40.2°
(E) 86.6°

45. The rate of growth of a certain organism varies jointly as the warmth of the sun and the square of the available food, and inversely as the number of enemies. If the growth rate remains constant when the warmth of the sun is cut in half and the number of enemies doubles, what can be said about the quantity of available food?

(A) It is doubled.
(B) It is 4 times as great.
(C) It is cut in half.
(D) It is divided by 4.
(E) It remains the same.

GO ON TO THE NEXT PAGE

46. If $f(x) = |x| + 2$, $g(y) = 3y - 2$, and $h(z) = f(g(z)) + g(z)$, the least value of $h(z)$ is

(A) 0
(B) 2
(C) 4
(D) –4
(E) A minimum value does not exist.

47. If x, $3x + 3$, $5x + 5$ are three consecutive terms of a geometric sequence, the sum of these three terms is

(A) –2.25
(B) –12.25
(C) –14.85
(D) –4.75
(E) –10

48. In the figure on the right, S is the set of points in the shaded region. Which of the following represents the set T consisting of all points $(x - y, y)$, where (x, y) is a point in S?

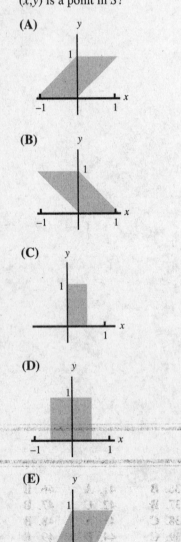

GO ON TO THE NEXT PAGE

49. If $p(x) = 3x^2 + 4x + 1$ and $p(a) = 9$, then $a =$

   (A) $-0.67 \pm 1.49i$
   (B) -0.82 or 2.66
   (C) 2.43 or -1.10
   (D) 0.82 or -2.16
   (E) -2.43 or 1.10

50. Mary, John, and Fran are on a trip. Mary drives during the first hour at an average speed of 50 miles per hour. John drives during the next 2 hours at an average speed of 48 miles per hour. Fran drives for the next 3 hours at an average speed of 52 miles per hour. They reach their destination after exactly 6 hours. Their mean speed (in miles per hour) was

   (A) 52
   (B) 50
   (C) $51\dfrac{1}{3}$
   (D) $50\dfrac{1}{3}$
   (E) $50\dfrac{2}{3}$

USE THIS SPACE FOR SCRATCH WORK

ANSWER KEY

1. A	6. B	11. A	16. E	21. C	26. B	31. E	36. B	41. A	46. B
2. B	7. B	12. C	17. A	22. B	27. C	32. A	37. B	42. C	47. B
3. B	8. B	13. C	18. B	23. B	28. D	33. D	38. C	43. A	48. B
4. C	9. B	14. A	19. E	24. C	29. E	34. D	39. C	44. B	49. E
5. B	10. B	15. A	20. D	25. B	30. B	35. C	40. D	45. A	50. D

ANSWER EXPLANATIONS

In these solutions the following notation is used:

  a: active—Calculator use is necessary or, at a minimum, extremely helpful.

  g: Graphing calculator is preferred.

  i: inactive—Calculator use is not helpful and may even be a hindrance.

1. i A The xy-plane is intersected when $z = 0$, and so the equation of the line is $3x + 4y = 60$. [5.5].

2. i B When coefficients are rational, the irrational roots come in conjugate pairs, but no relationship between the other roots is necessary. [2.4].

3. i B $3x + 4 = 2y + 4$. $\dfrac{3x}{2x} = \dfrac{2y}{2x}$. Therefore, $\dfrac{2y}{2x} \cdot \dfrac{y}{x} = \dfrac{3}{2}$. [5.6].

4. a C Use the change-of-base theorem to get
$$\log_7 \sqrt{3} = \frac{\log_{10} \sqrt{3}}{\log_{10} 7} \approx \frac{0.2386}{0.8451} \approx 0.28. \ [4.2].$$

5. a B $f(g(x)) = \sin(e^x)$, so $f(g(\pi)) = \sin(e^\pi) \approx -0.91$. [1.2].

6. i B $\sin \theta = \dfrac{b}{r}$. Therefore, $r \sin \theta = b$. [3.1].

7. g B With your calculator in radian mode, graph
$$y = \sqrt{5} \cos\left(x + \frac{\pi}{5}\right)$$
using ZTrig. Use CALC/ value = 0, to see that $y \approx 1.81$.

An alternative solution uses the fact that the y intercept is the y–coordinate when $x = 0$. Substituting 0 for x in the formula for y yields
$$y = \sqrt{5} \cos\frac{\pi}{5} \approx 1.81. \ [3.4].$$

8. a B From the 10 points, 3 must be chosen each time. Therefore, $\dbinom{10}{3} = \dfrac{10!}{3!7!} = 120$. [5.1].

9. a B Put your calculator in radian mode:
$$\frac{f(1,2)}{g(1,2)} = \frac{\tan 1 + \tan 2}{1 - \tan 1 \cdot \tan 2} \approx \frac{-0.6276}{4.4030} \approx -0.1425.$$
[1.5, 3.5].
<u>Alternative Solution</u>:
$$\frac{f(1,2)}{g(1,2)} = \frac{\tan 1 + \tan 2}{1 - \tan 1 \cdot \tan 2} = \tan 3 \approx -0.1425.$$

10. i B To recover the original function $y = f(x)$ from $y = f(x + 2) - 3$, x must equal -2 and must equal -3. This corresponds to a shift of 2 units left and 3 down. [5.5].

11. i A Slope of I is $-\dfrac{a}{b}$; slope of II is $\dfrac{a}{b}$; slope of III is $\dfrac{a}{b}$; slope of IV is $\dfrac{b}{a}$. I and IV are negative reciprocals. [2.2].

12. g C Plot the graphs of the answer choices to see that only choice C is symmetric about the origin and thus is an odd function.

An alternative solution relies on the fact that $f(-x) = -f(x)$ for an odd function. Of the choices given, $f(x) = x^3 + x$ is the only one that meets this condition: $f(-x) = (-x)^3 + (-x) = -x^3 - x = -(x^3 + x) = -f(x)$. [1.4].

13. i C $f(x)$ reduces to $x - 1$ with the restriction $x \neq -1$. Therefore, $f(i) = i - 1$. [1.2].

14. g A In order for $f^{-1}(x)$ to be a function, the domain of f must be such that for each value of y there is only one value of x. Graph f in the window $x\varepsilon[-30°, 270°]$ and $y\varepsilon[-1.5, 1.5]$ to capture the full range in the answer choices, and observe that A is the only answer choice that meets this condition. [1.3, 3.6].

15. i A Factor and reduce and then take the limit:
$$\frac{(2x - a)(x - a)}{(x + a)(x - a)}. \text{ Limit} = \frac{a}{2a} = \frac{1}{2}. \ [4.5].$$

16. i E $f(x) = \dfrac{x^2 - 1}{x(x - 1)}$. Since division by zero is undefined, $x \neq 0$ or 1. Therefore, the domain is all real numbers except 0 or 1. [1.1].

17. i A $\dfrac{\tan x}{\cot x} = \dfrac{\dfrac{\sin x}{\cos x}}{\dfrac{\cos x}{\sin x}} = \dfrac{\sin^2 x}{\cos^2 x} = \dfrac{1^2}{2^2} = \dfrac{1}{4}$. [3.1].

18. g   B   The supplement of an angle is 180°—the measure of the angle, so enter $\tan(108 - \sin^{-1}(.4321))$ to get the correct answer choice B.

19. i   E   The shaded region is the intersection of C and the union of A and B. [5.9].

20. g   D   Graph $y = \dfrac{x+3}{x}$ and $y = 5$ in the standard window and use CALC/intersect to find the point of intersection 0.75. Inspection of the graphs suggests that $\dfrac{x}{x+3}$ lies below 5 for all $x < 0$ and $x > 0.75$.

An alternative solution is to solve the equation $\dfrac{x+3}{x} = 5$ to get the boundary points $x = 0$ and $x = \dfrac{3}{4}$. Use test points in each of the three regions determined by these boundary points: $(-\infty, 0)$, $\left(0, \dfrac{3}{4}\right)$, and $\left(\dfrac{3}{4}, \infty\right)$. The inequality is true for test points in the first and third regions. [2.5].

21. g   C   Graph f using ZTrig and use CALC/max to find the maximum value of 3.46. Since inspecting the graph indicates that the maximum and minimum values of the function are opposite, the amplitude is also approximately 3.46.

An alternative solution is to multiply and divide f by $2\sqrt{3}$, making the coefficients of $\sin x$ and $\cos x$ trig values of special angles:

$$f(x) = 2\sqrt{3}\left(\frac{1}{2}\cos x + \frac{\sqrt{3}}{2}\sin x\right)$$

$$= 2\sqrt{3}(\sin 30° \cos x + \cos 30° \sin x)$$

$$= 2\sqrt{3}\sin(30° + x).$$

The amplitude is $2\sqrt{3} \approx 3.46$. [3.4].

22. i   B   If (x, y) represents any such point, $|y|$ represents the distance between that point and the x-axis. $d = \sqrt{(x-0)^2 + (y-0)^2} = \sqrt{x^2 + y^2}$. Therefore, $\sqrt{x^2 + y^2} = 2|y|$. Squaring both sides and simplifying gives $x^2 - 3y^2 = 0$. [4.1].

23. i   B   Substituting gives $x^2 + y^2 = x^2 - 4x + 4 + y^2 - 4y + 4$. Simplifying gives the equation of the line through p_1 and p_2: $x + y = 2$. The slope of this line is -1, and so the slope of the perpendicular is $+1$. [4.1].

24. i   C   Since $A = 1$ and $C = -1$ have opposite signs, the graph is either a hyperbola or its degenerate form—two intersecting lines. Complete the square to determine which: $x^2 + 4x + 4 - (y^2 + 6y + 9) = 12 + 4 - 9$, or $(x+2)^2 - (y+3)^2 = 7$. Since the right side is not zero, the graph is a hyperbola. [4.1].

25. i   B   Call the roots r_1 and r_2. Then $\dfrac{r_1}{r_2} = 3$, or $r_1 = 3r_2$. It follows that $(x - r_1)(x - r_2) = (x - 3r_2)(x - r_2) = x^2 - 4r_2 x + 3r_2^2 = 0$. Multiply both sides by 4 to get $4x^2 - 16r_2 x + 12r_2^2 = 0$. Equating coefficients with the given equation yields $12r_2^2 = 12$ and $-16r_2 = q$. Therefore, $r_2 = \pm 1$, so $q = \mp 16$, answer choice B. [2.3, 2.4].

26. i   B $(n + 1)! = (n + 1)n!$ and $(n + 2)! = (n + 2)(n + 1)!$. Therefore, the sum of the two ratios in the problem simplify to $\dfrac{1}{n+1} + \dfrac{1}{n+2} = \dfrac{n+2+n+1}{(n+1)(n+2)} = \dfrac{2n+3}{(n+1)(n+2)}$. [5.1].

27. a   C   Put your calculator in degree mode because the answer choices are in degrees. Law of cosines: $49 = 36 + 25 - 2 \cdot 6 \cdot 5 \cdot \cos C$. Therefore, $\cos C = 0.2$, which implies that $C = \text{Cos}^{-1}(0.2) \approx 78.5°$. [3.7].

28. a   D   Put your calculator in radian mode. Since $\sin A > 0$ and $\tan A < 0$, $\angle A$ must be in quadrant II. Reference angle for $\angle A = \text{Sin}^{-1} 0.8364 \approx 0.9907$. Therefore, $\angle A \approx -0.9907 \approx 2.15$. [3.5].

29. g   E   Put your calculator in parametric mode and graph $x = 3\sin t$ and $y = 3\sin t$ in the standard window. Observe that the graph is a line segment, then graph in ZSquare to see that the slope of the segment is 1.

An alternative solution is to eliminate the parameter, to get $y = x$. Since $-1 \le \sin t \le 1$, it follows that $-3 \le x, y \le 3$, so the graph is a line segment rather than a line. [4.6].

TIP: When you eliminate the parameter, you might be discarding restrictions that the parameter put on the value of x and y. Always check for restrictions on x and y in the original problem. Also, always look at all the answer choices to see possibilities that you had not thought of.

30. i   B $f(x - 3) = 0$ when $x - 3 = 3, -2,$ or 1. Therefore, $x = 6, 1,$ or 4. [2.4].

31. i   E   Use the definition of Δ carefully to get
$(x\Delta y)\Delta((x\Delta x)\Delta y) = (x - 2y)\Delta((x - 2x)\Delta y)$
$= (x - 2y)\Delta(-x - 2y)$
$= (x - 2y) - 2(-x - 2y)$
$= 3x + 2y$
[5.9].

32. g   A   Graph $x(x - 2)(x + 3)(x + 5)$ in the standard window. The graph crosses the x-axis at -5, $-3, 0,$ and 2 and is below it on the open intervals $(-5,-3)$ and $(0,2)$. These open intervals contain only the integers -4 and 1, so the correct answer choice is A.

An alternative solution is to solve the associated equation $x(x - 2)(x + 3)(x + 5) = 0$ to get $x = 0,2,-3,-5$ and test the inequality using points on the intervals $(-\infty,-5)$ $(-5,-3)$, $(-3,0)$, $(0,2)$, and $(2,\infty)$. The two intervals $(-5,-3)$ and $(0,2)$ satisfy the inequality and between them contain only two integers. [2.5].

33. g   D   Set your graphing calculator to degree mode and $a + bi$ mode. Since $(2\operatorname{cis}50°)^3 = 8\operatorname{cis}150°$, evaluate $8(\cos 150 + i\sin 150)$ to get the correct answer choice D. [4.7].

34. g   D   Graph $y = e^{12x}$ and $y = 10000$ using window settings of $x\varepsilon[0,1]$ and $y\varepsilon[0,20000]$. Use CALC/intersect to find the point of intersection 0.77 as the correct answer choice D.

An alternative solution is to substitute 10000 for A and 12 for t; then take the natural logarithm of both sides of the equation to get $\ln 1000 = 12k$.
Thus, $k = \dfrac{\ln 10000}{12} \approx 0.77$. [4.2].

35. g   C   Set your calculator to radian mode, and graph $2\sin 2x$ and $3\cos 2x$. Since $0 \le 2x \le \dfrac{\pi}{2}, 0 \le x \le \dfrac{\pi}{4}$, so use window settings of $x\varepsilon\left[0, \dfrac{\pi}{4}\right]$ and $y\varepsilon[-3,3]$. Use CALC/intersect to find the sole point of intersection 0.49.

An alternative solution is to divide both sides of the equation by $2\cos 2x$ to get $\dfrac{\sin 2x}{\cos 2x} = \tan$
$2x = \dfrac{3}{2}$. Calculate $\dfrac{\tan^{-1}3/2}{2} \approx 0.49$, the correct answer choice C. [3.5].

36. a   B   Complete the square:
$x^2 - 4x + y^2 - 5y + 6.25 + z^2 - 6z + 9 = 4 + 6.25 + 9 = 19.25$. Therefore, radius $= \sqrt{19.25} \approx 4.39$ [5.5].

37. i   B $f(x) = x^2 + 1, f(g(x)) = (g(x))^2 + 1 = 4x - 3$, $(g(x))^2 = 4x - 4, g(x) = \pm\sqrt{4x - 4} = \pm 2\sqrt{x - 1}$. Therefore, $2\sqrt{x - 1}$ is an expression for $g(x)$. [1.2].

38. g   C   The point closest to the origin is on the line through the origin and perpendicular to the given line. The slope of the given line is $\dfrac{7}{3}$, so the slope of the perpendicular is $-\dfrac{3}{7}$ and its equation is $y = -\dfrac{3}{7}x$. The desired point is the point of intersection of this line and the given line. Graph the two equations in the standard window, and use CALC/intersect to find the point of intersection $(1.33,-0.57)$.

An alternative solution is to solve the system $7x - 3y = 11$ and $y = -\dfrac{3}{7}x$ by substituting the expression for y in the second equation into the first equation: $7x - 3\left(-\dfrac{3}{7}x\right) = 11$, so $x = \dfrac{77}{58} \approx 1.33$ and $y = -\dfrac{33}{40} \approx -0.57$. [2.2].

39. g   C   Plot the graphs of $y = x + \sqrt{3x + 7}$ and $y = 7$ in the standard window. Observe that the domain of $y = x + \sqrt{3x + 7}$ is $x \ge -2.\overline{3}$, and that this function does not rise above $y = 7$ as long as $x \le 3$ if $x \le 3$. Thus, the correct answer choice is C.

An alternative solution is to solve the inequality $x + \sqrt{3x + 7} \le 7$. Solving the associated equation $x + \sqrt{3x + 7} = 7$ algebraically, establishing regions, and testing points can accomplish this. First note that the left side is only defined if $3x + 7 \ge 0$, or if $x \ge -\dfrac{7}{3} = -2.\overline{3}$. Subtract x from both sides of $x+\sqrt{3x+7}=7$ and square: $3x+7=49-14x$ $+x^2$. Then subtract $3x+7$ from both sides and factor: $0=x^2-17x+42=(x-14)(x-3)$, to yield the solutions $x=3$ and 14. Since $x>3$ does not solve the inequality, the solution set is $[-2.\overline{3},3]$. [2.5].

40. a   D   The diameter of the cylinder is the diagonal of the square. Therefore, a side of the square base is $4\sqrt{2}$. The area of the two square bases is 64, and the area of each of the four sides is $40\sqrt{2}$. Total surface area $= 160\sqrt{2}+64 \approx 290$. [5.5].

41. g   A   Choose any point on one of the lines, such as $(-1,-1)$ on the first line. Then use program DPL to find the distance 1.05 between the point and the second line. This is the (perpendicular) distance between the two lines.

An alternative solution is to use the formula for the distance between a point and a line, with $x=-1$, $y=-1$, $A=8$, $B=-14$, and $C=11$:

$$\text{distance} = \frac{8(-1)-14(-1)+11}{\sqrt{8^2+(-14)^2}} \approx 1.05$$

42. g   C   Calculate the common ratio as $\frac{2}{6}=\frac{1}{3}$. The first term is 6, so the nth term is $t_n=6\left(\frac{1}{3}\right)^{n-1}$. Use the sum and sequence features of your calculator to evaluate the sum of the 100 terms in the generated sequence:

LIST/MATH/sum(LIST/OPS/
seq($6(1/3)$^X, X, 0, 99)) = 9.

The range is 0 to 99 instead of 1 to 100 because the formula for t_n uses the exponent $n-1$.

An alternative solution is to use the formula for the sum of a geometric sequence:

$$S_n = \frac{t_1(1-r^n)}{1-r} = \frac{6\left(1-(1/3)^{100}\right)}{1-(1/3)} = \frac{6(1-0)}{2/3} = 9.$$
[5.4].

43. i   A ± 1 are the only numbers that are both within the range of csc and cos. If each equals 1, $2x+30=90°$ and $3y-15=0°$. This is not allowed since $0°$ is not positive. If each equals -1, $2x+30=270°$ and $3y-15=180°$. Thus, $x+y=185°$. [3.3].

44. a   B   The slope of a line equals the tangent of the angle, α, formed by the line and the positive x-axis. Thus, $\tan \angle 2 = \frac{3}{7}$, which implies that $\angle 2 = \text{Tan}^{-1}\left(\frac{3}{7}\right) \approx 23.2°$. Since $\tan \angle \alpha = -2$, $\tan \angle 1 = +2$, which implies that $\angle 1 = \text{Tan}^{-1}(2) \approx 63.4°$. Therefore, $\angle 3 = 180° - 23.2° - 63.4° \approx 93.4°$. [3.5].

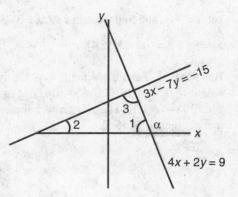

45. i   A $\frac{RE}{WF^2}=K$. Let $R=5$, $E=2$, $W=4$, $F=1$ (arbitrary choices to make the arithmetic easy). $K = \frac{5 \cdot 2}{4 \cdot 1} = \frac{5}{2}$. Changing the variables as indicated gives $\frac{5 \cdot 4}{2F^2} = \frac{5}{2} \cdot F^2 = 4$. Since F is the amount of food, $F=2$ only. The original value of F was doubled. [5.6].

Alternative Solution: $\frac{RE}{WF^2}=K$. With the changes, the formula becomes $\frac{R \cdot 2E}{\frac{1}{2}W \cdot F^2} = K$. Simplifying gives $\frac{4RE}{WF^2} = K$.

Therefore, F^2 must be 4 times as large as it was originally in order for the fraction to remain equal to K. Thus, the original value of F must be doubled.

46. g   B   Enter $|x|+2$ into Y_1, $3x-2$ into Y_2, and $Y_1(Y_2(x))$ into Y_3. De-select Y_1 and Y_2, and graph Y_3 in the standard window. Inspection of the graph leads to the correct answer choice B.

An alternative solution uses the definitions of the functions to determine that $h(z)=f(3z-2)+3z-2=|3z-2|+2+3z-2=|3z-2|+3z$. Two cases must be examined. If $3z-2 \geq 0$, $h(z) \geq 2$ because

$z \geq \dfrac{2}{3}$ and $h(z) = 6z - 2$. If $3z - 2 < 0$, $h(z) = 2 - 3z$ + $3z = 2$. Combining these two cases yields the correct answer choice B. [1.2, 4.3].

47. a   B   The ratios of the second to first terms and third to second terms must be equal, so $\dfrac{3x+3}{x} = \dfrac{5x+5}{3x+3} = \dfrac{5}{3}$. Cross multiply and solve to get $x = -\dfrac{9}{4}$, and find the sum $x + (3x = 3) + (5x + 5) = 9x + 8 = -12.25$. [5.4].

48. i   B   Analyze the transformation of the corner points: $(0,0) \rightarrow (0,0)$, $(1,0) \rightarrow (1,0)$, $(1,1) \rightarrow (0,1)$, and $(0,1) \rightarrow (-1,1)$. The only answer choice with these corners is B. [5.5].

49. g   E   Set one side of the equation to zero, and use program QUADFORM with $a = 3$, $b = 4$, $c = -8$ for the correct answer choice E.

An alternative solution is to substitute these values into the Quadratic Formula, evaluate, and use your calculator to obtain decimal values. [2.3].

50. i   D   Mary drove for $1 \times 50 = 50$ miles.
John drove for $2 \times 48 = 96$ miles.
Fran drove for $3 \times 52 = 156$ miles.

$$\text{Average (mean) speed} = \dfrac{\text{total distance}}{\text{total time}}$$
$$= \dfrac{302}{6} = 50\dfrac{1}{3} \text{ miles per hour.}$$
[5.8].

SELF-EVALUATION CHART FOR MODEL TEST 4

SUBJECT AREA	QUESTIONS	NUMBER OF RIGHT WRONG OMITTED

Mark correct answers with C, wrong answers with X, and omitted answers with O.

Algebra
(8 questions)
Review section

3	4	20	25	32	34	41	45
5.6	4.2	2.5	2.3	2.5	4.2	2.2	5.6

___ ___ ___

Solid geometry
(4 questions)
Review section

1	8	36	40
5.5	5.1	5.5	5.5

___ ___ ___

Coordinate geometry
(7 questions)
Review section

10	11	22	23	24	38	48
5.5	2.2	4.1	4.1	4.1	2.2	5.5

___ ___ ___

Trigonometry
(10 questions)
Review section

6	7	17	18	21	27	28	35	43	44
3.1	3.4	3.1	3.1	3.4	3.7	3.5	3.6	3.3	3.5

___ ___ ___

Functions
(12 questions)
Review section

2	5	9	12	13	14	16	30	37	39	46	49
2.4	1.2	1.5	1.4	1.2	1.3	1.1	2.4	1.2	2.5	1.2	2.3

___ ___ ___

Miscellaneous
(9 questions)
Review section

15	19	26	29	31	33	42	47	50
4.5	5.9	5.1	4.6	5.9	4.7	5.4	5.4	5.8

___ ___ ___

TOTALS   ___ ___ ___

Raw score = (number right) – ¼ (number wrong) = ____

Round your raw score to the nearest whole number = ____

Evaluate Your Performance
Model Test 4

Rating	Number Right
Excellent	41–50
Very good	33–40
Above average	25–32
Average	15–24
Below average	Below 15

A B C
A B C
A B C
A B C
A

ANSWER SHEET FOR MODEL TEST 5

Determine the correct answer for each question. Then, using a no. 2 pencil, blacken completely the oval containing the letter of your choice.

1. Ⓐ Ⓑ Ⓒ Ⓓ Ⓔ
2. Ⓐ Ⓑ Ⓒ Ⓓ Ⓔ
3. Ⓐ Ⓑ Ⓒ Ⓓ Ⓔ
4. Ⓐ Ⓑ Ⓒ Ⓓ Ⓔ
5. Ⓐ Ⓑ Ⓒ Ⓓ Ⓔ
6. Ⓐ Ⓑ Ⓒ Ⓓ Ⓔ
7. Ⓐ Ⓑ Ⓒ Ⓓ Ⓔ
8. Ⓐ Ⓑ Ⓒ Ⓓ Ⓔ
9. Ⓐ Ⓑ Ⓒ Ⓓ Ⓔ
10. Ⓐ Ⓑ Ⓒ Ⓓ Ⓔ
11. Ⓐ Ⓑ Ⓒ Ⓓ Ⓔ
12. Ⓐ Ⓑ Ⓒ Ⓓ Ⓔ
13. Ⓐ Ⓑ Ⓒ Ⓓ Ⓔ
14. Ⓐ Ⓑ Ⓒ Ⓓ Ⓔ
15. Ⓐ Ⓑ Ⓒ Ⓓ Ⓔ
16. Ⓐ Ⓑ Ⓒ Ⓓ Ⓔ
17. Ⓐ Ⓑ Ⓒ Ⓓ Ⓔ

18. Ⓐ Ⓑ Ⓒ Ⓓ Ⓔ
19. Ⓐ Ⓑ Ⓒ Ⓓ Ⓔ
20. Ⓐ Ⓑ Ⓒ Ⓓ Ⓔ
21. Ⓐ Ⓑ Ⓒ Ⓓ Ⓔ
22. Ⓐ Ⓑ Ⓒ Ⓓ Ⓔ
23. Ⓐ Ⓑ Ⓒ Ⓓ Ⓔ
24. Ⓐ Ⓑ Ⓒ Ⓓ Ⓔ
25. Ⓐ Ⓑ Ⓒ Ⓓ Ⓔ
26. Ⓐ Ⓑ Ⓒ Ⓓ Ⓔ
27. Ⓐ Ⓑ Ⓒ Ⓓ Ⓔ
28. Ⓐ Ⓑ Ⓒ Ⓓ Ⓔ
29. Ⓐ Ⓑ Ⓒ Ⓓ Ⓔ
30. Ⓐ Ⓑ Ⓒ Ⓓ Ⓔ
31. Ⓐ Ⓑ Ⓒ Ⓓ Ⓔ
32. Ⓐ Ⓑ Ⓒ Ⓓ Ⓔ
33. Ⓐ Ⓑ Ⓒ Ⓓ Ⓔ
34. Ⓐ Ⓑ Ⓒ Ⓓ Ⓔ

35. Ⓐ Ⓑ Ⓒ Ⓓ Ⓔ
36. Ⓐ Ⓑ Ⓒ Ⓓ Ⓔ
37. Ⓐ Ⓑ Ⓒ Ⓓ Ⓔ
38. Ⓐ Ⓑ Ⓒ Ⓓ Ⓔ
39. Ⓐ Ⓑ Ⓒ Ⓓ Ⓔ
40. Ⓐ Ⓑ Ⓒ Ⓓ Ⓔ
41. Ⓐ Ⓑ Ⓒ Ⓓ Ⓔ
42. Ⓐ Ⓑ Ⓒ Ⓓ Ⓔ
43. Ⓐ Ⓑ Ⓒ Ⓓ Ⓔ
44. Ⓐ Ⓑ Ⓒ Ⓓ Ⓔ
45. Ⓐ Ⓑ Ⓒ Ⓓ Ⓔ
46. Ⓐ Ⓑ Ⓒ Ⓓ Ⓔ
47. Ⓐ Ⓑ Ⓒ Ⓓ Ⓔ
48. Ⓐ Ⓑ Ⓒ Ⓓ Ⓔ
49. Ⓐ Ⓑ Ⓒ Ⓓ Ⓔ
50. Ⓐ Ⓑ Ⓒ Ⓓ Ⓔ

MODEL TEST

5

50 questions   1 hour

Tear out the preceding answer sheet. Decide which is the best choice by rounding your answer when appropriate. Blacken the corresponding space on the answer sheet. When finished, check your answers with those at the end of the test. For questions that you got wrong, note the sections containing the material that you must review. Also, if you do not fully understand how you arrived at some of the correct answers, you should review the appropriate sections. Finally, fill out the self-evaluation sheet on page 239 in order to pinpoint the topics that give you the most difficulty.

TEST DIRECTIONS

<u>Directions</u>: Decide which answer choice is best. If the exact numerical value is not one of the answer choices, select the closest approximation. Fill in the oval on the answer sheet that corresponds to your choice.

Notes:
(1) You will need to use a scientific or graphing calculator to answer some of the questions.
(2) You will have to decide whether to put your calculator in degree or radian mode for some problems.
(3) All figures that accompany problems are plane figures unless otherwise stated. Figures are drawn as accurately as possible to provide useful information for solving the problem, except when it is stated in a particular problem that the figure is not drawn to scale.
(4) Unless otherwise indicated, the domain of a function is the set of all real numbers for which the functional value is also a real number.

<u>Reference Information.</u> The following formulas are provided for your information.

Volume of a right circular cone with radius r and height h: $V = \frac{1}{3}\pi r^2 h$

Lateral area of a right circular cone if the base has circumference c and the slant height is l: $S = \frac{1}{2}cl$

Volume of a sphere of radius r: $V = \frac{4}{3}\pi r^3$

Surface area of a sphere of radius r: $S = 4\pi r^2$

Volume of a pyramid of base area B and height h: $V = \frac{1}{3}Bh$

---

1. The amplitude of function $f(x) = -\sin x \cdot \cos x$ is

   (A) 1
   (B) 2
   (C) $\dfrac{1}{2}$
   (D) $-\dfrac{1}{2}$
   (E) -1

2. What is the remainder when $3x^4 - 2x^3 - 20x^2 - 12$ is divided by $x + 2$?

   (A) -4
   (B) -28
   (C) -6
   (D) -36
   (E) -60

3. If $1 - \dfrac{1}{x} = 2 - \dfrac{2}{x}$, then $3 - \dfrac{3}{x} =$

   (A) $-\dfrac{1}{3}$

   (B) $\dfrac{1}{3}$

   (C) 3
   (D) -3
   (E) 0

4. The lines $3x - 4y + 8 = 0$ and $8x + 6y - 4 = 0$ intersect at point P. One angle formed at point P contains

   (A) $30°$
   (B) $45°$
   (C) $60°$
   (D) $90°$
   (E) Point P does not exist because the lines are parallel.

5. The domain of $f(x) = \log_{10}(\sin x)$ contains which of the following intervals?

   (A) $0 \le x \le \pi$
   (B) $-\dfrac{\pi}{2} \le x \le \dfrac{\pi}{2}$
   (C) $0 < x < \pi$
   (D) $-\dfrac{\pi}{2} < x < \dfrac{\pi}{2}$
   (E) $\dfrac{\pi}{2} < x < \dfrac{3\pi}{2}$

6. Which of the following is the ratio of the surface area of the sphere with radius r to its volume?

(A) $\dfrac{4}{\pi}$

(B) $\dfrac{r}{\pi}$

(C) $\dfrac{3}{r}$

(D) $\dfrac{r}{4}$

(E) $\dfrac{4}{r}$

USE THIS SPACE FOR SCRATCH WORK

7. If the two solutions of $x^2 - 9x + c = 0$ are complex conjugates, which of the following describes all possible values of c?

(A) $c = 0$

(B) $c \neq 0$

(C) $c > \dfrac{81}{4}$

(D) $c > 81$

(E) $c < 9$

8. If $\tan x = 3$, the numerical value of $\sqrt{\csc x}$ is

(A) 0.97
(B) 1.03
(C) 1.78
(D) 3.16
(E) 0.32

9. In the figure, the graph of $f(x)$ has two transformations performed on it. First it is rotated 180° about the origin, and then it is reflected about the x-axis. Which of the following is the equation of the resulting curve?

(A) $y = -f(x)$
(B) $y = f(x + 2)$
(C) $x = f(y)$
(D) $y = f(x)$
(E) none of the above

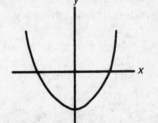

10. $\displaystyle \lim_{x \to \infty} \frac{3x^3 - 7x^2 + 2}{4x^2 - 3x - 1} =$

(A) $\dfrac{3}{4}$

(B) 0

(C) 3

(D) 1

(E) ∞

GO ON TO THE NEXT PAGE

11. As x increases from $-\dfrac{\pi}{4}$ to $\dfrac{3\pi}{4}$, the value of $\cos x$

   (A) always increases
   (B) always decreases
   (C) increases and then decreases
   (D) decreases and then increases
   (E) does none of the above

12. The vertical distance between the minimum and maximum values of the function $y = \left|-\sqrt{2}\,\sin\sqrt{3}x\right|$ is

   (A) 1.414
   (B) 2.828
   (C) 1.732
   (D) 3.464
   (E) 2.094

13. If the domain of $f(x) = -|x| + 2$ is $\{x: -1 \le x \le 3\}$, $f(x)$ has a minimum value when x equals

   (A) 0
   (B) –1
   (C) 1
   (D) 3
   (E) There is no minimum value.

14. If $f: (x,y) \to (x + y, y)$ for every pair (x,y) in the plane, for what point(s) is it true that $(x,y) \to (x,y)$?

   (A) $(0,0)$
   (B) $(-1,1)$
   (C) the set of points (x,y) such that $x = 0$
   (D) the set of points (x,y) such that $y = 0$
   (E) the set of points such that $x = y$

15. A positive rational root of the equation $4x^3 - x^2 + 16x - 4 = 0$ is

   (A) $\dfrac{1}{2}$
   (B) 1
   (C) $\dfrac{1}{4}$
   (D) 2
   (E) $\dfrac{3}{4}$

16. The norm of vector $\vec{V} = 3\vec{i} - \sqrt{2}\,\vec{j}$ is

   (A) 4.24
   (B) 2.45
   (C) 3.61
   (D) 3.32
   (E) 1.59

GO ON TO THE NEXT PAGE

17. If five coins are flipped and all the different ways they could fall are listed, how many elements of this list will contain more than three heads?

(A) 16
(B) 10
(C) 5
(D) 6
(E) 32

18. The negation of the statement "For all sets, there is one subset" is

(A) for all sets, there is not one subset
(B) for no sets, there is one subset
(C) for some sets, there is not one subset
(D) for some sets, there is one subset
(E) for no sets, there is not one subset

19. The graph of the curve represented by $\begin{Bmatrix} x = \sec\theta \\ y = \cos\theta \end{Bmatrix}$ is

(A) a line
(B) a hyperbola
(C) an ellipse
(D) a line segment
(E) a portion of a hyperbola

20. Point (3,2) lies on the graph of the inverse of $f(x) = 2x^3 + x + A$. The value of A is

(A) 15
(B) –15
(C) 18
(D) 54
(E) –54

21. If $f(x) = ax^2 + bx + c$ and $f(1) = 3$ and $f(-1) = 3$, then $a + c$ equals

(A) –3
(B) 0
(C) 3
(D) 6
(E) 2

22. In $\triangle ABC$, $\angle B = 42°$, $\angle C = 30°$, and $AB = 100$. The length of BC is

(A) 47.6
(B) 66.9
(C) 190.2
(D) 133.8
(E) none of the above

GO ON TO THE NEXT PAGE

23. If $4\sin x + 3 = 0$ on $0 \le x < 2\pi$, then $x =$

   (A) 5.435
   (B) 0.848
   (C) 3.990 or 5.435
   (D) 0.848 or 5.435
   (E) −0.848

24. The primary period of the function $f(x) = 2 \cdot \cos^2 3x$ is

   (A) 2π

   (B) $\dfrac{\pi}{3}$

   (C) π

   (D) $\dfrac{\pi}{2}$

   (E) 3π

25. In $a + bi$ form, the reciprocal of $2 + 6i$ is

   (A) $-\dfrac{1}{16} + \dfrac{3}{16}i$

   (B) $\dfrac{1}{16} + \dfrac{3}{16}i$

   (C) $\dfrac{1}{20} - \dfrac{3}{20}i$

   (D) $\dfrac{1}{20} + \dfrac{3}{20}i$

   (E) none of the above

26. A central angle of two concentric circles is $\dfrac{3\pi}{14}$. The area of the large sector is twice the area of the small sector. What is the ratio of the lengths of the radii of the two circles?

   (A) 0.50:1
   (B) 0.71:1
   (C) 0.25:1
   (D) 1:1
   (E) 0.67:1

27. If the region bounded by the lines $y = -\dfrac{4}{3}x + 4$, $x = 0$, and $y = 0$ is rotated about the y-axis, the volume of the figure formed is

   (A) 18.8
   (B) 37.7
   (C) 56.5
   (D) 84.8
   (E) 113.1

28. If there are known to be 4 broken transistors in a box of 12, and 3 transistors are drawn at random, what is the probability that none of the 3 is broken?

(A) 0.375
(B) 0.255
(C) 0.250
(D) 0.750
(E) 0.556

29. In order for the inverse of $f(x) = \sin 2x$ to be a function, the domain of f can be limited to

(A) $-\dfrac{\pi}{2} \le x \le \dfrac{\pi}{2}$

(B) $0 \le x \le \dfrac{\pi}{2}$

(C) $\dfrac{\pi}{4} \le x \le \dfrac{3\pi}{4}$

(D) $\dfrac{\pi}{2} \le x \le \pi$

(E) $0 \le x \le \pi$

30. R varies as the square of z and inversely as the cube of T. If z is tripled and T is doubled, the value of R is

(A) multiplied by 3
(B) multiplied by $\dfrac{9}{8}$
(C) multiplied by 8
(D) divided by 3
(E) divided by $\dfrac{2}{3}$

31. Two roots of $x^3 + 3x^2 + Kx - 12 = 0$ are unequal but have the same absolute value. The value of K is

(A) 4
(B) -4
(C) 6
(D) -6
(E) -9

32. If n is an integer, what is the remainder when $3x^{2n+3} - 4x^{2n+2} + 5x^{2n+1} - 8$ is divided by $x + 1$?

(A) -4
(B) -10
(C) 0
(D) -20
(E) The remainder cannot be determined.

USE THIS SPACE FOR SCRATCH WORK

GO ON TO THE NEXT PAGE

33. Four men, A, B, C, and D, line up in a row. What is the probability that man A is at either end of the row?

(A) $\dfrac{1}{2}$

(B) $\dfrac{1}{3}$

(C) $\dfrac{1}{4}$

(D) $\dfrac{1}{6}$

(E) $\dfrac{1}{12}$

34. $\displaystyle\sum_{i=3}^{10} 5 =$

(A) 260
(B) 50
(C) 40
(D) 5
(E) none of these

35. The graph of $y^4 - 3x^2 + 7 = 0$ is symmetric with respect to which of the following?

   I. the x-axis
  II. the y-axis
 III. the origin

(A) only I
(B) only II
(C) only III
(D) only I and II
(E) I, II, and III

36. In a group of 30 students, 20 take French, 15 take Spanish, and 5 take neither language. How many students take both French and Spanish?

(A) 0
(B) 5
(C) 10
(D) 15
(E) 20

37. If $f(x) = \dfrac{x+1}{x^2+1}$ and $g(x) = \dfrac{x^2+1}{x+1}$, find the slope of the line that passes through points $\left(\sqrt{5}, g\left(\sqrt{5}\right)\right)$ and $\left(\sqrt{2}, f\left(\sqrt{2}\right)\right)$.

(A) 0.06
(B) 2.13
(C) −0.63
(D) −1.26
(E) 1.28

USE THIS SPACE FOR SCRATCH WORK

GO ON TO THE NEXT PAGE

38. The plane whose equation is $2x + 3y + 5z = 35$ forms a pyramid in the first octant with the coordinate planes. Its volume is

(A) 190.6
(B) 238.2
(C) 285.8
(D) 381.1
(E) 566.8

39. Solve the equation $\sin 15x + \cos 15x = 0$. What is the sum of the three smallest positive solutions?

(A) $\dfrac{\pi}{20}$

(B) $\dfrac{\pi}{3}$

(C) $\dfrac{7\pi}{20}$

(D) $\dfrac{21\pi}{20}$

(E) $\dfrac{21\pi}{4}$

40. Given the set of data 1, 1, 2, 2, 2, 3, 3, x, y, where x and y represent two different integers. If the mode is 2, which of the following statements must be true?

(A) If $x = 1$ or 3, then y must $= 2$.
(B) Both x and y must be > 3.
(C) Either x or y must $= 2$.
(D) It does not matter what values x and y have.
(E) Either x or y must $= 3$, and the other must $= 1$.

41. If $f(x) = \sqrt{2x + 3}$ and $g(x) = x^2$, for what value(s) of x does $f(g(x)) = g(f(x))$?

(A) 5.45
(B) –0.55
(C) 0.46
(D) –0.55 and 5.45
(E) 0.46 and 6.46

42. If $3x - x^2 \geq 2$ and $y^2 + y \leq 2$, then

(A) $-1 \leq xy \leq 2$
(B) $-2 \leq xy \leq 2$
(C) $-4 \leq xy \leq 4$
(D) $-4 \leq xy \leq 2$
(E) 1, 2, and 4 only

43. In $\triangle ABC$, if $\sin A = \dfrac{1}{3}$ and $\sin B = \dfrac{1}{4}$, $\sin C =$

(A) 0.14
(B) 0.58
(C) 0.56
(D) 3.15
(E) 2.51

GO ON TO THE NEXT PAGE

44. The solution set of $\dfrac{|x-1|}{x} > 2$ is

(A) $0 < x < \dfrac{1}{3}$

(B) $x < \dfrac{1}{3}$

(C) $x > \dfrac{1}{3}$

(D) $\dfrac{1}{3} < x < 1$

(E) $x > 0$

45. If (x,y) represents a point on the graph of $y = 2x + 1$, which of the following could be a portion of the graph of the set of points (x^2,y)?

(A)
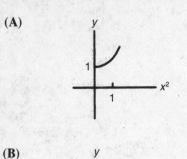

(B)
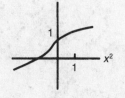

(C)
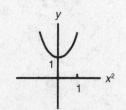

(D)
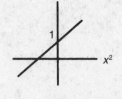

(E)

46. In the figure, the bases, *ABC* and *DEF*, of the right prism are equilateral triangles of side *s*. The altitude of the prism *BE* is *h*. If a plane cuts the figure through points *A*, *C*, and *E*, two solids, *EABC*, and *EACFD*, are formed. What is the ratio of the volume of *EABC* to the volume of *EACFD*?

(A) $\dfrac{1}{2}$

(B) $\dfrac{1}{3}$

(C) $\dfrac{\sqrt{3}}{3}$

(D) $\dfrac{\sqrt{3}}{4}$

(E) $\dfrac{1}{4}$

47. A new machine can produce *x* widgets in *y* minutes, while an older one produces *u* widgets in *w* hours. If the two machines work together, how many widgets can they produce in *t* hours?

(A) $t\left(\dfrac{x}{60y} + \dfrac{u}{w}\right)$

(B) $t\left(\dfrac{60x}{y} + \dfrac{u}{w}\right)$

(C) $60t\left(\dfrac{x}{y} + \dfrac{u}{w}\right)$

(D) $t\left(\dfrac{y}{60x} + \dfrac{w}{u}\right)$

(E) $t\left(\dfrac{x}{y} + \dfrac{60u}{w}\right)$

48. The length of the major axis of the ellipse $3x^2 + 2y^2 - 6x + 8y - 1 = 0$ is

(A) $\sqrt{3}$

(B) $2\sqrt{3}$

(C) $\sqrt{6}$

(D) $2\sqrt{6}$

(E) 4

GO ON TO THE NEXT PAGE

USE THIS SPACE FOR SCRATCH WORK

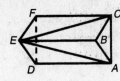

49. If a coordinate system is devised so that the positive y-axis makes an angle of 60° with the positive x-axis, what is the distance between the points with coordinates $(4,-3)$ and $(5,1)$?

(A) 4.12
(B) 4.58
(C) 3.87
(D) 3.61
(E) 7.14

50. If $x - 7$ divides $x^3 - 3k^3x^2 - 13x - 7$, then $k =$

(A) 1.34
(B) 1.20
(C) 5.04
(D) 4.63
(E) 1.72

USE THIS SPACE FOR SCRATCH WORK

ANSWER KEY

1. C	6. C	11. C	16. D	21. C	26. B	31. B	36. C	41. B	46. A
2. B	7. C	12. A	17. D	22. C	27. B	32. D	37. E	42. D	47. B
3. E	8. B	13. D	18. C	23. C	28. B	33. A	38. B	43. C	48. D
4. D	9. D	14. D	19. E	24. B	29. C	34. C	39. C	44. A	49. B
5. C	10. E	15. C	20. B	25. C	30. B	35. E	40. A	45. E	50. B

ANSWER EXPLANATIONS

In these solutions the following notation is used:

a: active—Calculator use is necessary or, at a minimum, extremely helpful.

g: Graphing calculator is preferred.

i: inactive—Calculator use is not helpful and may even be a hindrance.

1. g   C   Plot the graph of $y = -\sin x \cos x$ in the Ztrig window. Inspection of the graph indicates that the maximum is 0.5 and the minimum is -0.5. Therefore, the amplitude is 0.5.

An alternative solution uses the identity $\sin 2x = 2 \sin x \cos x$. Therefore, $-\sin x \cos x = \frac{1}{2} \sin 3x$, which has an amplitude of $\frac{1}{2}$. [3.4].

2. g   B   Let $f(x) = 3x^4 - 2x^3 - 20x^2 - 12$ and recall that $f(-2)$ is equal to the remainder upon division of $f(x)$ by $x + 2$. Enter $f(x)$ into Y_1, return to the Home Screen, and enter $Y_1(-2)$ to get the correct answer choice.

An alternative solution is to use synthetic division to find the remainder. [2.4].

$$
\begin{array}{r|rrrrr}
-2 & 3 & -2 & -20 & 0 & -12 \\
   &   & -6 & 16 & 8 & -16 \\
\hline
   & 3 & -8 & -4 & 8 & -28 = \text{remainder}
\end{array}
$$

3. i   E   Solve the equation by adding $\frac{2}{x} - 1$ to both sides and getting $\frac{2}{x} - \frac{1}{x} = 2 - 1$ or $\frac{1}{x} = 1$, so $x = 1$. Therefore, $3 - \frac{3}{x} = 3 - 3 = 0$. [5.9].

4. i   D   The slope of the first line is $\frac{3}{4}$, and of the second line is $-\frac{4}{3}$. Slopes are negative reciprocals, and so lines are perpendicular. [2.2].

5. i   C   Since the domain of $\log_{10}$ is positive numbers, then the domain of f consists of values of x for which $\sin x$ is positive. This is only true for $0 < x < \pi$. [4.2, 3.1].

6. i   C $\dfrac{\text{Surface area of sphere}}{\text{Volume of sphere}} = \dfrac{4\pi r^2}{\frac{4}{3}\pi r^2} = \dfrac{4 \cdot 3}{\frac{4}{3} r \cdot 3}$

$$= \frac{12}{4r} = \frac{3}{r}. \quad [5.5].$$

7. g   C   Plot the graph of $y = x^2 - 9x$ in the standard window and observe that you must extend the window in the negative y direction to capture the vertex of the parabola. Since this vertex must lie above the x-axis for the solutions to be complex conjugates, c must be bigger than |minimum| $= 20.25 = \dfrac{81}{4}$. (The minimum is found using CALC/minimum.)

An alternative solution is to use the fact that for the solutions to be complex conjugates, the discriminant $b^2 - 4ac = 81 - 4c < 0$, or $c > \dfrac{81}{4}$.

8. g   B   Since $\tan x = 3$, x could be in the first or third quadrants. Since, however, $\sqrt{\csc x}$ is only defined when $\csc x \geq 0$, we need only consider x in the first quadrant. Thus, we can enter

$$\sqrt{\left(\frac{1}{(\sin(\tan^{-1} 3))}\right)}$$ to get the correct answer choice B.

9. i   D   The two transformations put the graph right back where it started. [5.5].

10. g   E   Enter $(3x^3 - 7x^{2+2})/(4x^2 - 3x - 1)$ into Y_1. Enter TBLSET and set TblStart $= 110$ and ΔTbl $= 10$. Then enter TABLE and scroll down to larger and larger x values until you are convinced that Y_1 grows without bound.

An alternative solution is to divide the numerator and denominator by x^3 and then let $x \to \infty$. The numerator approaches 3 while the denominator approaches 0, so the whole fraction grows without bound. [4.5].

11. g   C   Plot the graph of $y = \cos x$ in an $x \varepsilon \left[-\dfrac{\pi}{4}, \dfrac{3\pi}{4}\right]$ and $y \varepsilon [-1, 1]$ window and inspect the graph to arrive at the correct answer choice C.

An alternative solution is to sketch the unit circle and observe that the x-coordinate of a point increases as it moves from $-\dfrac{\pi}{4}$ to 0 and then decreases as it moves from 0 to $\dfrac{3\pi}{4}$. [3.4].

12. g   A   Plot the graph of $y = \left| -\sqrt{2} \sin \sqrt{3x} \right|$ using Ztrig. The minimum value of the function is clearly zero, and you can use CALC/maximum to establish 1.414 as the maximum value.

This function inside the absolute value is sinusoidal with amplitude $\sqrt{2} \approx 1.414$. The absolute value eliminates the bottom portion of the sinusoid, so this is the vertical distance between the maximum and the minimum as well. [3.4].

13. g   D   Plot the graph of $y = -|x| + 2$ on an $x \varepsilon [-1,3]$ and $y \varepsilon [-3,3]$ window. Examine the graph to see that its minimum value is achieved when $x = 3$.

An alternative solution is to realize that y is smallest when x is largest because of the negative absolute value. [4.3].

14. i   D   By definition, the transformation f takes x into $x + y$. If x is taken into x, then y must be zero. Conversely, f takes any point $(x,0)$ into itself. [1.5].

15. g   C   Plot the graph of $y = 4x^3 - x^2 + 16x - 4$ in the standard window and zoom in once to get a clearer picture of the location of the zero. Use CALC/zero to determine that the zero is at $x = 0.25$.

An alternative solution is to use the Rational Roots Theorem to determine that the only possible rational roots are $\pm 1, \pm 2, \pm 4, \pm \dfrac{1}{2}, \dfrac{1}{4}$. Synthetic division with these values in turn eventually will yield the correct answer choice.

Another alternative solution is to observe that the left side of the equation can be factored: $4x^3 - x^2 + 16x - 4 = x^2(4x - 1) + 4(4x - 1) = (4x - 1)(x^2 + 1) = 0$. Since $x^2 + 1$ can never equal zero, the only solution is $x = \dfrac{1}{4}$. [2.4].

16. a   D $|\vec{V}| = \sqrt{3^2 + \left(-\sqrt{2}\right)^2} = \sqrt{9 + 2} = \sqrt{11} \approx 3.32$. [5.5].

17. a   D   "More than 3" implies 4 or 5, and so the number of elements is $\dbinom{5}{4} + \dbinom{5}{5} = 6$. [5.1].

$$\dbinom{5}{4} = \frac{5!}{4!1!} = \frac{120}{24} = 5$$

$$\dbinom{5}{5} = \frac{5!}{5!0!} = \frac{120}{120} = 1$$

Therefore, $\dbinom{5}{4} + \dbinom{5}{5} = 5 + 1 = 6$.

18. i   C   By the definition of negation, the negation of the given statements is Choice C. [5.7].

19. g   E   In parametric mode, plot the graph of $x = \sec t$ and $y = \cos t$ in the standard window to see that it looks like a portion of a hyperbola. You should verify that it is a portion of a hyperbola rather than the whole hyperbola by noting that $y = \cos t$ implies $-1 \le y \le 1$.

An alternative solution is to use the fact that secant and cosine are reciprocals so that elimination of the parameter t yields the equation $xy = 1$. This is the equation of a hyperbola and again, since $y = \cos t$ implies $-1 \le y \le 1$, the correct answer is E. [4.6].

20. i   B   If $(3,2)$ lies on the inverse of f, $(2,3)$ lies on f. Substituting in f gives $2 \cdot 2^3 + 2 + A = 3$. Therefore, $A = -15$. [1.3].

21. i   C   Substitute 1 for x to get $a + b + c = 3$. Substitute -1 for x to get $a - b + c = 3$. Add these two equations to get $a + b = 3$. [2.3].

22. a   C $\angle A = 108°$. Law of sines: $\dfrac{a}{\sin 108°} = \dfrac{100}{\sin 30°}$. Therefore, $a = 200 \sin 108° = 200 \sin 72° \approx 190.2$. [3.7].

23. g   C   Plot the graph of $4 \sin x + 3$ in radian mode in an $x \varepsilon [0, 2\pi]$ and $y \varepsilon [-2,8]$ window. Use CALC/zero twice to find the correct answer choice.

An alternative solution is to solve the equation for $\sin x$ and use your calculator, in radian mode, to evaluate $\sin^{-1}\left(-\dfrac{3}{4}\right)$. This yields the value -0.848, the Arcsine of $-\dfrac{3}{4}$. Since this value is not between 0 and 2π, you must find the value of x in the required interval that has the same terminal side ($2\pi - 0.848 = 5.435$) as well as the third quadrant angle that has the same reference angle ($\pi + .848 = 3.990$). [3.6].

24. g **B** Plot the graph of $y = 2\cos(3x)^2$ using Ztrig, and trace one cycle to find its length of approximately 1.05, approximately $\dfrac{\pi}{3}$.

An alternative solution uses the double angle formula for cosine: $\cos 2u = 2\cos^2 u - 1$, with $u = 3x$, to write $f(x) = \cos 6x + 1$. The primary period of this function is $\dfrac{2\pi}{6} = \dfrac{\pi}{3}$. [3.5].

25. i **C** $\dfrac{1}{2+6i} \cdot \dfrac{2-6i}{2-6i} = \dfrac{2-6i}{4-(-36)} = \dfrac{1}{20} - \dfrac{3}{20}i$.

[4.7].

26. a **B** The central angle $\dfrac{3\pi}{14}$ is not necessary.

$\dfrac{1}{2}R^2\theta = \dfrac{1}{2}r^2\theta \cdot 2$, and so $\dfrac{1}{2} = \dfrac{r^2}{R^2}$. Therefore,

$\dfrac{r}{R} = \dfrac{\sqrt{2}}{2} \approx \dfrac{1.414}{2} \approx 0.71{:}1$. [3.2].

Tips: (1) Volumes of similar figures are proportional to the cube of corresponding linear measures.
(2) Areas of similar figures are proportional to the square of corresponding linear measures.

---

27. a **B** The line cuts the x-axis at 3 and the y-axis at 4 to form a right triangle that, when rotated about the y-axis, forms a cone with radius 3 and altitude 4. Volume $= \dfrac{1}{3}\pi r^2 h = \dfrac{1}{3}\pi(9)(4) = 12\pi \approx$ 37.7. [5.5].

28. a **B** There are $\binom{8}{3} = 56$ ways to select 3 good transistors. There are $\binom{12}{3} = 220$ ways to select any 3 transistors. $P(3 \text{ good ones}) = \dfrac{56}{220} = \dfrac{14}{55} \approx$ 0.255. [5.3].

An alternative solution is to note that since there are 4 broken transistors, there must be 8 good ones. $P(\text{first pick is good}) = \dfrac{8}{12}$. Of the remaining 11 transistors, 7 are good, and so $P(\text{second pick is good}) = \dfrac{7}{11}$. Finally, $P(\text{third pick is good}) = \dfrac{6}{10}$. Therefore, $P(\text{all three are good}) =$

$\dfrac{8}{12} \cdot \dfrac{7}{11} \cdot \dfrac{6}{10} = \dfrac{14}{55} \approx 0.255$.

29. g **C** Plot the graph of $y = \sin 2x$ using Ztrig, and recall that for the inverse of a function to be a function, any horizontal line can only cross the graph of the function in one point. Using TRACE, you can see that this only occurs when $\dfrac{\pi}{4} \le x \le \dfrac{3\pi}{4}$.

An alternative solution is to use your knowledge of the sine graph and the fact that $\sin 2x$ has period $\dfrac{2\pi}{2} = \pi$. The function $\sin 2x$ increases from 0 to 1 on $\left[0, \dfrac{\pi}{4}\right]$, decreases from 1 to 0 on $\left[\dfrac{\pi}{4}, \dfrac{\pi}{2}\right]$, from 0 to -1 on $\left[\dfrac{\pi}{2}, \dfrac{3\pi}{4}\right]$, and finally from -1 back to 0 on $\left[\dfrac{3\pi}{4}, \pi\right]$. Each y value between 1 and -1 is achieved exactly once between $\dfrac{\pi}{4}$ and $\dfrac{3\pi}{4}$. [1.3, 3.4, 3.6].

30. i **B** $\dfrac{RT^3}{Z^2} = K$. Let $R = 1$, $T = 2$, $Z = 3$ (arbitrary choices) to get $K = \dfrac{8}{9}$. The new values are $Z = 9$, $T = 4$, and so $\dfrac{R(64)}{81} = \dfrac{8}{9}$. Therefore, $R = \dfrac{9}{8}$, and the original value of R has been multiplied by $\dfrac{9}{8}$. [4.5].

An alternative solution is to take the original equation, $\dfrac{RT^3}{Z^2} = K$ and make the indicated changes: $\dfrac{R(2T)^3}{(3Z)^2} = \dfrac{8RT^3}{9Z^2}$. In order for this fraction to remain equal to K, the value of R must be multiplied by $\dfrac{9}{8}$.

31. i **B** The sum of the roots, $r + (-r) + p$, equals $-\dfrac{3}{1} = -3$. Therefore, the third root, p, equals -3. Substituting -3 for x in the equation gives $-27 + 27 + 3K - 12 = 0$. $K = -4$. [2.4].

32. i **D** Substitute -1 for x. -1 to an even exponent $= 1$, and to an odd exponent $= -1$. $3(-1) - 4(1) + 5(-1) - 8 = -20$. [2.4].

33. i   A   In half of the arrangements man A is at one end. Therefore, p(man A is in an end seat) $= \dfrac{1}{2}$. [5.4].

Alternative Solution: There are $_4P_4 = 24$ ways the four men can be arranged in a line. If A is at the front of the row, there are $_3P_3 = 6$ ways the remaining 3 men can be arranged. Similarly, if A is at the end of the row, the other 3 men can be arranged in 6 ways. Therefore, A is at one end of the row for 12 of the arrangements. Thus, the probability that A is at one of the ends is $\dfrac{12}{24} = \dfrac{1}{2}$.

34. i   C   The summation indicates that 5 be summed 8 times (when $i = 3, 4, \ldots, 11$), so the sum is $8 \times 5 = 40$. [5.4].

35. g   E   Graph $y = \sqrt[4]{3x^2 + 7}$ and $y = -\sqrt[4]{3x^2 + 7}$ in a standard window. Observe the graph is symmetrical with respect to the x-axis, the y-axis, and the origin.

The equation is unchanged if x is replaced by $-x$; if y is replaced by $-y$, or if both replacements take place, so all three symmetries are present. [1.4].

36. i   C   From the Venn diagram below you get the following equations:

$$a + b + c + d = 30 \qquad (1)$$
$$b + c \quad\;\; = 20 \qquad (2)$$
$$c + d = 15 \qquad (3)$$
$$a \quad\quad\;\; = 5$$

Subtract equation (2) from equation (1): $a + d = 10$. Since $a = 5$, $d = 5$. Substituting 5 for d in equation (3) leaves $c = 10$. [5.9].

All students

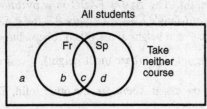

37. g   E   Enter $\dfrac{x+1}{x^2+1}$ into Y_1 and $\dfrac{x^2+1}{x+1}$ into Y_2. Return to the Home Screen and enter $\dfrac{Y_2\left(\sqrt{5}\right) - Y_1\left(\sqrt{2}\right)}{\sqrt{5} - \sqrt{2}}$ to get the correct answer choice E. [1.2, 2.2].

38. a   B   The plane cuts the x-axis at 17.5, the y-axis at 11.7, and the z-axis at 7. The base is a right tri-

angle with area $\approx \dfrac{1}{2}(17.5)(11.7) \approx 102.1$. $V = \dfrac{1}{3}Bh$ $\approx \dfrac{1}{3}(102.1)(7) \approx 238.2$. [5.5].

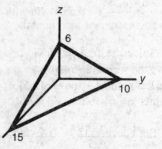

39. g   C   Graph $y = \sin 15x + \cos 15x$ in an $x \varepsilon\left[0, \dfrac{\pi}{5}\right]$ and $y \varepsilon[-2,2]$ window. Use CALC/zero to find the far left zero, return to the Home Screen, and store it in A. Then use CALC/zero to find the middle zero, return to the Home Screen, and store it in B. Finally, use CALC/zero to find the far right zero, return to the Home Screen, and enter A + B + X to get 1.099. ... Since π is just bigger than 3, answer choice C is the only close candidate. Evaluate the decimal approximation to $\dfrac{7\pi}{20}$ to verify.

An alternative solution is to rewrite the equation as $\tan 15x = -1$. The three smallest positive solutions for $15x$ are $\dfrac{3\pi}{4}, \dfrac{7\pi}{4}, \dfrac{11\pi}{4}$, so the three smallest positive solutions for x are $\dfrac{3\pi}{60}, \dfrac{7\pi}{60}, \dfrac{11\pi}{60}$, and their sum is $\dfrac{7x}{20}$. [3.5].

40. i   A   The number of 2s must exceed the number of other values. Some of the choices can be eliminated. B: one integer could be 2. C: x or y could be 2, but not necessarily. D and E: if $x = 3$ and $y = 1$, there will be no mode. Therefore, Choice A is the answer. [5.9].

41. g   B   Enter $\sqrt{2x+3}$ into Y_1, x^2 into Y_2, $Y_1(Y_2(X))$ into Y_3, and $Y_2(Y_1(X))$ into Y_4. De-select Y_1 and Y_2, and graph Y_3 and Y_3 in an $x\varepsilon[-1.5,2]$ and $y\varepsilon[0,5]$ window (because $2x+3 \geq 0$). Use CALC/intersect to find the correct answer choice B.

An alternative solution is to evaluate $f(g(x)) = \sqrt{2x^3 + 3}$ and $g(f(x)) = 2x + 3$, set the two equal, square both sides, and solve the resulting quadratic: $2x^2 + 3 = (2x + 3)^2 = 4x^2 + 12x + 9$ or $x^2 + 6x + 3 = 0$. The Quadratic Formula yields

$x = \dfrac{-6+\sqrt{24}}{2}$ or $x = \dfrac{-6-\sqrt{24}}{2}$. However, the second is not in the domain of g, so

$x = \dfrac{-6+\sqrt{24}}{2} \approx -0.55$ is the only solution. [1.2].

42. i **D** Solve the first inequality to get $1 \leq x \leq 2$. Solve the second inequality to get $-2 \leq y \leq 1$. The smallest product xy possible is -4, and the largest product xy possible is $+2$. [2.5].

43. a **C** Sin $C = \sin(180° - (A + B)) = \sin(A + B) =$

$\sin A \cdot \cos B + \cos A \cdot \sin B = \dfrac{1}{3} \cdot \dfrac{\sqrt{15}}{4} + \dfrac{\sqrt{8}}{3} \cdot \dfrac{1}{4}$

$= \dfrac{\sqrt{15}+2\sqrt{2}}{12} \approx$

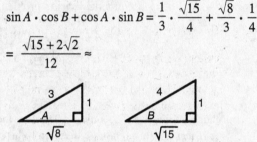

An alternative solution is to draw $\triangle ABC$ and drop a perpendicular, CM, from C to side AB. Since, in a right triangle, $\sin A = \dfrac{\text{opposite side}}{\text{hypotenuse}} = \dfrac{1}{3}$ and

$\sin B = \dfrac{1}{4}$, the lengths of segments AC, CM, and

CB are indicated. Using the Pythagorean theorem gives $AM = \sqrt{8}$ and $MB = \sqrt{15}$, and so $AB = \sqrt{8} + \sqrt{15}$. By the law of sines in $\triangle ABC$,

$\dfrac{\sin C}{\sqrt{8}+\sqrt{15}} = \dfrac{\sin A}{4} = \dfrac{\frac{1}{3}}{4}$. Therefore,

$\sin C = \dfrac{1}{12}\left(\sqrt{8}+\sqrt{15}\right) \approx \dfrac{2\sqrt{2}+\sqrt{15}}{12} \approx 0.56.$

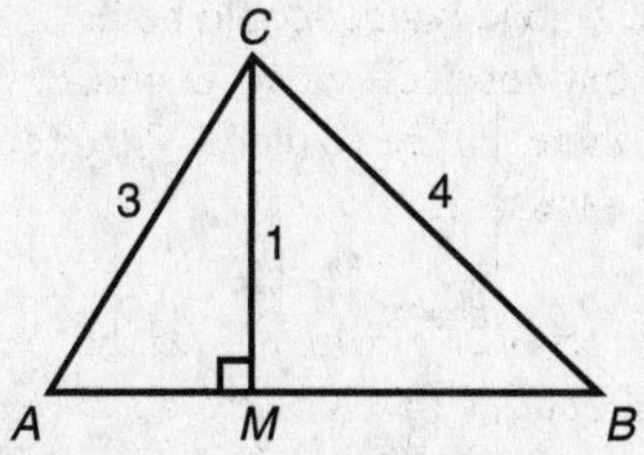

(triangle diagram with vertex C at top, base A–M–B, segments labeled 3, 1, 4)

TIP: Always look at the answer choices. You may find a clue as you work through your solution. In this case, when you find that $AB = \sqrt{8} + \sqrt{15}$, which equals $2\sqrt{2} + \sqrt{15}$, you might conclude that Choice C is the correct answer, thus saving much work.

44. g **A** Graph $y = \dfrac{|x-1|}{x}$ and $y = 2$ in the standard window. There is a vertical asymptote at $x = 0$, and the curve is above the horizontal $y = 2$ just to the right of 0. Answer choice A is the only possibility, but to be sure, use CALC/intersect to determine that the point of intersection is at $x = \dfrac{1}{3}$.

An alternative solution is to use the associated equation $\dfrac{|x-1|}{x} = 2$ to find boundary values and then test points. Since the left side is undefined when $x = 0$, zero is one boundary value. Multiply both sides by x to get $|x - 1| = 2x$ and analyze the two cases for absolute value. If $x - 1 \geq 0$, then $x - 1 = 2x$ so $x = -1$, which is impossible because $x \geq 1$. Therefore, $x - 1 < 0$, so $1 - x = 2x$ or $x = \dfrac{1}{3}$ is the other boundary value. Testing points in the intervals $(-\infty, 0)$, $\left(0, \dfrac{1}{3}\right)$, and $\left(\dfrac{1}{3}, \infty\right)$ yields the correct answer choice A. [4.3, 2.5].

45. i **E** From the table of values below, point (x, y) can be seen to lie on the graph of $y = 2x + 1$, and the set of points (x^2, y) lies on the graph we are looking for. Plotting a few points gives a rough sketch that indicates the answer is Choice E. [5.5].

x	-2	-1	0	1	2	3
x^2	4	1	0	1	4	9
y	-3	-1	1	3	5	7

46. i **A** The volume of the prism is area of base times height. The figure $EABC$ is a pyramid with base triangle ABC and height BE, the same as the base and height of the prism. The volume of the pyramid is $\dfrac{1}{3}$ (base times height), $\dfrac{1}{3}$ the volume of the prism. Therefore, the other solid, $EACFD$, is $\dfrac{2}{3}$ the volume of the prism. The ratio of the volumes is $\dfrac{1}{2}$. [5.5].

47. i **B** Since the new machine produces x widgets in y minutes, it can produce $\dfrac{60x}{y}$ widgets per hour.

The old machine produces $\dfrac{u}{w}$ widgets per hour.

Adding these and multiplying by t yields the correct answer choice B.

48. i D Get the center axis form of the equation by completing the square: $3x^2 - 6x + 2y^2 + 8y = 1$

$3(x^2 - 2x + 1) + 2(y^2 + 4y + 4) = 1 + 3 + 8 = 12$,

which leads to $\dfrac{3(x-1)^2}{12} + \dfrac{2(y+2)^2}{12} = 1$ and

finally to $\dfrac{(x-1)^2}{4} + \dfrac{(y+2)^2}{6} = 1$. Thus, half the

major axis is $\sqrt{6}$, making the major axis $2\sqrt{6}$. [4.1]

49. a B From the figure it can be seen that the lines drawn parallel to the axes and the line through the two points form a triangle with one angle of $120°$ and adjacent sides of 4 and 1. From the law of cosines,

$$d^2 = 1 + 16 - 2 \cdot 1 \cdot 4 \cos 120°$$
$$= 17 + 4$$

Thus, the distance between the two points $d = \sqrt{21} \approx 4.58$. [5.9].

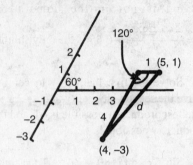

50. a B Set up as a synthetic division problem:

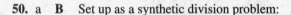

$7\rfloor$	1	$-3k^3$	-13	-7
		7	$49-21k^3$	7
	1	$7-3k^3$	$36-21k^3$	**0**

The remainder is 0 because $x - 7$ divides the polynomial. For this synthetic division to be correct,

$36 - 21k^3 = 1$, or $k^3 = \dfrac{36}{21}$, so $k \approx 1.20$. [2.4].

SELF-EVALUATION CHART FOR MODEL TEST 5

SUBJECT AREA	QUESTIONS	NUMBER OF RIGHT WRONG OMITTED

Mark correct answers with C, wrong answers with X, and omitted answers with O.

Algebra
(9 questions)

3	4	7	25	30	31	42	44	47

Review section

4.5	2.2	2.3	4.7	4.5	2.4	2.5	4.3	5.9

___ ___ ___

Solid geometry
(4 questions)

6	27	38	46

Review section

5.5	5.5	5.5	5.5

___ ___ ___

Coordinate geometry
(6 questions)

9	19	35	45	48	49

Review section

5.5	4.6	1.4	5.5	4.1	5.9

___ ___ ___

Trigonometry
(10 questions)

1	8	11	12	22	23	24	26	39	43

Review section

3.4	3.1	3.4	3.4	3.7	3.6	3.5	3.2	3.5	3.8

___ ___ ___

Functions
(12 questions)

2	5	13	14	15	20	21	29	32	37	41	50

Review section

2.4	1.1	4.3	5.5	2.4	1.3	2.3	1.3	2.4	1.2	1.2	2.4

___ ___ ___

Miscellaneous
(9 questions)

10	16	17	18	28	33	34	36	40

Review section

4.5	5.5	5.1	5.7	5.3	5.4	5.4	5.9	5.9

___ ___ ___

TOTALS   ___ ___ ___

Raw score = (number right) − ¼ (number wrong) = _____

Round your raw score to the nearest whole number = _____

Evaluate Your Performance
Model Test 5

Rating	Number Right
Excellent	41–50
Very good	33–40
Above average	25–32
Average	15–24
Below average	Below 15

ANSWER SHEET FOR MODEL TEST 6

Determine the correct answer for each question. Then, using a no. 2 pencil, blacken completely the oval containing the letter of your choice.

1. Ⓐ Ⓑ Ⓒ Ⓓ Ⓔ	18. Ⓐ Ⓑ Ⓒ Ⓓ Ⓔ	35. Ⓐ Ⓑ Ⓒ Ⓓ Ⓔ
2. Ⓐ Ⓑ Ⓒ Ⓓ Ⓔ	19. Ⓐ Ⓑ Ⓒ Ⓓ Ⓔ	36. Ⓐ Ⓑ Ⓒ Ⓓ Ⓔ
3. Ⓐ Ⓑ Ⓒ Ⓓ Ⓔ	20. Ⓐ Ⓑ Ⓒ Ⓓ Ⓔ	37. Ⓐ Ⓑ Ⓒ Ⓓ Ⓔ
4. Ⓐ Ⓑ Ⓒ Ⓓ Ⓔ	21. Ⓐ Ⓑ Ⓒ Ⓓ Ⓔ	38. Ⓐ Ⓑ Ⓒ Ⓓ Ⓔ
5. Ⓐ Ⓑ Ⓒ Ⓓ Ⓔ	22. Ⓐ Ⓑ Ⓒ Ⓓ Ⓔ	39. Ⓐ Ⓑ Ⓒ Ⓓ Ⓔ
6. Ⓐ Ⓑ Ⓒ Ⓓ Ⓔ	23. Ⓐ Ⓑ Ⓒ Ⓓ Ⓔ	40. Ⓐ Ⓑ Ⓒ Ⓓ Ⓔ
7. Ⓐ Ⓑ Ⓒ Ⓓ Ⓔ	24. Ⓐ Ⓑ Ⓒ Ⓓ Ⓔ	41. Ⓐ Ⓑ Ⓒ Ⓓ Ⓔ
8. Ⓐ Ⓑ Ⓒ Ⓓ Ⓔ	25. Ⓐ Ⓑ Ⓒ Ⓓ Ⓔ	42. Ⓐ Ⓑ Ⓒ Ⓓ Ⓔ
9. Ⓐ Ⓑ Ⓒ Ⓓ Ⓔ	26. Ⓐ Ⓑ Ⓒ Ⓓ Ⓔ	43. Ⓐ Ⓑ Ⓒ Ⓓ Ⓔ
10. Ⓐ Ⓑ Ⓒ Ⓓ Ⓔ	27. Ⓐ Ⓑ Ⓒ Ⓓ Ⓔ	44. Ⓐ Ⓑ Ⓒ Ⓓ Ⓔ
11. Ⓐ Ⓑ Ⓒ Ⓓ Ⓔ	28. Ⓐ Ⓑ Ⓒ Ⓓ Ⓔ	45. Ⓐ Ⓑ Ⓒ Ⓓ Ⓔ
12. Ⓐ Ⓑ Ⓒ Ⓓ Ⓔ	29. Ⓐ Ⓑ Ⓒ Ⓓ Ⓔ	46. Ⓐ Ⓑ Ⓒ Ⓓ Ⓔ
13. Ⓐ Ⓑ Ⓒ Ⓓ Ⓔ	30. Ⓐ Ⓑ Ⓒ Ⓓ Ⓔ	47. Ⓐ Ⓑ Ⓒ Ⓓ Ⓔ
14. Ⓐ Ⓑ Ⓒ Ⓓ Ⓔ	31. Ⓐ Ⓑ Ⓒ Ⓓ Ⓔ	48. Ⓐ Ⓑ Ⓒ Ⓓ Ⓔ
15. Ⓐ Ⓑ Ⓒ Ⓓ Ⓔ	32. Ⓐ Ⓑ Ⓒ Ⓓ Ⓔ	49. Ⓐ Ⓑ Ⓒ Ⓓ Ⓔ
16. Ⓐ Ⓑ Ⓒ Ⓓ Ⓔ	33. Ⓐ Ⓑ Ⓒ Ⓓ Ⓔ	50. Ⓐ Ⓑ Ⓒ Ⓓ Ⓔ
17. Ⓐ Ⓑ Ⓒ Ⓓ Ⓔ	34. Ⓐ Ⓑ Ⓒ Ⓓ Ⓔ	

MODEL TEST

Tear out the preceding answer sheet. Decide which is the best choice by rounding your answer when appropriate. Blacken the corresponding space on the answer sheet. When finished, check your answers with those at the end of the test. For questions that you got wrong, note the sections containing the material that you must review. Also, if you do not fully understand how you arrived at some of the correct answers, you should review the appropriate sections. Finally, fill out the self-evaluation sheet on page 259 in order to pinpoint the topics that give you the most difficulty.

TEST DIRECTIONS

Directions: Decide which answer choice is best. If the exact numerical value is not one of the answer choices, select the closest approximation. Fill in the oval on the answer sheet that corresponds to your choice.

Notes:
(1) You will need to use a scientific or graphing calculator to answer some of the questions.
(2) You will have to decide whether to put your calculator in degree or radian mode for some problems.
(3) All figures that accompany problems are plane figures unless otherwise stated. Figures are drawn as accurately as possible to provide useful information for solving the problem, except when it is stated in a particular problem that the figure is not drawn to scale.
(4) Unless otherwise indicated, the domain of a function is the set of all real numbers for which the functional value is also a real number.

Reference Information. The following formulas are provided for your information.

Volume of a right circular cone with radius r and height h: $V = \frac{1}{3}\pi r^2 h$

Lateral area of a right circular cone if the base has circumference c and the slant height is l: $S = \frac{1}{2}cl$

Volume of a sphere of radius r: $V = \frac{4}{3}\pi r^3$

Surface area of a sphere of radius r: $S = 4\pi r^2$

Volume of a pyramid of base area B and height h: $V = \frac{1}{3}Bh$

---

1. If point (a,b) lies on the graph of function f, which of the following points must lie on the graph of the inverse f?

   (A) (a,b)
   (B) $(-a,b)$
   (C) $(a,-b)$
   (D) (b,a)
   (E) $(-b,-a)$

2. Harry had grades of 70, 80, 85, and 80 on his quizzes. If all quizzes have the same weight, what grade must he get on his next quiz so that his average will be 80?

   (A) 85
   (B) 90
   (C) 95
   (D) 100
   (E) more than 100

3. Which of the following is an asymptote of $f(x) = \dfrac{x^2 + 3x + 2}{x + 2} \cdot \tan \pi x$?

   (A) $x = 2$
   (B) $x = 1$
   (C) $x = -2$
   (D) $x = -1$
   (E) $x = \dfrac{1}{2}$

4. A trace of the plane $5x - 2y + 3z = 10$ is

   (A) $5x + 2y = 10$
   (B) $3z = 2y$
   (C) $2y + 3z = 10$
   (D) $5x + 3z = 10$
   (E) $2y = 5x + 10$

5. The sum of the roots of $3x^3 + 4x^2 - 4x = 0$ is

   (A) $\dfrac{4}{3}$
   (B) 0
   (C) $-\dfrac{4}{3}$
   (D) 4
   (E) $-\dfrac{3}{4}$

6. If $f(x) = x - \dfrac{1}{x}$, then $f(a) + f\left(\dfrac{1}{a}\right) =$

   (A) 0

   (B) $2a - \dfrac{2}{a}$

   (C) $a - \dfrac{1}{a}$

   (D) $\dfrac{a^4 - a^2 + 1}{a(a^2 - 1)}$

   (E) 1

7. If $f(x) = x^4 - 4x^3 + 6x^2 - 4x + 2$, then $f(2) - f\left(\sqrt{2}\right) =$

   (A) 0.97
   (B) 1.42
   (C) 0.86
   (D) 1.73
   (E) 1.03

8. If $f(x) \geq 0$ for all x, then $f(2 - x)$ is

   (A) ≥ 0
   (B) ≥ 2
   (C) ≥ -2
   (D) ≤ 2
   (E) ≤ 0

9. How many four-digit numbers can be formed from the numbers 0, 2, 4, 8 if no digit is repeated?

   (A) 24
   (B) 18
   (C) 64
   (D) 36
   (E) 27

10. If $x - 1$ is a factor of $x^2 + ax - 4$, then a has the value

    (A) 4
    (B) 3
    (C) 2
    (D) 1
    (E) none of the above

11. If 10 coins are to be flipped and the first 5 all come up heads, what is the probability that exactly 3 more heads will be flipped?

    (A) 0.3125
    (B) 0.0439
    (C) 0.6000
    (D) 0.1172
    (E) 0.1250

GO ON TO THE NEXT PAGE

12. If $i = \sqrt{-1}$ and n is a positive integer, which of the following statements is FALSE?

    (A) $i^{4n} = 1$
    (B) $i^{4n+1} = -i$
    (C) $i^{4n+2} = -1$
    (D) $i^{n+4} = i^n$
    (E) $i^{4n+3} = -i$

13. If $\log_r 3 = 7.1$, then $\log_r \sqrt{3} =$

    (A) $\sqrt[5]{7.1}$
    (B) 2.66
    (C) 3.55
    (D) $\dfrac{\sqrt{3}}{r}$
    (E) $\dfrac{7.1}{r}$

14. If $f(x) = 4x^2$ and $g(x) = f(\sin x) + f(\cos x)$, then $g(23°)$ is

    (A) 1
    (B) 4
    (C) 4.29
    (D) 5.37
    (E) 8

15. What is the sum of the roots of the equation $\left(x - \sqrt{2}\right)\left(x^2 - \sqrt{3}x + \pi\right) = 0$?

    (A) 1.414
    (B) 3.15
    (C) −0.318
    (D) −0.315
    (E) 4.56

16. Which of the following equations has (have) graphs consisting of two perpendicular lines?

    I. $xy = 0$
    II. $|y| = |x|$
    III. $|xy| = 1$

    (A) only I
    (B) only II
    (C) only III
    (D) only I and II
    (E) I, II, and III

17. A line, m, is parallel to a plane, X, and is 6 inches from X. The set of points that are 6 inches from m and 1 inch from X form

    (A) a line parallel to m
    (B) two lines parallel to m
    (C) four lines parallel to m
    (D) one point
    (E) the empty set

GO ON TO THE NEXT PAGE

18. In the figure below, if $VO = VY$, what is the slope of segment VO?

(A) $\sqrt{3}$

(B) $-\sqrt{3}$

(C) $\dfrac{\sqrt{3}}{2}$

(D) $-\dfrac{\sqrt{3}}{2}$

(E) Cannot be determined from the given information.

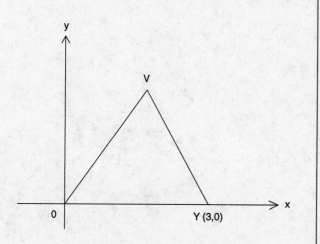

19. A cylindrical bar of metal has a base radius of 2 and a height of 9. It is melted down and reformed into a cube. A side of the cube is

(A) 2.32
(B) 4.84
(C) 97.21
(D) 3.84
(E) 113.10

20. The graph of $y = (x + 2)(2x - 3)$ can be expressed as a set of parametric equations. If $x = 2t - 2$ and $y = f(t)$, the $f(t) =$

(A) $2t(4t - 5)$
(B) $(2t - 2)(4t - 7)$
(C) $2t(4t - 7)$
(D) $(2t - 2)(4t - 5)$
(E) $2t(4t + 1)$

21. If points $\left(\sqrt{2}, y_1\right)$ and $\left(-\sqrt{2}, y_2\right)$ lie on the graph of $y = x^3 + ax^2 + bx + c$, and $y_1 - y_2 = 3$, then $b =$

(A) 1.473
(B) −0.939
(C) −2.167
(D) −0.354
(E) 1.061

GO ON TO THE NEXT PAGE

22. Which one of the following is NOT a fifth root of 1?

    (A) $1(\cos 0 + i \cdot \sin 0)$
    (B) $1(\cos 72° + i \cdot \sin 72°)$
    (C) $1(\cos 154° + i \cdot \sin 154°)$
    (D) $1(\cos 216° + i \cdot \sin 216°)$
    (E) $1(\cos 288° + i \cdot \sin 288°)$

23. If a and b are real numbers, with $a > b$ and $|a| < |b|$, then

    (A) $a > 0$
    (B) $a < 0$
    (C) $b > 0$
    (D) $b < 0$
    (E) none of the above

24. If $[x]$ is defined to represent the greatest integer less than or equal to x, and $f(x) = \left| x - [x] - \frac{1}{2} \right|$, the maximum value of $f(x)$ is

    (A) 1
    (B) 2
    (C) 0
    (D) $\frac{1}{2}$
    (E) -1

25. $\lim\limits_{x \to 2} \dfrac{x^3 - 8}{x^2 - 4} =$

    (A) 0
    (B) 2
    (C) 3
    (D) 1
    (E) ∞

26. A right circular cone whose base radius is 4 is inscribed in a sphere of radius 5. What is the ratio of the volume of the cone to the volume of the sphere?

    (A) $0.333 : 1$
    (B) $0.864 : 1$
    (C) $0.222 : 1$
    (D) $0.288 : 1$
    (E) $0.256 : 1$

27. If $x_0 = 1$ and $x_{n+1} = \sqrt[3]{2x_n}$, then $x_3 =$

    (A) 1.260
    (B) 1.412
    (C) 1.361
    (D) 1.408
    (E) 1.396

28. The y-intercept of $y = \left| \sqrt{2} \csc 3\left(x + \dfrac{\pi}{5}\right) \right|$ is

(A) 0.22
(B) 1.49
(C) 4.58
(D) 0.67
(E) 1.41

29. If the center of the circle $x^2 + y^2 + ax + by + 2 = 0$ is point $(4, -8)$, then $a + b =$

(A) −4
(B) 4
(C) 8
(D) −8
(E) 24

30. If $p(x) = 3x^2 + 9x + 7$ and $p(a) = 2$, then $a =$

(A) only 0.736
(B) only −2.264
(C) 0.736 or 2.264
(D) 0.736 or −2.264
(E) −0.736 or −2.264

31. If i is a root of $x^4 + 2x^3 - 3x^2 + 2x - 4 = 0$, the product of the real roots is

(A) 0
(B) −2
(C) 2
(D) 4
(E) −4

32. If $\sin A = \dfrac{3}{5}, 90° \le A \le 180°, \cos B = \dfrac{1}{3},$ and $270° \le B \le 360°, \sin(A + B) =$

(A) −0.333
(B) 0.733
(C) −0.832
(D) 0.954
(E) −0.554

33. If x varies directly as t and t varies inversely as the square of y, what is the relationship between x and y?

(A) x varies as y^2
(B) x varies inversely as y^2
(C) y varies as x^2
(D) y varies inversely as x^2
(E) No variation between x and y can be determined.

GO ON TO THE NEXT PAGE

34. If sec 1.4 = x, find the value of csc(2 Arctan x).

   (A) 0.33
   (B) 3.03
   (C) 1.00
   (D) 1.06
   (E) 0.87

35. The graph of $|y - 1| = |x + 1|$ forms an X. The two branches of the X intersect at a point whose coordinates are

   (A) (1,1)
   (B) (−1,1)
   (C) (1,−1)
   (D) (−1,−1)
   (E) (0,0)

36. For what value of x between 0° and 360° does $\cos 2x = 2 \cos x$?

   (A) 68.5° or 291.5°
   (B) only 68.5°
   (C) 103.9° or 256.1°
   (D) 90° or 270°
   (E) 111.5° or 248.5°

37. For what value(s) of x will the graph of the function $f(x) = \sin \sqrt{B - x^2}$ have a maximum?

   (A) $\dfrac{\pi}{2}$

   (B) $\sqrt{B - \dfrac{\pi}{2}}$

   (C) $\sqrt{B - \left(\dfrac{\pi}{2}\right)^2}$

   (D) $\pm\sqrt{B - \dfrac{\pi}{2}}$

   (E) $\pm\sqrt{B - \left(\dfrac{\pi}{2}\right)^2}$

38. For each positive integer n, let S_n = the sum of all positive integers less than or equal to n. Then S_{51} equals

   (A) 50
   (B) 1326
   (C) 1275
   (D) 1250
   (E) 51

39. If the graphs of $3x^2 + 4y^2 - 6x + 8y - 5 = 0$ and $(x-2)^2 = 4(y+2)$ are drawn on the same coordinate system, at how many points do they intersect?

(A) 0
(B) 1
(C) 2
(D) 3
(E) 4

40. $\text{Log}_x 2 = \log_3 x$ is satisfied by two values of x. Their sum equals

(A) 0
(B) 1.73
(C) 2.35
(D) 2.81
(E) 3.14

41. In the expansion of $(2b^2 - 3b^{-3})^n$, if the fifth term does not contain a factor of b, what is the value of n?

(A) 6
(B) 10
(C) $\dfrac{15}{2}$
(D) $\dfrac{25}{2}$
(E) 9

42. If $\dfrac{3\sin 2\theta}{1 - \cos 2\theta} = \dfrac{1}{2}$ and $0° \le \theta \le 180°$, then $\theta =$

(A) $0°$
(B) $0°$ or $180°$
(C) $80.5°$
(D) $0°$ or $80.5°$
(E) $99.5°$

43. If $f(x,y) = 2x^2 - y^2$ and $g(x) = 2^x$, which one of the following is equal to 2^{2x}?

(A) $f(x,g(x))$
(B) $f(g(x),x)$
(C) $f(g(x),g(x))$
(D) $f(g(x),0)$
(E) $g(f(x,x))$

44. Two positive numbers, a and b, are in the sequence 4, a, b, 12. The first three numbers form a geometric sequence, and the last three numbers form an arithmetic sequence. The difference $b - a$ equals

(A) 1

(B) $1\frac{1}{2}$

(C) 2

(D) $2\frac{1}{2}$

(E) 3

45. A sector of a circle has an arc length of 2.4 feet and an area of 14.3 square feet. How many degrees are in the central angle?

(A) 63.4°

(B) 20.2°

(C) 14.3°

(D) 11.5°

(E) 12.9°

46. The y-coordinate of one focus of the ellipse $36x^2 + 25y^2 + 144x - 50y - 731 = 0$ is

(A) 7.81

(B) 1

(C) 4.32

(D) −2

(E) 3.32

47. In the figure, $ABCD$ is a square. M is the point one-third of the way from B to C. N is the point one-half of the way from D to C. Then $\theta =$

(A) 36.9°

(B) 45.0°

(C) 50.8°

(D) 30.0°

(E) 36.1°

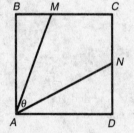

48. For what positive value(s) of $x \le 180°$ does $\tan 2x = 2 \cot 2x$?

(A) 54.7°

(B) 25° and 155°

(C) 27.4° and 117.4°

(D) 27.4°, 62.6°, 117.4°, and 152.6°

(E) none of the above

49. Under which of the following conditions is $\dfrac{x(x-y)}{y}$ negative?

(A) $x < y < 0$
(B) $y < x < 0$
(C) $0 < y < x$
(D) $x < 0 < y$
(E) All of the above.

50. The binary operation * is defined over the set of real numbers to be $a*b = \begin{cases} a\sin\dfrac{b}{a} & \text{if } a > b \\ b\cos\dfrac{a}{b} & \text{if } a < b \end{cases}$. What is the value of 2 * (5 * 3)?

(A) 4.01
(B) 3.65
(C) 1.84
(D) 2.79
(E) 2.14

ANSWER KEY

1. D	6. A	11. A	16. D	21. B	26. E	31. E	36. E	41. B	46. C
2. A	7. A	12. B	17. B	22. C	27. E	32. D	37. E	42. C	47. B
3. E	8. A	13. C	18. E	23. D	28. B	33. B	38. B	43. C	48. D
4. D	9. B	14. B	19. B	24. D	29. C	34. B	39. C	44. E	49. A
5. C	10. B	15. B	20. C	25. C	30. E	35. B	40. D	45. D	50. E

ANSWER EXPLANATIONS

In these solutions the following notation is used:

a: active—Calculator use is necessary or, at a minimum, extremely helpful.

g: Graphing calculator is preferred.

i: inactive—Calculator use is not helpful and may even be a hindrance.

1. i D Since inverse functions are symmetric about the line $y = x$, if point (a,b) lies on f, point (b,a) must lie on f^{-1}. [1.3].

2. a A Average $= \dfrac{70 + 80 + 85 + 80 + x}{5} = 80.$
Therefore, $x = 85$. [5.8].

3. g E Plot the graph $y = (x^2 + 3x + 2)/(x + 2)\tan \pi x$ in an $x\varepsilon[-2.5, 2.5]$ and $y\varepsilon[-5, 5]$ window and use TRACE to approximate the location of asymptotes. The only answer choice that could be an asymptote occurs at $x = \dfrac{1}{2}$.

An alternative solution is to factor the numerator of $f(x)$ and observe that the factor $x + 2$ divides out: $f(x) = \dfrac{(x+2)(x+1)}{x+2} \tan \pi x = (x+1)\tan \pi x.$ The only asymptotes occur because of $\tan \pi x$, when πx is a multiple of $\dfrac{\pi}{2}$. Setting $\pi x = \dfrac{\pi}{2}$ yields $x = \dfrac{1}{2}$. [4.5].

4. i D The equation of a trace of a plane can be found by letting one of the variables $= 0$. By inspection, only Choice D is a trace. [5.5].

5. g C Plot the graph of $y = 3x^3 + 4x^2 - 4x$ in the standard window and use CALC/zero to find the three zeros of this function. Sum these three values to get the correct answer choice.

An alternative solution is first to observe that x factors out, so that $x = 0$ is one zero. The other factor is a quadratic, so the sum of its zeros is $-\dfrac{b}{a} = -\dfrac{4}{3}$. [2.3].

6. i A $f(a) = a - \dfrac{1}{a}$. $f\left(\dfrac{1}{a}\right) = \dfrac{1}{a} - a$. Therefore, $f(a) + f\left(\dfrac{1}{a}\right) = 0$. [1.2].

7. g A Enter the function $f(x)$ into Y_1 and return to the Home Screen. Enter $Y_1(2) - Y_1\left(\sqrt{2}\right)$ to get the correct answer choice.

An alternative solution is to use a scientific calculator to evaluate the function at $x = 2$ and at $x = \sqrt{2}$, and then subtract. [1.2].

8. i A The $f(2 - x)$ just shifts and reflects the graph horizontally; it does not have any vertical effect on the graph. Therefore, regardless of what is substituted for x, $f(x) \geq 0$. [1.2].

9. i B Only 3 of the numbers can be used in the thousands place, 3 are left for the hundreds place, 2 for the tens place, and only one for the units place. $3 \cdot 3 \cdot 2 \cdot 1 = 18$. [5.1].

10. i B Substituting 1 for x gives $1 + a - 4 = 0$, and so $a = 3$. [2.3].

11. a A The first 5 flips have no effect on the last 5 flips, so the problem becomes "What is the probability of getting exactly 3 heads in the flip of 5 coins?" $\binom{5}{3} = 10$ outcomes contain 3 heads out of a total of $2^5 = 32$ possible outcomes. $P(3H) = \dfrac{5}{16} \approx 0.3125$. [5.3].

12. i B $i^{4n} = 1$; $i^{4n+1} = i$; $i^{4n+2} = -1$; $i^{4n+3} = -i$; $i^{4n+4} = (i^{4n})(i^4) = (1)(1)$. [4.7, 4.2].

13. i C Since $\sqrt{x} = x^{1/2}$, $\log_r \sqrt{3} = \log_r 3^{1/2} = \dfrac{1}{2}\log_r 3 = 3.55$. [4.2].

14. i B $g(x) = f(\sin x) + f(\cos x) = 4\sin^2 x + 4\cos^2 x = 4(\sin^2 x + \cos^2 x) = 4(1) = 4$. [1.2, 3.5]

15. a B This is a tricky problem. If you just plot the graph of $y = \left(x - \sqrt{2}\right)\left(x^2 - \sqrt{3x} + \pi\right)$, you will see only one real zero, at approximately 1.414 $(\approx \sqrt{2})$. This is because the zeros of the quadratic factor are imaginary. Since, however, they are imaginary conjugates, their sum is real—namely twice the real part. Therefore, graphing the function, using CALC/zero to find the zeros, and summing them will give you the wrong answer.

To get the correct answer, you must use the fact that the sum of the zeros of the quadratic factor is $-\dfrac{b}{a} = \sqrt{3} \approx 1.732$. Since the zero of the linear factor is $\sqrt{2} \approx 1.414$, the sum of the zeros is about $1.732 + 1.414 \approx 3.15$. [2.3].

16. i D Graph I consists of the lines $x = 0$ and $y = 0$, which are the coordinate axes and are therefore perpendicular. Graph II consists of $y = |x|$ and $y = -|x|$, which are at $\pm 45°$ to the coordinate axes and are therefore perpendicular. Graph III consists of the hyperbolas $xy = 1$ and $xy = -1$. Therefore, the correct answer choice is D.

There are two reasons why a graphing calculator solution is not recommended here. One is that equations, not functions, are given, and solving these equations so that they can be graphed involves two branches each. The other reason is that even with graphs, you would have to make judgments about perpendicularity. At a minimum, this would require you to graph the equations in a square window. [4.3].

17. i B Points 6 in. from m form a cylinder, with m as axis, which is tangent to plane X. Points 1 in. from X are two planes parallel to X, one above and one below X. The cylinder intersects only one of the planes in two lines parallel to m. [5.5].

18. i E Since the y-coordinate of the point V could be at any height, the slope of VO could be any value. [5.5].

19. a B Volume of cylinder $= \pi r^2 h = 36\pi =$ volume of cube $= s^3$. Therefore, $s = \sqrt[3]{36\pi} \approx 4.84$. [5.5].

20. i C Substitute $2t - 2$ for x. [4.6].

21. a B $y_1 = 2^{3/2} + 2a + \sqrt{2}b + c$ and $y_2 = -(2)^{3/2} + 2a - 2\sqrt{2}b + c$. So, $y_1 - y_2 = (2^{3/2} + 2^{3/2}) + 2\sqrt{2}b = 3$. Therefore, $5.65685 + 2.828b \approx 3$ and $b \approx \dfrac{3 - 5.65685}{2.8284} \approx -0.939$. [2.4].

22. i The n nth roots of any complex number can be represented by n vectors symmetrically drawn about the origin $\dfrac{360°}{n}$ apart. The vectors for the

fifth roots of any complex number should be $\dfrac{360°}{5} = 72°$ apart. The only answer that disrupts this pattern is Choice C. [4.7]

23. i D Here, a could be either positive or negative. However, b must be negative. [4.3, 2.5].

24. g D Plot the graph of $y = \text{abs}(x - \text{int}(x) - 1/2)$ in an $x\varepsilon[-5,5]$ and $y\varepsilon[-2,2]$ window and observe that the maximum value is $\dfrac{1}{2}$.

An alternative solution is to sketch a portion of the graph by hand and observe the maximum value. [4.3, 4.4].

25. g C Plot the graph $y = x^3 - 8/x^2 - 4$ in the standard window. Using CALC/value, observe that y is not defined when $x = 2$. Therefore, enter 1.999 for x and observe that y is approximately equal to 3.

An alternative solution is to factor the numerator and denominator, divide out $x - 2$, and substitute the limiting value 2 into the resulting expression:

$$\lim_{x \to 2} \frac{x^3 - 8}{x^2 - 4} = \lim_{x \to 2} \frac{(x-2)(x^2 + 2x + 2)}{(x-2)(x+2)}$$

$$= \lim_{x \to 2} \frac{(x^2 + 2x + 4)}{x + 2} = \frac{4 + 4 + 4}{2 + 2} = 3.$$

[4.5].

26. a E A sketch will help you see that the height of the cone is $5 + 3 = 8$.

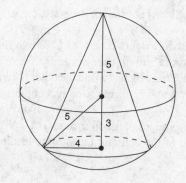

The volume of the cone is $V_c = \frac{1}{3}\pi r^2 h = \frac{1}{3}\pi(16)(8)$, and the volume of the sphere is $V_s = \frac{4}{3}\pi r^3 = \frac{4}{3}\pi(125)$. The desired ratio is $V_c : V_s = 0.256 : 1$. [5.5].

27. g   E   Enter 1 into your calculator. Then enter $\sqrt[3]{2\text{Ans}}$ three times, to accomplish three iterations that result in x_3 to get the correct answer choice E.

An alternative solution is to use the formula to evaluate x_1, x_2, and x_3, in turn. [5.9]

28. g   B   With your calculator in radian mode, plot the graph of $y = \text{abs}\left(\sqrt{2}\left(1/\sin\left(x + \pi/5\right)\right)\right)$ in an $x\varepsilon[-1,1]$ and $y\varepsilon[0,5]$ window. Use TRACE to determine that the y-intercept is approximately 1.49.

An alternative solution is to substitute 0 for x and evaluate y. [3.4].

29. i   C   The equation of the circle is $(x - 4)^2 + (y + 8)^2 = r^2$. Multiplying out indicates that $a = -8$ and $b = 16$, and so $a + b = 8$. [4.1].

30. g   E   Since $p(a) = 2$, $3a^2 + 9a + 5 = 0$. Solve this quadratic by using your program for the Quadratic Formula to get the correct answer choice.

An alternative solution is to substitute 3 for a, 9 for b, and 5 for c in the Quadratic Formula and evaluate. [2.3].

31. g   E   Since i is a root of the equation, so is $-i$. Plot the graph of $y = x^4 + 2x^3 - 3x^2 + 2x - 4$ in the standard window. Use CALC/zero to find one of the roots, and store the answer (X) in A. Then use CALC/zero again to find the other root, and multiply A and X to get the correct choice.

The product of all four roots is $(-1)^4$ times the ratio of the constant term to the leading coeffi-

cient, or -4. Since i is a root, so is $-i$, and their product is $-i^2 = 1$. Therefore, the product of the two real roots is -4. [2.3].

32. a   D   From $\sin^2 x + \cos^2 x = 1$ and $90° \le A \le 180°$; $\sin A = \frac{3}{5}$ implies that $\cos A = -\frac{4}{5}$; from $270° \le B \le 360°$; $\cos B = \frac{1}{3}$ implies that $\sin B = -\frac{2\sqrt{2}}{3}$.

Sin $(A + B) = \sin A \cdot \cos B + \cos A \cdot \sin B = \frac{3 + 8\sqrt{2}}{15} \approx 0.954$. [3.5].

33. i   B $\frac{x}{t} = K$ and $ty^2 = C$, where K and C are constants. Solve the first equation for t: $t = \frac{x}{K}$. Substituting for t in the second equation gives $\frac{xy^2}{K} = C$. Thus, $xy^2 = CK$, which indicates that x varies inversely as y^2. [5.6].

34. a   B   Since $\sec 1.4 = x$, $\cos 1.4 = \frac{1}{x}$ or $x = \frac{1}{\cos 1.4}$. Therefore, $\csc(2 \text{Arctan} x) = 1/\sin(2\tan^{-1}(1/\cos 1.4)) \approx 3.03$ [3.6].

35. g   B   The equation $|y - 1| = |x + 1|$ defines two functions: $y = \pm|x + 1| + 1$. Plot these graphs in the standard window and observe that they intersect at $(-1,1)$.

An alternative solution is to recall that the important point of an absolute value occurs when the expression within the absolute value sign equals zero. The important point of this absolute value problem occurs when $y - 1 = 0$ and $x - 1 = 0$, i.e., at $(-1,1)$. [4.3].

36. g   E   With your calculator in degree mode, plot the graphs of $y = \cos 2x$ and $y = 2\cos x$ in an $x\varepsilon[0,360]$ and $y\varepsilon[-2,2]$ window. Use CALC/intersect to find the correct answer choice E.

An alternative solution is to use the identity $\cos 2x = 2\cos^2 x - 1$ and solve the quadratic $2\cos^2 x - 2\cos x - 1 = 0$, getting $\cos x = \frac{2 \pm \sqrt{12}}{4} = \frac{1 \pm \sqrt{3}}{2}$.

Since $\frac{1+\sqrt{3}}{2} > 1$, solve $\cos x = \frac{1-\sqrt{3}}{2} \approx -0.3660$. Use your calculator to evaluate $\cos^{-1}(-0.3660) = 111.5°$. Then find the third quadrant angle that has the same reference angle ($180° - 111.5° = 68.5°$)—namely, $180° + 68.5° = 248.5°$. [3.5].

37. i E Since $\sin\theta$ has a maximum at $\theta = \frac{\pi}{2}$, $\sqrt{B-x^2} = \frac{\pi}{2}$. Thus, $B - x^2 = \left(\frac{\pi}{2}\right)^2$ and $x^2 = B - \left(\frac{\pi}{2}\right)^2$. Therefore, $x = \pm\sqrt{B - \left(\frac{\pi}{2}\right)^2}$. [3.4].

38. g B Enter LIST/MATH/sum(LIST/OPS/seq(X,X,1,51)) to compute the desired sum.

An alternative solution is to observe that the sequence is arithmetic with $t_1 = 1$ and $d = 1$. Using the formula for the sum of the first n terms of an arithmetic sequence, $S_{51} = \frac{51}{2}(2 + 50 \cdot 1) = 51 \cdot 26 = 1326$. [5.4].

39. g C Complete the square in the first equation to get $3(x-1)^2 + 4(y+1)^2 = 12$. Solving this equation for y yields $y = \pm\sqrt{\frac{12 - 3(x-1)^2}{4}} - 1$. Solving for y in the second equation, $y = \frac{(x-2)^2}{4} - 2$.

Plot the graphs of these three equations in the standard window to see that the graphs intersect in two places.

After completing the square in the first equation, divide by 12 to get the standard form equation of an ellipse, $\frac{(x-1)^2}{4} + \frac{(y+1)^2}{3} = 1$. The second equation is the standard form of a parabola. Sketch these two equations and observe the number of points of intersection. [4.1]

40. a D Let $y = \log_x 2 = \log_3 x$. Converting to exponential form gives $x^y = 2$ and $3^y = x$. Substitute to get $3^{y^2} = 2$, which can be converted into

$y^2 = \frac{\log 2}{\log 3} \approx 0.6309$. Thus, $y \approx \pm 0.7943$. Therefore, $3^{0.7943} = x \approx 2.393$ or $3^{-0.7943} = x \approx 0.4178$. Therefore, the sum of two x's is 2.81. [4.2].

41. i B The exponent on $(-3b^{-3})$ in the fifth term is 4. The exponent on $(2b^2)$ in the fifth term is $n - 4$. $(b^2)^{n-4} \cdot (b^{-3})^4 = b^0 \cdot 2(n-4) + (-3)4 = 0$. $2n - 8 - 12 = 0$, $2n = 20$, and $n = 10$. [5.2].

42. g C With your calculator in degree mode, plot the graphs of $y = 1/2$ and $y = (3\sin(2x))/(1 - \cos(2x))$ and in the Ztrig window. Use CALC/intersect to find the one point of intersection in the specified interval, at 80.5°.

An alternative solution is to cross-multiply the original equation and use the double angle formulas for sine and cosine, to get

$6\sin 2\theta = 1 - \cos 2\theta$

$12\sin\theta\cos\theta = 1 - (1 - 2\sin^2\theta) = 2\sin^2\theta$

$6\sin\theta\cos\theta - \sin^2\theta = 0$

$\sin\theta(6\cos\theta - \sin\theta) = 0$

Therefore, $\sin\theta = 0$ or $\tan\theta = 6$. It follows that $\theta = 0°, 180°$, or $80.5°$. The first two solutions make the denominator of the original equation equal to zero, so 80.5° is the only solution. [3.5]

43. i C To get 2^{2x}, either one term must be zero in $f(x,y)$ or both must contain 2^x. Choice D gives $2 \cdot 2^{2x}$, which is wrong. The only other possibility is Choice C: $f(g(x),g(x)) = 2(2^x)^2 - (2^x)^2 = 2^{2x}$. [1.2, 1.5, 4.2].

44. i E From the geometric sequence, $b = a\left(\frac{a}{4}\right)$. From the arithmetic sequence, $2b - a = 12$ since $r = \frac{a}{4}$ and $d = b - a$. Substituting gives $2\left(\frac{a^2}{4}\right) - a = 12$. Solving gives $a = 6$ or -4. Eliminate -4 since a is given to be positive. Substituting the 6 gives $2b - 6 = 12$, giving $b = 9$. Therefore, $b - a = 3$. [5.4].

45. a D Since $s = r\theta^R$, then $2.4 = r\theta$, which implies that $r = \frac{2.4}{\theta}$. $A = \frac{1}{2}r^2\theta$, and so $14.3 = \frac{1}{2}r^2\theta$, which implies that $r^2 = \frac{28.6}{\theta}$. Therefore, $\left(\frac{2.4}{\theta}\right)^2 = \frac{28.6}{\theta}$, which implies that $2.4^2 = 28.6\theta$. Therefore, $\theta = \frac{5.76}{28.6} \approx 0.2014^R \approx 11.5°$. [3.2].

46. a   C   Complete the square and put the equation of the ellipse in standard form:

$$36x^2 + 25y^2 + 144x - 50y - 731$$
$$= 36(x+2)^2 + 25(y-1)^2 - 900$$

$$\frac{(x+2)^2}{25} + \frac{(y-1)^2}{36} = 1$$

The center of the ellipse is at $(-2,1)$, with $a^2 = 36$ and $b^2 = 25$, and the major axis is parallel to the y axis. Each focus is $c = \sqrt{a^2 - b^2}$ units above and below the center. Therefore, the y coordinates of the foci are $1 \pm \sqrt{11} \approx 4.32$ and -2.32. [4.1].

47. a   B   Because you are bisecting one side and trisecting another side, it is convenient to let the length of the sides be a number divisible by both 2 and 3. Let $AB = AD = 6$. Thus $BM = 2$, $MC = 4$, and $CN = ND = 3$. Let $\angle NAD = x$, so that, using right triangle NAD, $\tan x = \dfrac{3}{6} = 0.5$, which implies that $x = \tan^{-1} 0.5 \approx 26.6°$. Let $\angle MAB = y$, so that, using right triangle MAB, $\tan y = \dfrac{2}{6}$, which implies that $y = \text{Tan}^{-1} \dfrac{1}{3} \approx 18.4°$. Therefore, $\theta \approx 90° - 26.6° - 18.4° \approx 45°$. [3.7].

48. g   D   With your calculator in degree mode, plot the graphs of $y = \tan(2x)$ and $y = 2/(\tan(2x))$ in an $x\varepsilon[0,180]$ and $y\varepsilon[-4,4]$ window. Since there are 4 points of intersection, D is the only possible correct answer choice.

An alternative solution is to observe that $\tan 2x = \dfrac{2}{\tan 2x}$ implies $\tan^2 2x = 2$, or $\tan 2x = \pm\sqrt{2}$. The reference angle for $2x$ is 54.7°, and since we are looking for $x\varepsilon(0°,180°)$, we must look for $2x\varepsilon(0°,360°)$ with this reference angle. Since $\tan 2x = \pm\sqrt{2}$, there is a value of $2x$ in each quadrant, so there are four solutions. [3.5].

49. i   A   You must check each answer choice, one at a time. In A, $x < 0$, $y < 0$, $x - y < 0$, so the expression is negative. In B, $x < 0$, $y < 0$, $x - y > 0$, so the expression is positive. At this point you know that the correct answer choice must be A. [5.9].

50. a   E   Put your calculator in radian mode:

$5 * 3 = 5 \sin \dfrac{3}{5} \approx 2.823$. $2 * (5 * 3) = 2 * 2.823 \approx$ $2.823 \cos \dfrac{2}{2.823} \approx 2.14$. [5.9].

SELF-EVALUATION CHART FOR MODEL TEST 6

SUBJECT AREA	QUESTIONS	NUMBER OF RIGHT WRONG OMITTED		

Mark correct answers with C, wrong answers with X, and omitted answers with O.

Algebra
(9 questions)
Review section

5	12	15	20	21	23	25	33	40
2.3	4.7	2.3	4.6	2.4	4.3	4.5	5.6	4.2

___ ___ ___

Solid geometry
(4 questions)
Review section

4	17	19	26
5.5	5.5	5.5	5.5

___ ___ ___

Coordinate geometry
(5 questions)
Review section

16	18	29	35	39
4.3	5.5	4.1	4.3	4.1

___ ___ ___

Trigonometry
(10 questions)
Review section

22	28	32	34	36	37	42	45	47	48
4.7	3.4	3.5	3.6	3.5	3.4	3.5	3.2	3.7	3.5

___ ___ ___

Functions
(13 questions)
Review section

1	3	6	7	8	10	13	14	24	30	31	43	46
1.3	4.5	1.2	1.2	1.2	2.3	4.2	1.2	4.3	2.3	2.4	1.2	4.1

___ ___ ___

Miscellaneous
(9 questions)
Review section

2	9	11	27	38	41	44	49	50
5.8	5.1	5.3	5.9	5.4	5.2	5.4	5.9	5.9

___ ___ ___

TOTALS   ___ ___ ___

Raw score = (number right) – $\frac{1}{4}$ (number wrong) = ____

Round your raw score to the nearest whole number = ____

Evaluate Your Performance
Model Test 6

Rating	Number Right
Excellent	41–50
Very good	33–40
Above average	25–32
Average	15–24
Below average	Below 15

ANSWER SHEET FOR MODEL TEST 7

Determine the correct answer for each question. Then, using a no. 2 pencil, blacken completely the oval containing the letter of your choice.

1. Ⓐ Ⓑ Ⓒ Ⓓ Ⓔ	18. Ⓐ Ⓑ Ⓒ Ⓓ Ⓔ	35. Ⓐ Ⓑ Ⓒ Ⓓ Ⓔ
2. Ⓐ Ⓑ Ⓒ Ⓓ Ⓔ	19. Ⓐ Ⓑ Ⓒ Ⓓ Ⓔ	36. Ⓐ Ⓑ Ⓒ Ⓓ Ⓔ
3. Ⓐ Ⓑ Ⓒ Ⓓ Ⓔ	20. Ⓐ Ⓑ Ⓒ Ⓓ Ⓔ	37. Ⓐ Ⓑ Ⓒ Ⓓ Ⓔ
4. Ⓐ Ⓑ Ⓒ Ⓓ Ⓔ	21. Ⓐ Ⓑ Ⓒ Ⓓ Ⓔ	38. Ⓐ Ⓑ Ⓒ Ⓓ Ⓔ
5. Ⓐ Ⓑ Ⓒ Ⓓ Ⓔ	22. Ⓐ Ⓑ Ⓒ Ⓓ Ⓔ	39. Ⓐ Ⓑ Ⓒ Ⓓ Ⓔ
6. Ⓐ Ⓑ Ⓒ Ⓓ Ⓔ	23. Ⓐ Ⓑ Ⓒ Ⓓ Ⓔ	40. Ⓐ Ⓑ Ⓒ Ⓓ Ⓔ
7. Ⓐ Ⓑ Ⓒ Ⓓ Ⓔ	24. Ⓐ Ⓑ Ⓒ Ⓓ Ⓔ	41. Ⓐ Ⓑ Ⓒ Ⓓ Ⓔ
8. Ⓐ Ⓑ Ⓒ Ⓓ Ⓔ	25. Ⓐ Ⓑ Ⓒ Ⓓ Ⓔ	42. Ⓐ Ⓑ Ⓒ Ⓓ Ⓔ
9. Ⓐ Ⓑ Ⓒ Ⓓ Ⓔ	26. Ⓐ Ⓑ Ⓒ Ⓓ Ⓔ	43. Ⓐ Ⓑ Ⓒ Ⓓ Ⓔ
10. Ⓐ Ⓑ Ⓒ Ⓓ Ⓔ	27. Ⓐ Ⓑ Ⓒ Ⓓ Ⓔ	44. Ⓐ Ⓑ Ⓒ Ⓓ Ⓔ
11. Ⓐ Ⓑ Ⓒ Ⓓ Ⓔ	28. Ⓐ Ⓑ Ⓒ Ⓓ Ⓔ	45. Ⓐ Ⓑ Ⓒ Ⓓ Ⓔ
12. Ⓐ Ⓑ Ⓒ Ⓓ Ⓔ	29. Ⓐ Ⓑ Ⓒ Ⓓ Ⓔ	46. Ⓐ Ⓑ Ⓒ Ⓓ Ⓔ
13. Ⓐ Ⓑ Ⓒ Ⓓ Ⓔ	30. Ⓐ Ⓑ Ⓒ Ⓓ Ⓔ	47. Ⓐ Ⓑ Ⓒ Ⓓ Ⓔ
14. Ⓐ Ⓑ Ⓒ Ⓓ Ⓔ	31. Ⓐ Ⓑ Ⓒ Ⓓ Ⓔ	48. Ⓐ Ⓑ Ⓒ Ⓓ Ⓔ
15. Ⓐ Ⓑ Ⓒ Ⓓ Ⓔ	32. Ⓐ Ⓑ Ⓒ Ⓓ Ⓔ	49. Ⓐ Ⓑ Ⓒ Ⓓ Ⓔ
16. Ⓐ Ⓑ Ⓒ Ⓓ Ⓔ	33. Ⓐ Ⓑ Ⓒ Ⓓ Ⓔ	50. Ⓐ Ⓑ Ⓒ Ⓓ Ⓔ
17. Ⓐ Ⓑ Ⓒ Ⓓ Ⓔ	34. Ⓐ Ⓑ Ⓒ Ⓓ Ⓔ	

MODEL TEST

7

50 questions   1 hour

Tear out the preceding answer sheet. Decide which is the best choice by rounding your answer when appropriate. Blacken the corresponding space on the answer sheet. When finished, check your answers with those at the end of the test. For questions that you got wrong, note the sections containing the material that you must review. Also, if you do not fully understand how you arrived at some of the correct answers, you should review the appropriate sections. Finally, fill out the self-evaluation sheet on page 277 in order to pinpoint the topics that give you the most difficulty.

TEST DIRECTIONS

<u>Directions</u>: Decide which answer choice is best. If the exact numerical value is not one of the answer choices, select the closest approximation. Fill in the oval on the answer sheet that corresponds to your choice.

Notes:
(1) You will need to use a scientific or graphing calculator to answer some of the questions.
(2) You will have to decide whether to put your calculator in degree or radian mode for some problems.
(3) All figures that accompany problems are plane figures unless otherwise stated. Figures are drawn as accurately as possible to provide useful information for solving the problem, except when it is stated in a particular problem that the figure is not drawn to scale.
(4) Unless otherwise indicated, the domain of a function is the set of all real numbers for which the functional value is also a real number.

<u>Reference Information.</u> The following formulas are provided for your information.

Volume of a right circular cone with radius r and height h: $V = \dfrac{1}{3}\pi r^2 h$

Lateral area of a right circular cone if the base has circumference c and the slant height is l: $S = \dfrac{1}{2}cl$

Volume of a sphere of radius r: $V = \dfrac{4}{3}\pi r^3$

Surface area of a sphere of radius r: $S = 4\pi r^2$

Volume of a pyramid of base area B and height h: $V = \dfrac{1}{3}Bh$

---

1. $2^{2/3} + 2^{4/3} =$

    (A) 1.6
    (B) 1.9
    (C) 3.2
    (D) 4.0
    (E) 4.1

2. In three dimensions, what is the set of all points for which $x = 0$?

    (A) the origin
    (B) a line parallel to the x-axis
    (C) the yz-plane
    (D) a plane containing the x-axis
    (E) the x-axis

3. Expressed with positive exponents only, $\dfrac{ab^{-1}}{a^{-1} - b^{-1}}$ is equivalent to

    (A) $\dfrac{a^2}{a-b}$

    (B) $\dfrac{a^2}{a-1}$

    (C) $\dfrac{b-a}{ab}$

    (D) $\dfrac{a^2}{b-a}$

    (E) $\dfrac{1}{a-b}$

4. If $f(x) = \sqrt[3]{x}$ and $g(x) = x^3 + 8$, find $(f \circ g)(3)$.

    (A) 5
    (B) 3.3
    (C) 50.5
    (D) 11
    (E) 35

5. $x > \sin x$ for

    (A) all $x > 0$
    (B) all $x < 0$
    (C) all x for which $x \neq 0$
    (D) all x
    (E) all x for which $-\dfrac{\pi}{2} < x < 0$

6. The sum of the zeros of $f(x) = 3x^2 - 5$ is

    (A) 1.3
    (B) 1.8
    (C) 1.7
    (D) 3.3
    (E) 0

GO ON TO THE NEXT PAGE

7. If x varies inversely as the cube root of y, and if $x = 3$ and $y = 4$, the constant of variation is

(A) 1.9
(B) 4.8
(C) 0.04
(D) 192
(E) 0.53

8. In the figure, c equals

(A) 1
(B) xy
(C) $\dfrac{x}{y}$
(D) $\dfrac{y}{x}$
(E) -1

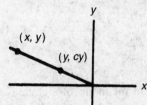

9. The plane $ax + by + cz = 12$ intersects the x-axis at $(2,0,0)$, the y-axis at $(0,-3,0)$, and the z-axis at $(0,0,-4)$. $a + b + c =$

(A) 13
(B) 7
(C) 6
(D) 0
(E) -1

10. $P(x) = x^5 + x^4 - 2x^3 - x - 1$ has at most n positive zeros. Then $n =$

(A) 0
(B) 1
(C) 2
(D) 3
(E) 5

11. If $x^5y^8z^{-3} = \dfrac{8x^4}{y^{-8}z^3}$, then $x =$

(A) $\dfrac{1}{8}$

(B) $\dfrac{1}{4}$

(C) 8
(D) $8y^2z^2$
(E) $8y^{16}z^6$

GO ON TO THE NEXT PAGE

12. If $f(x)$ is a linear function and $f(2) = 1$ and $f(4) = -2$, then $f(x) =$

(A) $-\dfrac{3}{2}x + 4$

(B) $\dfrac{3}{2}x - 2$

(C) $-\dfrac{3}{2}x + 2$

(D) $\dfrac{3}{2}x - 4$

(E) $-\dfrac{2}{3}x + \dfrac{7}{3}$

USE THIS SPACE FOR SCRATCH WORK

13. The length of the radius of a circle is one-half the length of an arc of the circle. How large is the central angle that intercepts that arc?

(A) $60°$
(B) $120°$
(C) π^R
(D) 1^R
(E) 2^R

14. If $f(x) = 2^x + 1$, then $f^{-1}(7) =$

(A) 2.8
(B) 3
(C) 3.6
(D) 2.4
(E) 2.6

15. The statement "If it is green, then it is a turkey" is true. Which of the following is (are) also true?

I. It is not green; therefore it is not a turkey.
II. It is a turkey; therefore it is green.
III. It is not a turkey; therefore it is not green.

(A) only I
(B) only II
(C) only III
(D) only I and III
(E) only II and III

16. The 71st term of 30, 27, 24, 21, . . . , is

(A) 180
(B) −183
(C) −180
(D) 240
(E) 5325

GO ON TO THE NEXT PAGE

17. If $0 < x < \dfrac{\pi}{2}$ and $\tan 5x = 3$, to the nearest tenth, what is the value of $\tan x$?

   (A) 0.5
   (B) 0.4
   (C) 0.3
   (D) 0.2
   (E) 0.1

USE THIS SPACE FOR SCRATCH WORK

18. If $4.05^p = 5.25^q$, what is the value of $\dfrac{p}{q}$?

   (A) −0.11
   (B) 0.11
   (C) 1.30
   (D) 1.19
   (E) 1.67

19. A cylinder has a base radius of 2 and a height of 9. To the nearest whole number, by how much does the lateral area exceed the sum of the areas of the two bases?

   (A) 101
   (B) 96
   (C) 88
   (D) 81
   (E) 75

20. If $\cos 67° = \tan x°$, then $x =$

   (A) 0.4
   (B) 6.8
   (C) 21
   (D) 29.3
   (E) 7.8

21. $P(x) = x^3 + 18x - 30$ has a zero in the interval

   (A) (0, 0.5)
   (B) (0.5, 1)
   (C) (1, 1.5)
   (D) (1.5, 2)
   (E) (2, 2.5)

22. The lengths of the sides of a triangle are 23, 32, and 37. To the nearest degree, what is the value of the largest angle?

   (A) 83°
   (B) 122°
   (C) 128°
   (D) 142°
   (E) 71°

23. A point, p, is 5 inches from a plane, X. The set of points 13 inches from p and 2 inches from X forms

(A) a circle parallel to X with a radius <13
(B) two circles parallel to X with radii >10
(C) a sphere with radius <13
(D) two lines parallel to X
(E) two planes parallel to X

24. Two cards are drawn from a regular deck of 52 cards. What is the probability that both will be 7s?

(A) 0.149
(B) 0.012
(C) 0.005
(D) 0.009
(E) 0.04

25. If $\sqrt{y} = 3.216$, then $\sqrt{10y} =$

(A) 321.6
(B) 32.16
(C) 10.17
(D) 5.67
(E) 4.23

26. What is the domain of the function $f(x) = \log \sqrt{2x^2 - 15}$?

(A) $-7.5 < x < 7.5$
(B) $x < -7.5$ or $x > 7.5$
(C) $x < -2.7$ or $x > 2.7$
(D) $x < -3.2$ or $x > 3.2$
(E) $x < 1.9$ or $x > 1.9$

27. The sum of the roots of $3x - 7x^{-1} + 3 = 0$ is

(A) $\dfrac{7}{3}$
(B) 1
(C) $-\dfrac{7}{3}$
(D) -1
(E) $\dfrac{3}{7}$

28. Let S be the sum of the first n terms of the arithmetic sequence 3, 7, 11, . . . , and let T be the sum of the first n terms of the arithmetic sequence 8, 10, 12, ... For $n > 1$, $S = T$ for

(A) no value of n
(B) one value of n
(C) two values of n
(D) three values of n
(E) four values of n

GO ON TO THE NEXT PAGE

29. On the interval $\left[-\dfrac{\pi}{4},\dfrac{\pi}{4}\right]$, the function $f(x) = \sqrt{1+\sin^2 x}$ has a maximum value of

(A) 0.78
(B) 1
(C) 1.1
(D) 1.2
(E) 1.4

30. A point has rectangular coordinates (3,4). The polar coordinates are $(5,\theta)$. What is the value of θ?

(A) 37°
(B) 30°
(C) 51°
(D) 53°
(E) 60°

31. If $f(x) = x^2 - 4$, for what real number values of x will $f(f(x)) = 0$?

(A) 2.4
(B) ±2.4
(C) 2 or 6
(D) ±1.4 or ±2.4
(E) no values

32. If $f(x) = x \log x$ and $g(x) = 10^x$, then $g(f(2)) =$

(A) 24
(B) 17
(C) 4
(D) 2
(E) 0.6

33. If $f(x) = x^{\sqrt{x}}$, then $f\left(\sqrt{2}\right) =$

(A) 1.6
(B) 2.7
(C) 1.5
(D) 2.0
(E) 1.4

34. The figure at right shows the graph of 5^x. What is the sum of the areas of the triangles?

(A) 32,550
(B) 16,225
(C) 2604
(D) 1302
(E) 651

USE THIS SPACE FOR SCRATCH WORK

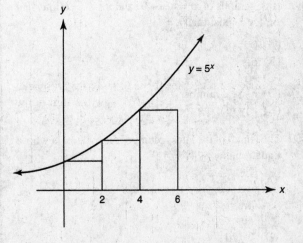

GO ON TO THE NEXT PAGE

35. (p,q) is called a *lattice point* if p and q are both integers. How many lattice points lie in the area between the two curves $x^2 + y^2 = 9$ and $x^2 + y^2 - 6x + 5 = 0$?

   (A) 0
   (B) 1
   (C) 2
   (D) 3
   (E) 4

36. If $\sin A = \dfrac{3}{5}$, $90° < A < 180°$, $\cos B = \dfrac{1}{3}$, and $270° < B < 360°$, the value of $\sin (A + B)$ is

   (A) –0.33
   (B) 0.73
   (C) –0.83
   (D) –0.55
   (E) 0.95

37. For all real numbers x, $f(2x) = x^2 - x + 3$. An expression for $f(x)$ in terms of x is

   (A) $2x^2 - 2x + 3$
   (B) $4x^2 - 2x + 3$
   (C) $\dfrac{x^2}{4} - \dfrac{x}{2} + 3$
   (D) $\dfrac{x^2}{2} - \dfrac{x}{2} + 3$
   (E) $x^2 - x + 3$

38. For what value(s) of k is $x^2 - kx + k$ divisible by $x - k$?

   (A) only 0
   (B) only 0 or $-\dfrac{1}{2}$
   (C) only 1
   (D) any value of k
   (E) no value of k

39. If the graphs of $x^2 = 4(y + 9)$ and $x + ky = 6$ intersect on the x-axis, then $k =$

   (A) 0
   (B) 6
   (C) –6
   (D) no real number
   (E) any real number

40. The length of the latus rectum of the hyperbola whose equation is $x^2 - 4y^2 = 16$ is

   (A) 1
   (B) $\sqrt{20}$
   (C) 16
   (D) $2\sqrt{20}$
   (E) 2

USE THIS SPACE FOR SCRATCH WORK

GO ON TO THE NEXT PAGE

41. If $f_n = \begin{cases} \dfrac{f_{n-1}}{2} & \text{when } f_{n-1} \text{ is an even number} \\ 3 \cdot f_{n-1} + 1 & \text{when } f_{n-1} \text{ is an odd number} \end{cases}$

and $f_1 = 3$, then $f_5 =$

(A) 1
(B) 2
(C) 4
(D) 8
(E) 16

42. How many distinguishable rearrangements of the letters in the word CONTEST start with the two vowels?

(A) 120
(B) 60
(C) 10
(D) 5
(E) None of these

43. The amount of heat received by a body varies inversely as the square of its distance from the heat source. In comparison, how much heat is received by a body that is 3 times as far from the heat source?

(A) 3 times as much
(B) 9 times as much
(C) $\dfrac{1}{3}$ as much
(D) $\dfrac{1}{9}$ as much
(E) $\dfrac{1}{6}$ as much

44. How many positive integers are there in the solution set of $\dfrac{x}{x-2} > 5$?

(A) 0
(B) 2
(C) 4
(D) 5
(E) an infinite number

45. During the year 1995 the price of ABC Company stock increased by 125%, and during the year 1996 the price of the stock increased by 80%. Over the period from January 1, 1995, through December 31, 1996, by what percentage did the price of ABC Company stock rise?

(A) 103%
(B) 205%
(C) 305%
(D) 405%
(E) 505%

GO ON TO THE NEXT PAGE

46. If $x_0 = 3$ and $x_{n+1} = x_n \sqrt{x_n + 1}$, then $x_3 =$

(A) 15.9
(B) 31.7
(C) 173.9
(D) 44.9
(E) 65.2

47. When the smaller root of the equation $3x^2 + 4x - 1 = 0$ is subtracted from the larger root, the result is

(A) 0.7
(B) 1.8
(C) 2.0
(D) 1.3
(E) −1.3

48. Each of a group of 50 students studies either French or Spanish but not both, and either math or physics but not both. If 16 students study French and math, 26 study Spanish, and 12 study physics, how many study both Spanish and physics?

(A) 5
(B) 6
(C) 8
(D) 4
(E) 10

49. If x, y, and z are positive, with $xy = 24$, $xz = 48$, and $yz = 72$, then $x + y + z =$

(A) 22
(B) 36
(C) 50
(D) 62
(E) 96

50. $\operatorname{Sin}^{-1}(\cos 100°) =$

(A) 1.0
(B) 1.4
(C) 0.2
(D) −1.4
(E) −0.2

USE THIS SPACE FOR SCRATCH WORK

ANSWER KEY

1. E	6. E	11. C	16. C	21. C	26. C	31. D	36. E	41. D	46. E
2. C	7. B	12. A	17. C	22. A	27. D	32. C	37. C	42. A	47. B
3. D	8. D	13. E	18. D	23. B	28. B	33. C	38. A	43. D	48. D
4. B	9. E	14. E	19. C	24. C	29. D	34. D	39. E	44. A	49. A
5. A	10. B	15. C	20. C	25. C	30. D	35. D	40. E	45. C	50. E

ANSWER EXPLANATIONS

In these solutions the following notation is used:

- a: active—Calculator use is necessary or, at a minimum, extremely helpful.
- g: Graphing calculator is preferred.
- i: inactive—Calculator use is not helpful and may even be a hindrance.

1. a E $10^{2/3} \approx 1.59$. $10^{4/3} \approx 2.52$. Therefore, $10^{2/3} + 10^{4/3} \approx 1.59 + 2.52 \approx 4.1$. [4.2].

2. i C When $x = 0$, y and z can be any value. Therefore, any point in the yz-plane is a possible member of the set. [5.5].

3. i D $\dfrac{\dfrac{a}{b}}{\dfrac{1}{a} - \dfrac{1}{b}} \cdot \dfrac{ab}{ab} = \dfrac{a^2}{b-a}$. [4.2].

4. a B $f \circ g(3) = f(g(3)) = \sqrt[3]{3^3 + 8} = \sqrt[3]{35} \approx 3.27106 = 3.3$ [1.2].

5. g A Plot the graph of $y = x - \sin x$ and observe that the graph lies above the x-axis for all $x > 0$.

An alternative solution is to sketch the graphs of $y = \sin x$ and $y = x$ by hand and ascertain that the latter lies above the former for all $x > 0$. You might want to test this by calculating y for each function for some positive x values near 0. [3.4].

6. g E Plot the graph of $y = 3x^2 - 5$ in the standard window. The symmetry about the y-axis indicates that the zeros are opposites and therefore sum to zero.

An alternative solution is to use the fact that the zeros of a quadratic function sum to $-\dfrac{b}{a} = 0$. [2.3].

7. a B Since x varies inversely with the cube root of y, $x = \dfrac{k}{\sqrt[3]{y}}$, or $k = x\sqrt[3]{y}$. Substituting the given values for x and y yields the correct answer choice. [5.6].

8. i D The slope of the line through (x,y) and $(0,0) = \dfrac{y}{x}$. The slope of the line through (y,cy) and $(0,0) = \dfrac{cy}{y} = c$. The two slopes are equal, and so $c = \dfrac{y}{x}$. [2.2].

9. i E Substituting the coordinates of the three points into the equation gives $2a = 12$, $-3b = 12$, and $-4c = 12$. $a + b + c = 6 - 4 - 3 = -1$. [5.5].

10. i B By Descartes' rule of signs, the one sign change in $P(x)$ implies there is exactly one positive real zero. [2.4].

11. i C $x^5 y^8 z^{-3} = \dfrac{8x^4}{y^{-8}z^3} = 8x^4 y^8 z^{-3}$. Dividing out common factors leaves $x = 8$. [5.9].

12. i A The slope of $f(x) = \dfrac{-2-1}{4-2} = \dfrac{-3}{2}$. Using the point-slope form, $f(x) - 1 = \dfrac{-3}{2}(x - 2)$. Therefore, $f(x) = \dfrac{-3}{2}x + 4$. [2.2].

13. i E $s = r\theta$. $2r = r\theta$. $\theta = 2^R$. [3.2].

14. g E Plot the graphs of $y = 2^x + 1$ in the standard window. Since $f^{-1}(7)$ is the value of x such that $f(x) = 7$, also plot the graph of $y = 7$. The correct answer choice is the point where these two graphs intersect.

An alternative solution is to solve the equation $2^x + 1 = 7$: $x = \log_2 6 = \dfrac{\log 6}{\log 2} \approx 2.3$. [4.2].

15. i   C   III is the contrapositive of the given true statement, and so III is true also. [5.7].

16. g   C   Use LIST/seq to construct the sequence as $seq(30 - 3x,x,0,70)$ and store this sequence in a list (e.g., L_1). On the Home Screen, enter $L_1(71)$ for the 71st term, -180.

An alternative solution is to use the formula for the nth term of an arithmetic sequence: $t_{71} = 30 - (70)(3) = -180$. [5.4].

17. a   C   Put your calculator in radian mode: $5x = Tan^{-1} 3 \approx 1.249$. $x \approx 0.2498$. Therefore, $\tan x = 0.255139 \approx 0.3$. [3.1].

18. a   D   Take $\log_{4.05}$ of both sides of the equation, getting $p = \log_{4.05} 5.25^q$, or $p = q\log_{4.05} 5.25$. Dividing both sides by q and changing to base 10 yields

$$\frac{p}{q} = \frac{\log 5.25}{\log 4.05} \approx 1.19 .\ [4.2].$$

19. a   C   Area of one base $= \pi r^2 = 4\pi$.
Lateral area $= 2\pi rh = 36\pi$.
Lateral area $-$ two bases $= 36\pi - 8\pi = 28\pi \approx 87.96 \approx 88$. [5.5, 6.5].

20. a   C   Put your calculator in degree mode: $\cos 67° \approx 0.3907$ $Tan^{-1} (0.3907) \approx 21$. [3.1].

21. g   C   Plot the graph of $y = x^3 + 18x - 30$ in an $x\varepsilon[0,3]$ and $y\varepsilon[-5,5]$ window, and use CALC/zero to locate a zero at $x \approx 1.48$.

An alternative solution is to use your calculator to identify a change of sign in $P(x)$ when moving from the left to the right side of an interval. Checking the interval in each answer choice in this fashion yields the correct answer choice C. [2.4]

22. a   A   The angle opposite the 37 side (call it $\angle A$) is the largest angle. By the law of cosines, $37^2 = 23^2 + 32^2 - 2(23)(32) \cos A$.

$$Cos A = \frac{37^2 - 23^2 - 32^2}{-2(23)(32)} \approx \frac{-184}{-1472} = 0.125.$$

Therefore, $A = Cos^{-1} (0.125) \approx 83°$. [3.7].

23. i   B   The set of points 13 inches from p is a sphere with center at p and radius 13 inches. The set of points 2 inches from X consists of two planes parallel to X, one above and one below X. Since the radius, 13, is greater than $5 + 2$, the two sets intersect in two circles. [5.5].

24. a   C   There are four 7s in a deck, and so P(first draw is a 7) $= \dfrac{4}{52} = \dfrac{1}{13}$. There are now only three 7s among the remaining 51 cards, and so P(second draw is a 7) $= \dfrac{3}{51} = \dfrac{1}{17}$. Therefore, P(both draws are 7s) $= \dfrac{1}{13} \cdot \dfrac{1}{17} = \dfrac{1}{221} \approx 0.00452 \approx 0.005$. [5.3].

25. a   C   Since $\sqrt{y} = 3.216$, $y \approx 10.34$, and $10y \approx 103.4$, so $\sqrt{10y} \approx 10.17$. [4.2].

26. g   C   Plot the graph of $y = \log \sqrt{2x^2 - 15}$ in the standard window and zoom in once. Use the TRACE to determine that there are no y values on the graph between the approximate x values of -2.7 and 2.7.

An alternative solution is to use the fact that the domain of the square root function is $x \geq 0$ and its range is $y \geq 0$. However, since the domain of the log function is $x > 0$, the domain of the function f must satisfy $2x^2 - 15 > 0$, or $x^2 > 7.5$. The approximate solution to this inequality is the correct answer choice C. [2.5]

27. g   D   Plot the graph of $y = 3x - 7x^{-1} + 3$ in a standard window. Inspect the graph to see that the zeros are approximately -2 and 1. Therefore, their sum is approximately -1.

An alternative solution is to use the fact that the roots of $y = 3x - 7x^{-1} + 3 = 0$ are the same as the roots of $y = 3x^2 + 3x - 7 = 0$. The sum of the roots of the latter equation is $= -\dfrac{b}{a} = \dfrac{-3}{3} = -1$.

28. i **B** $S = \frac{n}{2}[6 + (n-1)4] = T = \frac{n}{2}[16 + (n-1)2]$.
Solving for n gives 6. [5.4].

29. g **D** Plot the graph of $y = \sqrt{(1 + \sin(x)^2)}$ in an

$x\varepsilon\left[-\frac{\pi}{4}, \frac{\pi}{4}\right]$ and $y\varepsilon[-2, 2]$ window, and observe
that the maximum value of y occurs at the end-
points. Return to the Home Screen and calculate

$Y_1\left(\frac{\pi}{4}\right)$ for the correct answer choice.

An alternative solution uses the fact that on the

interval $\left[-\frac{\pi}{4}, \frac{\pi}{4}\right]$, the maximum value of

$|\sin x| = \frac{\sqrt{2}}{2}$. Therefore, the maximum value of

$f(x)$ is $\sqrt{1 + \frac{1}{2}} = \sqrt{\frac{3}{2}} \approx 1.2$. [3.4].

30. a **D** $\sin\theta = \frac{4}{5}$.

Therefore, $\theta = \sin^{-1}\left(\frac{4}{5}\right) \approx 53°$.

[4.7, 3.6].

31. g **D** Plot the graph of $f(f(x))$ in the standard window by entering $Y_1 = x^2 - 4$, $Y_2 = Y_1(x)$ and
de-selecting Y_1. The graph has 4 zeros located
symmetrically about the y-axis, making answer
choice D the only possible one.

An alternative solution is to evaluate $f(f(x)) = (x^2 - 4)^2 - 4 = 0$ and solve by setting $x^2 - 4 = \pm 2$, or
$x^2 = 6$ or 2, so $x = \pm\sqrt{6}$ or $\pm\sqrt{2}$. Again, D is the
only answer choice with 4 solutions. [1.2, 2.3].

32. i **C** $f(2) = 2\log 2 = \log 2^2 = \log 4$.
$g(\log 4) = 10^{\log 4} = 4$. [4.2].

33. a **C** $f(\sqrt{2}) = (\sqrt{2})^{\sqrt{\sqrt{2}}} \approx (\sqrt{2})^{1.19} \approx 1.5$. [4.2].

34. a **D** The width of each rectangle is 2 and the
heights are $5^0 = 1$, $5^2 = 25$, and $5^4 = 625$. Therefore, the total area is $2(1 + 25 + 625) = 1302$. [4.2].

35. g **D** Plot the graphs of $y = \pm\sqrt{9 - x^2}$ and

$y = \pm\sqrt{-x^2 + 6x - 5}$ in the standard window, but
with FORMAT set to GridOn. The "grid" consists exactly of the lattice points. ZOOM/ZBox
around the area enclosed by the two graphs, and
count the number of lattice points in that area to
be 3. The points $(1,0)$ and $(3,0)$ appear close to
the boundary, but a mental check finds that $(1,0)$
is on the boundary of the second curve, while $(3,0)$
is on the boundary of the first. [4.1].

36. a **E** First observe the facts that A is in Quadrant II
and B is in Quadrant IV imply that $A + B$ is in
Quadrant I or II, so that $(A + B) > 0$. With your
calculator either in degree or radian mode, enter
$\sin(\sin^{-1}(3/5) + \cos^{-1}(1/3))$ to get the correct
answer choice.

An alternative solution is to use the formula for
the sin of a sum of two angles. Using $\sin^2 x + \cos^2 x = 1$ and the fact that A is in Quadrant I, together

with $\sin A = \frac{3}{4}$, implies $\cos A = -\frac{4}{5}$. Similarly,

$\cos B = \frac{1}{3}$ and B in Quadrant IV implies $\sin B = $

$-\frac{2\sqrt{2}}{3}$. Substituting these values into sin

$(A + B) = \sin A \cos B + \cos A \sin B$ yields the correct answer choice. [3.5].

37. i **C**

$f(x) = f\left(2 \cdot \frac{x}{2}\right) = \left(\frac{x}{2}\right)^2 - \frac{x}{2} + 3 = \frac{x^2}{4} - \frac{x}{2} + 3$.

[1.2].

38. i   A   Using the factor theorem, substitute k for x and set the result equal to zero. Then $k^2 - k^2 + k = 0$, and $k = 0$. [2.4, 2.3].

39. i   E   If the graphs intersect on the x-axis, the value of y must be zero. Since the value of y is zero, it does not matter what k is. [2.3].

40. i   E $a^2 = 16$. $b^2 = 4$. Latus rectum $= \dfrac{2b^2}{a} = 2$. [4.1].

41. i   D

n	1	2	3	4	5
f_n	3	10	5	16	8

[5.4].

42. a   A   There are 5 consonants, CNTST, but the two Ts are indistinguishable, so there are $\dfrac{5!}{2} = 60$ ways of arranging these. There are two ways of arranging the 2 vowels in the front. Therefore, there are $2 \cdot 60 = 120$ distinguishable arrangements. [5.1].

43. i   D $H_1 d_1^{\,2} = H_2 d_2^{\,2}$. $d_2 = 3d$. $H_1 d_1^{\,2} = H_2(9d_1^{\,2})$. Therefore, $H_2 = \dfrac{1}{9} H_1$. [5.6].

44. g   A   Plot the graph of $y = \dfrac{x}{x-2} - 5$ in the standard window and observe the vertical asymptote at $x = 2$. $y > \dfrac{x}{x-2} - 5$ where this graph lies above the x-axis, and there are no integer values of x in this interval.

An alternative solution is to solve the equation $\dfrac{x}{x-2} = 5(x \neq 2)$: $x = 5x - 10$, or $x = \dfrac{5}{2}$. If you test values in the intervals $(-\infty, 2)$, $\left(2, \dfrac{5}{2}\right)$, and $\left(\dfrac{5}{2}, \infty\right)$ you find that only the middle interval satisfies the inequality, but it contains no integers. [2.5].

45. a   C   Let the starting price of the stock be $100. During the first year a 125% increase means a $125 increase to $225. During the second year an 80% increase of the $225 stock price means a $180 increase to $405. Thus, over the 2-year period the price increased $305 from the original $100 starting price. Therefore, the price increased 305%. [5.9].

46. g   E   Enter 3 into your calculator. Then enter Ans $\sqrt{\text{Ans} + 1}$ three times to accomplish three iterations that result in x_3 and the correct answer choice.

An alternative solution is to use the formula to evaluate x_1, x_2, and x_3 in turn. [5.9].

47. g   B   Use your calculator program for the Quadratic Formula to find the roots of the equation. Then subtract the smaller root (-1.54) from the larger (0.22) and round to get the correct answer choice.

An alternative solution is to substitute the values of a, b, and c into the Quadratic Formula to find the algebraic solutions; then subtract the smaller from the larger to find the decimal approximation to the answer:

$$x = \frac{-4 \pm \sqrt{16 + 12}}{6} = \frac{-2 \pm \sqrt{7}}{3};$$

$$\frac{-2 + \sqrt{7}}{3} - \frac{-2 - \sqrt{7}}{3} = \frac{2\sqrt{7}}{3} \approx 1.8. \text{ [2.3].}$$

48. i   D

	F	S
P	c	a
M	16	b

$a + b + c + 16 = 50$, $a + b = 26$, $a + c = 12$. Subtracting the first two equations and then the first and third gives $c = 8$, $b = 22$, and $a = 4$. Four students take both Spanish and physics. [5.9].

49. i   A   First evaluate the ratio $\dfrac{xy}{xz} = \dfrac{y}{z} = \dfrac{24}{48} = \dfrac{1}{2}$. Cross-multiply to get $z = 2y$, and substitute in the third equation: $yz = y(2y) = 2y^2 = 72$. Therefore, $y = 6$, $z = 12$, and $x = 4$, and $x + y + z = 22$. [5.9].

50. g   E   With your calculator in degree mode, evaluate $\sin^{-1}(\cos(100)) = -10°$ directly. To change to radians, return your calculator to radian mode and key in $-10°$ (using ANGLE/°). This will return $-0.17\ldots$.

An alternative solution is to convert $-10°$ to radians by multiplying by $\dfrac{\pi}{180°}$. [3.2].

SELF-EVALUATION CHART FOR MODEL TEST 7

SUBJECT AREA	QUESTIONS	NUMBER OF RIGHT WRONG OMITTED

Mark correct answers with C, wrong answers with X, and omitted answers with O.

Algebra
(8 questions)
Review section

1	3	7	25	26	27	44	47
4.2	4.2	5.6	4.2	2.5	2.3	2.5	2.3

___ ___ ___

Solid geometry
(4 questions)
Review section

2	9	19	23
5.5	5.5	5.5	5.5

___ ___ ___

Coordinate geometry
(5 questions)
Review section

8	12	35	39	40
2.2	2.2	5.9	2.3	4.1

___ ___ ___

Trigonometry
(9 questions)
Review section

5	13	17	20	22	29	30	36	50
3.4	3.2	3.1	3.1	3.7	3.4	3.6	3.5	3.6

___ ___ ___

Functions
(13 questions)
Review section

4	6	10	14	18	21	31	32	33	34	37	38	46
1.2	2.3	2.4	4.2	4.2	2.4	1.2	4.2	4.2	4.2	1.2	2.4	5.4

___ ___ ___

Miscellaneous
(11 questions)
Review section

11	15	16	24	28	41	42	43	45	48	49
5.9	5.7	5.4	5.3	5.4	5.3	5.1	5.6	5.9	5.9	5.9

TOTALS ___ ___ ___

Raw score = (number right) $-\frac{1}{4}$ (number wrong) = _____

Round your raw score to the nearest whole number = _____

Evaluate Your Performance
Model Test 7

Rating	Number Right
Excellent	41–50
Very good	33–40
Above average	25–32
Average	15–24
Below average	Below 15

ANSWER SHEET FOR MODEL TEST 8

Determine the correct answer for each question. Then, using a no. 2 pencil, blacken completely the oval containing the letter of your choice.

1. Ⓐ Ⓑ Ⓒ Ⓓ Ⓔ
2. Ⓐ Ⓑ Ⓒ Ⓓ Ⓔ
3. Ⓐ Ⓑ Ⓒ Ⓓ Ⓔ
4. Ⓐ Ⓑ Ⓒ Ⓓ Ⓔ
5. Ⓐ Ⓑ Ⓒ Ⓓ Ⓔ
6. Ⓐ Ⓑ Ⓒ Ⓓ Ⓔ
7. Ⓐ Ⓑ Ⓒ Ⓓ Ⓔ
8. Ⓐ Ⓑ Ⓒ Ⓓ Ⓔ
9. Ⓐ Ⓑ Ⓒ Ⓓ Ⓔ
10. Ⓐ Ⓑ Ⓒ Ⓓ Ⓔ
11. Ⓐ Ⓑ Ⓒ Ⓓ Ⓔ
12. Ⓐ Ⓑ Ⓒ Ⓓ Ⓔ
13. Ⓐ Ⓑ Ⓒ Ⓓ Ⓔ
14. Ⓐ Ⓑ Ⓒ Ⓓ Ⓔ
15. Ⓐ Ⓑ Ⓒ Ⓓ Ⓔ
16. Ⓐ Ⓑ Ⓒ Ⓓ Ⓔ
17. Ⓐ Ⓑ Ⓒ Ⓓ Ⓔ

18. Ⓐ Ⓑ Ⓒ Ⓓ Ⓔ
19. Ⓐ Ⓑ Ⓒ Ⓓ Ⓔ
20. Ⓐ Ⓑ Ⓒ Ⓓ Ⓔ
21. Ⓐ Ⓑ Ⓒ Ⓓ Ⓔ
22. Ⓐ Ⓑ Ⓒ Ⓓ Ⓔ
23. Ⓐ Ⓑ Ⓒ Ⓓ Ⓔ
24. Ⓐ Ⓑ Ⓒ Ⓓ Ⓔ
25. Ⓐ Ⓑ Ⓒ Ⓓ Ⓔ
26. Ⓐ Ⓑ Ⓒ Ⓓ Ⓔ
27. Ⓐ Ⓑ Ⓒ Ⓓ Ⓔ
28. Ⓐ Ⓑ Ⓒ Ⓓ Ⓔ
29. Ⓐ Ⓑ Ⓒ Ⓓ Ⓔ
30. Ⓐ Ⓑ Ⓒ Ⓓ Ⓔ
31. Ⓐ Ⓑ Ⓒ Ⓓ Ⓔ
32. Ⓐ Ⓑ Ⓒ Ⓓ Ⓔ
33. Ⓐ Ⓑ Ⓒ Ⓓ Ⓔ
34. Ⓐ Ⓑ Ⓒ Ⓓ Ⓔ

35. Ⓐ Ⓑ Ⓒ Ⓓ Ⓔ
36. Ⓐ Ⓑ Ⓒ Ⓓ Ⓔ
37. Ⓐ Ⓑ Ⓒ Ⓓ Ⓔ
38. Ⓐ Ⓑ Ⓒ Ⓓ Ⓔ
39. Ⓐ Ⓑ Ⓒ Ⓓ Ⓔ
40. Ⓐ Ⓑ Ⓒ Ⓓ Ⓔ
41. Ⓐ Ⓑ Ⓒ Ⓓ Ⓔ
42. Ⓐ Ⓑ Ⓒ Ⓓ Ⓔ
43. Ⓐ Ⓑ Ⓒ Ⓓ Ⓔ
44. Ⓐ Ⓑ Ⓒ Ⓓ Ⓔ
45. Ⓐ Ⓑ Ⓒ Ⓓ Ⓔ
46. Ⓐ Ⓑ Ⓒ Ⓓ Ⓔ
47. Ⓐ Ⓑ Ⓒ Ⓓ Ⓔ
48. Ⓐ Ⓑ Ⓒ Ⓓ Ⓔ
49. Ⓐ Ⓑ Ⓒ Ⓓ Ⓔ
50. Ⓐ Ⓑ Ⓒ Ⓓ Ⓔ

MODEL TEST

8

50 questions   1 hour

Tear out the preceding answer sheet. Decide which is the best choice by rounding your answer when appropriate. Blacken the corresponding space on the answer sheet. When finished, check your answers with those at the end of the test. For questions that you got wrong, note the sections containing the material that you must review. Also, if you do not fully understand how you arrived at some of the correct answers, you should review the appropriate sections. Finally, fill out the self-evaluation sheet on page 296 in order to pinpoint the topics that give you the most difficulty.

TEST DIRECTIONS

<u>Directions</u>: Decide which answer choice is best. If the exact numerical value is not one of the answer choices, select the closest approximation. Fill in the oval on the answer sheet that corresponds to your choice.

Notes:
(1) You will need to use a scientific or graphing calculator to answer some of the questions.
(2) You will have to decide whether to put your calculator in degree or radian mode for some problems.
(3) All figures that accompany problems are plane figures unless otherwise stated. Figures are drawn as accurately as possible to provide useful information for solving the problem, except when it is stated in a particular problem that the figure is not drawn to scale.
(4) Unless otherwise indicated, the domain of a function is the set of all real numbers for which the functional value is also a real number.

<u>Reference Information.</u> The following formulas are provided for your information.

Volume of a right circular cone with radius r and height h: $V = \frac{1}{3}\pi r^2 h$

Lateral area of a right circular cone if the base has circumference c and the slant height is l: $S = \frac{1}{2}cl$

Volume of a sphere of radius r: $V = \frac{4}{3}\pi r^3$

Surface area of a sphere of radius r: $S = 4\pi r^2$

Volume of a pyramid of base area B and height h: $V = \frac{1}{3}Bh$

---

1. In the diagram, the circle has a radius of 1 and center at the origin. If the point F represents a complex number, $a + bi$, which of the other points could represent the conjugate of F?

    (A) A
    (B) B
    (C) C
    (D) D
    (E) E

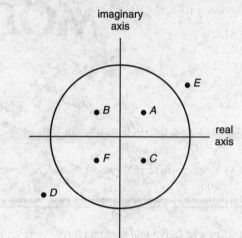

2. For what values of x and y is $|x - y| \leq |y - x|$?

    (A) $x < y$
    (B) $y < x$
    (C) $x > 0$ and $y < 0$
    (D) for no value of x and y
    (E) for all values of x and y

3. If (a,b) is a solution of the system of equations $\begin{cases} 2x - y = 7 \\ x + y = 8 \end{cases}$, then the difference, $a - b$, equals

    (A) 0
    (B) 2
    (C) 4
    (D) −12
    (E) −10

4. If $f(x) = x - 1$, $g(x) = 3x$, and $h(x) = \dfrac{5}{x}$, then $f^{-1}(g(h(5))) =$

    (A) 4
    (B) 2
    (C) $\dfrac{5}{12}$
    (D) $\dfrac{5}{6}$
    (E) $\dfrac{1}{2}$

5. A sphere is inscribed in a cube. The ratio of the volume of the sphere to the volume of the cube is

    (A) 1:2
    (B) 0.52:1
    (C) 1:3.1
    (D) 0.24:1
    (E) 0.79:1

GO ON TO THE NEXT PAGE

6. The equation $\sec^2 x - \tan x - 1 = 0$ has n solutions between $10°$ and $350°$. Then $n =$

   (A) 0
   (B) 1
   (C) 2
   (D) 3
   (E) 4

7. The nature of the roots of the equation $3x^4 + 4x^3 + x - 1 = 0$ is

   (A) three positive real roots and one negative real root
   (B) three negative real roots and one positive real root
   (C) one negative real root and three complex roots
   (D) one positive real root, one negative real root, and two complex roots
   (E) two positive real roots, one negative real root, and one complex root

8. For what value(s) of k is $x^2 + 3x + k$ divisible by $x + k$?

   (A) only 0
   (B) only 0 or 2
   (C) only 0 or -4
   (D) no value of k
   (E) any value of k

9. What number should be added to each of the three numbers 3, 11, and 27 so that the resulting numbers form a geometric sequence?

   (A) 2
   (B) 3
   (C) 4
   (D) 5
   (E) 6

10. What is the equation of the set of points that are 5 units from point (2,3,4)?

    (A) $2x + 3y + 4z = 5$
    (B) $x^2 + y^2 + z^2 - 4x - 6y - 8z = 25$
    (C) $(x-2)^2 + (y-3)^2 + (z-4)^2 = 25$
    (D) $x^2 + y^2 + z^2 = 5$
    (E) $\dfrac{x}{2} + \dfrac{y}{3} + \dfrac{z}{4} = 5$

11. If $3x^{3/2} = 4$, then $x =$

    (A) 1.1
    (B) 1.2
    (C) 1.3
    (D) 1.4
    (E) 1.5

GO ON TO THE NEXT PAGE

12. If $f(x) = x^3 - 4$, then the inverse of $f =$

(A) $-x^3 + 4$

(B) $\sqrt[3]{x+4}$

(C) $\sqrt[3]{x-4}$

(D) $\dfrac{1}{x^3 - 4}$

(E) $\dfrac{4}{\sqrt[3]{x}}$

13. If f is an odd function and $f(a) = b$, which of the following must also be true?

  I. $f(a) = -b$
  II. $f(-a) = b$
  III. $f(-a) = -b$

(A) only I

(B) only II

(C) only III

(D) only I and II

(E) only II and III

14. For all θ, $\tan\theta + \cos\theta + \tan(-\theta) + \cos(-\theta) =$

(A) 0

(B) $2\tan\theta$

(C) $2\cos\theta$

(D) $2(\tan\theta + \cos\theta)$

(E) 2

15. The period of the function $f(x) = k \cos kx$ is $\dfrac{\pi}{2}$. The amplitude of f is

(A) 2

(B) $\dfrac{1}{2}$

(C) 1

(D) $\dfrac{1}{4}$

(E) 4

16. If $f(x) = \dfrac{x+2}{(x-2)(x^2-4)}$, its graph will have

(A) one horizontal and three vertical asymptotes

(B) one horizontal and two vertical asymptotes

(C) one horizontal and one vertical asymptote

(D) zero horizontal and one vertical asymptote

(E) zero horizontal and two vertical asymptotes

USE THIS SPACE FOR SCRATCH WORK

GO ON TO THE NEXT PAGE

17. At a distance of 100 feet, the angle of elevation from the horizontal ground to the top of a building is 42°. The height of the building is

(A) 67 feet
(B) 74 feet
(C) 90 feet
(D) 110 feet
(E) 229 feet

18. A sphere has a surface area of 36π. Its volume is

(A) 113
(B) 339
(C) 201
(D) 84
(E) 905

19. A pair of dice is tossed 10 times. What is the probability that no 7s or 11s ever appear?

(A) 0.08
(B) 0.09
(C) 0.11
(D) 0.16
(E) 0.24

20. The lengths of two sides of a triangle are 50 inches and 63 inches. The angle opposite the 63-inch side is 66°. How many degrees are in the largest angle of the triangle?

(A) 72°
(B) 68°
(C) 71°
(D) 67°
(E) 66°

21. If the following instructions are followed, what number will be printed in line 6?

1. Let $A = 1$.
2. Let $x = 4$.
3. Let A be replaced by the sum of A and x.
4. Increase the value of x by 3.
5. If $x < 9$ go back to step 3.
   If $x \geq 9$ go to step 6.
6. Print the value of A.

(A) 7
(B) 10
(C) 12
(D) 0
(E) 9

USE THIS SPACE FOR SCRATCH WORK

GO ON TO THE NEXT PAGE

22. What is the period of the graph of the function $y = \dfrac{\sin x}{1 + \cos x}$?

   (A) 2π
   (B) π
   (C) $\dfrac{\pi}{2}$
   (D) $\dfrac{\pi}{4}$
   (E) 4π

23. For what values of k are the roots of the equation $kx^2 + 4x + k = 0$ real and unequal?

   (A) $0 < k < 2$
   (B) $|k| < 2$
   (C) $|k| > 2$
   (D) $k > 2$
   (E) $-2 < k < 0$ or $0 < k < 2$

24. A point moves in a plane so that its distance from the origin is always twice its distance from point $(1,1)$. All such points form

   (A) a line
   (B) a circle
   (C) a parabola
   (D) an ellipse
   (E) a hyperbola

25. If $f(x) = 3x^2 + 24x - 53$, find the negative value of $f^{-1}(0)$.

   (A) -58.8
   (B) -9.8
   (C) -1.8
   (D) -8.2
   (E) -0.2

26. The operation # is defined by the equation $a \# b = \dfrac{a}{b} - \dfrac{b}{a}$. What is the value of k if $3 \# k = k \# 2$?

   (A) ± 2.0
   (B) ± 3.0
   (C) ± 6.0
   (D) ± 2.4
   (E) ± 5.5

27. If $7^{x-1} = 6^x$, find x.

   (A) 0.08
   (B) -13.2
   (C) 0.22
   (D) 12.6
   (E) 0.52

USE THIS SPACE FOR SCRATCH WORK

GO ON TO THE NEXT PAGE

28. A red box contains eight items, of which three are defective, and a blue box contains five items, of which two are defective. An item is drawn at random from each box. What is the probability that one item is defective and one is not?

(A) $\dfrac{5}{8}$

(B) $\dfrac{19}{40}$

(C) $\dfrac{17}{32}$

(D) $\dfrac{17}{20}$

(E) $\dfrac{9}{40}$

29. If $(\log_3 x)(\log_5 3) = 3$, find x.

(A) 5
(B) 25
(C) 125
(D) 81
(E) 9

30. If $f(x) = \sqrt{x}$, $g(x) = \sqrt[3]{x+1}$, and $h(x) = \sqrt[4]{x+2}$, then $f(g(h(2))) =$

(A) 1.2
(B) 8.5
(C) 1.4
(D) 2.9
(E) 4.7

31. In $\triangle ABC$, $\angle A = 45°$, $\angle B = 30°$, and $b = 8$. Side $a =$

(A) 12
(B) 16
(C) 6.5
(D) 11
(E) 14

32. The equations of the asymptotes of the graph of $4x^2 - 9y^2 = 36$ are

(A) $y = x$ and $y = -x$
(B) $y = 0$ and $x = 0$
(C) $y = \dfrac{2}{3}x$ and $y = -\dfrac{2}{3}x$
(D) $y = \dfrac{3}{2}x$ and $y = -\dfrac{3}{2}x$
(E) $y = \dfrac{4}{9}x$ and $y = -\dfrac{4}{9}x$

GO ON TO THE NEXT PAGE

33. If $g(x - 1) = x^2 + 2$, then $g(x) =$

   (A) $x^2 - 2x + 3$
   (B) $x^2 + 2x + 3$
   (C) $x^2 + 2$
   (D) $x^2 - 2$
   (E) $x^2 - 3x + 2$

34. If $f(x) = 3x^3 - 2x^2 + x - 2$, then $f(i) =$

   (A) $-2i - 4$
   (B) $4i$
   (C) $4i - 4$
   (D) $-2i$
   (E) 0

35. If the hour hand of a clock moves k radians in 48 minutes, $k =$

   (A) 2.4
   (B) 5
   (C) 0.3
   (D) 0.4
   (E) 0.5

36. If the longer diagonal of a rhombus is 10 and the large angle is 100°, what is the area of the rhombus?

   (A) 45
   (B) 40
   (C) 37
   (D) 42
   (E) 50

37. Let $f(x) = \sqrt{x^3 - 4x}$ and $g(x) = 3x$. The sum of all values of x for which $f(x) = g(x)$ is

   (A) 0
   (B) 9
   (C) 9.4
   (D) 8
   (E) -8.5

38. How many subsets does a set with n elements have?

   (A) n^2
   (B) 2^n
   (C) $\dbinom{2n}{n}$
   (D) n
   (E) $n!$

GO ON TO THE NEXT PAGE

39. Given the statement "All vacationers are tourists," which conclusion follows logically?

   (A) If the Browns are tourists, then they are vacationers.
   (B) If the Smiths are not tourists, then they are not vacationers.
   (C) If the Polks are not vacationers, then they are not tourists.
   (D) All tourists are vacationers.
   (E) Some tourists are not vacationers.

40. For what positive value of n are the zeros of $P(x) = 5x^2 + nx + 12$ in ratio 2:3?

   (A) 0.42
   (B) 15.8
   (C) 1.32
   (D) 4.56
   (E) 25

41. If Arcsin $x = 2$ Arccos x, then $x =$

   (A) 0.5
   (B) 0.9
   (C) 0
   (D) ±0.9
   (E) ±0.5

42. A man piles 150 toothpicks in layers so that each layer has one less toothpick than the layer below. If the top layer has three toothpicks, how many layers are there?

   (A) 15
   (B) 17
   (C) 20
   (D) 148
   (E) 11,322

43. If the circle $x^2 + y^2 - 2x - 6y = r^2 - 10$ is tangent to the line $5x + 12y = 60$, the value of r is

   (A) 3.2
   (B) 1.5
   (C) 1.1
   (D) 1.7
   (E) 13

44. If $a_0 = 0.4$ and $a_{n+1} = 2|a_n| - 1$, then $a_5 =$

   (A) −0.2
   (B) 0.2
   (C) −0.6
   (D) 0.6
   (E) 0.4

USE THIS SPACE FOR SCRATCH WORK

GO ON TO THE NEXT PAGE

45. The line passing through $(1,4,-2)$ and $(2,1,4)$ can be represented by the set of equations

(A) $x = -1 + 2d$
   $y = 3 + d$
   $z = 2 + 4d$

(B) $x = 1 + 2d$
   $y = 4 + d$
   $z = -2 + 4d$

(C) $x = 2 + d$
   $y = 1 + 4d$
   $z = 4 - 2d$

(D) $x = 1 + d$
   $y = 4 - 3d$
   $z = -2 + 6d$

(E) $x = 1 - d$
   $y = -3 - 4d$
   $z = 2 + 2d$

46. As $n \rightarrow \infty$, find the limit of the product
$(\sqrt[3]{3})(\sqrt[6]{3})(\sqrt[12]{3}) \cdots (\sqrt[3 \cdot 2^n]{3})$.

(A) 2.3
(B) 1.9
(C) 2.2
(D) 2.0
(E) 2.1

47. The function $f(x) = 4x^3 - px^2 + qx - 2p$ crosses the x-axis at three points, 4, 7, and t. Find t.

(A) 0.73
(B) 0.93
(C) -0.79
(D) 0.64
(E) 0.85

48. If the length of the diameter of a circle is equal to the length of the major axis of the ellipse whose equation is $x^2 + 4y^2 - 4x + 8y - 28 = 0$, to the nearest whole number, what is the area of the circle?

(A) 113
(B) 64
(C) 28
(D) 254
(E) 452

GO ON TO THE NEXT PAGE

USE THIS SPACE FOR SCRATCH WORK

49. The force of the wind on a sail varies jointly as the area of the sail and the square of the wind velocity. On a sail of area 50 square yards, the force of a 15-mile-per-hour wind is 45 pounds. Find the force on the sail if the wind increases to 45 miles per hour.

(A) 135 pounds
(B) 405 pounds
(C) 450 pounds
(D) 225 pounds
(E) 675 pounds

50. If the riser of each step in the drawing is 6 inches and the tread is 8 inches, what is the value of |*AB*|?

(A) 40 inches
(B) 43.9 inches
(C) 46.6 inches
(D) 48.3 inches
(E) 50 inches

USE THIS SPACE FOR SCRATCH WORK

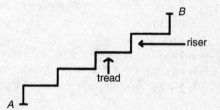

ANSWER KEY

1. B	6. D	11. B	16. C	21. C	26. D	31. D	36. D	41. B	46. E
2. E	7. D	12. B	17. C	22. A	27. D	32. C	37. C	42. A	47. E
3. B	8. B	13. C	18. A	23. E	28. B	33. B	38. B	43. B	48. A
4. A	9. D	14. C	19. A	24. B	29. C	34. D	39. B	44. B	49. B
5. B	10. C	15. E	20. B	25. B	30. A	35. D	40. B	45. D	50. B

ANSWER EXPLANATIONS

In these solutions the following notation is used:

a: active—Calculator use is necessary or, at a minimum, extremely helpful.
g: Graphing calculator is preferred.
i: inactive—Calculator use is not helpful and may even be a hindrance.

1. i   B   If $F = a + bi$, its conjugate $a - bi$ must be at point B. [4.7].

2. i   E   Since $|x - y|$ and $|y - x|$ both represent the distance between x and y, they must be equal, and they are equal for any values of x and y. [4.3].

3. i   B   Adding the equations gives $3x = 15$. $x = 5$ and $y = 3$. $a - b = 2$. [2.2, 5.9].

4. i   A $h(5) = 1$. $g(1) = 3$. Interchange x and y to find that $f^{-1}(x) = x + 1$, and so $f^{-1}(3) = 4$. [1.2, 1.3].

5. a   B   Diameter of sphere = side of cube.
Volume of sphere = $\frac{4}{3}\pi r^3$.
Volume of cube = $s^3 = (2r)^3 = 8r^3$.

$$\frac{\text{Volume of sphere}}{\text{Volume of cube}} = \frac{\frac{4}{3}\pi r^3}{8r^3} = \frac{\pi}{6} \approx \frac{3.14}{6} \approx \frac{0.52}{1}.$$

6. g   D   With your calculator in degree mode, plot the graph of $y = (1/\cos(x)^2) - \tan(x) - 1$ in an $x\varepsilon[10,350]$ and $y\varepsilon[-3,3]$ window and observe that the graph crosses the x-axis 3 times.

An alternative solution is to use the identity $\sec^2 x = 1 + \tan^2 x$, so the equation becomes $\tan^2 x - \tan x = 0$. Factoring and setting each factor to zero yields $\tan x = 0$ or $\tan x - 1 = 0$. On the interval $[10°, 350°]$, $\tan x = 0$ when $x = 180°$, and $\tan x = 1$ when $x = 45°$ or $225°$. [3.5].

7. g   D   Plot the graph of $y = 3x^4 + 4x^3 + x - 1$ in the standard window. Observe that the graph crosses the x-axis twice—once at a positive x value and once at a negative one. Since the function is a degree 4 polynomial, there are 4 roots, so the other two must be complex conjugates. [2.4].

8. i   B   If $P(x)$ represents the polynomial, $P(-k) = k^2 - 3k + k = 0$. $k(k - 2) = 0$. $k = 0, 2$. [2.4, 2.3].

9. i   D   If x is the number, then to have a geometric sequence, $\frac{11+x}{3+x} = \frac{27+x}{11+x}$. Cross-multiplying yields $121 + 22x + x^2 = 81 + 30x + x^2$, and subtracting x^2 and solving gives the desired solution. [5.4].

10. i   C   The set of points represents a sphere with equation $(x - 2)^2 + (y - 3)^2 + (z - 4)^2 = 5^2$. [5.5].

11. g   B   Plot the graph of $y = 3x^{3/2} - 4$ in the standard window and use CALC/zero to find the correct answer choice.

An alternative solution is to divide the equation by 3 to get $x^{3/2} = \frac{4}{3}$ and then raise both sides to the $\frac{2}{3}$ power: $x = \left(\frac{4}{3}\right)^{2/3} \approx 1.2$. [4.2].

12. i   B   Let $y = f(x) = x^3 - 4$. To get the inverse, interchange x and y and solve for y. $x = y^3 - 4$. $y = \sqrt[3]{x + 4}$. [1.3].

13. i   C   Use the definition of an odd function. If $f(a) = b$, then $f(-a) = -b$. Only III is true. [1.4].

14. i   C   Tangent is an odd function, so $\tan(-\theta) = -\tan\theta$, and cosine is an even function, so $\cos(-\theta) = \cos\theta$. Therefore, the sum in the problem is $2\cos\theta$. [1.4, 3.1].

15. i   E   Period = $\frac{2\pi}{k} = \frac{\pi}{2}$. $k = 4$. Amplitude = 4. [3.4].

16. g   C   Plot the graph of $y = \frac{x+2}{(x-2)(x^2-4)}$ in the standard window and observe one vertical asymptote ($x = 2$) and one horizontal asymptote ($y = 0$).

17. a   C   Tan $42° = \frac{x}{100}$.
$x = 100 \tan 42 \approx 90$.
[3.1].

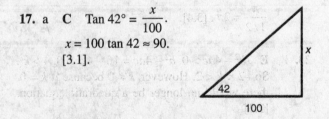

18. a **A** $4\pi r^2 = 36\pi$. $r^2 = 9$. $r = 3$. $V = \dfrac{4}{3}\pi r^3 = 36\pi \approx$
≈ 113. [5.5].

19. a **A** There are six ways to get a 7 and two ways to get an 11 on two dice, and so there are 28 ways to get anything else. Therefore, $P(\text{no 7 or 11}) =$

$P(\text{always getting something else}) = \left(\dfrac{28}{36}\right)^{10} \approx$

$(0.7777)^{10} \approx 0.08$. [5.3].

20. a **B** Use the law of sines: $\dfrac{\sin B}{50} = \dfrac{\sin 66°}{63}$. $\sin B =$

$\dfrac{50\sin 66}{63} \approx \dfrac{45.68}{63} \approx 0.725$. $B = \text{Sin}^{-1}(0.725) \approx$
$46°$ and $\angle A = 180 - 46 - 66 = 68°$. [3.7].

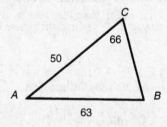

21. i **C** Keep track of the values of A and x:

Step	1	2	3	4	3	4
A	1		5		12	
x		4		7		10

Print 12 in line 6. [5.9].

22. g **A** With your calculator in radian mode, plot the graph of $y = \sin x/(1 + \cos x)$ and observe that the period is 2π.

An alternative solution is to use the identity $\tan\dfrac{x}{2} = \dfrac{\sin x}{1 + \cos x}$ and the fact that the period of

$\tan x$ is π to deduce the period of $\tan\dfrac{x}{2}$ as

$\dfrac{\pi}{1/2} = 2\pi$. [3.4].

23. i **E** $b^2 - 4ac > 0$. $b^2 - 4ac = 16 - 4k^2 > 0$. $4 > k^2$. So $-2 < k < 2$. However, $k \neq 0$ because if $k = 0$, there would no longer be a quadratic equation. [2.3].

24. i **B** If (x, y) represents any of the points,
$\sqrt{(x - 0)^2 + (y - 0)^2} = 2\sqrt{(x - 1)^2 + (y - 1)^2}$
$x^2 + y^2 = 4(x^2 - 2x + y^2 - 2y + 2)$. $3x^2 + 3y^2 - 8x - 8y + 8 = 0$. All such points form a circle. [4.1].

25. g **B** Find $f^{-1}(0)$ means to find a value of x that makes $3x^2 + 24x - 53 = 0$. Use your program for the Quadratic Formula to determine the correct answer choice.

Use the Quadratic Formula to evaluate the solutions and then use your calculator to find the decimal approximation to the negative solution. [1.3, 2.3].

26. g **D** Plot the graphs of $y = \dfrac{3}{x} - \dfrac{x}{3}$ and $y = \dfrac{x}{2} - \dfrac{2}{x}$ in the standard wdow, and use CALC/intersect to find the points of intersection ± 2.4.

$3\#k = \dfrac{3}{k} - \dfrac{k}{3} = \dfrac{9 - k^2}{3k}$ and $k\#2 = \dfrac{k}{2} - \dfrac{2}{k} = $

$\dfrac{k^2 - 4}{2k}$. Setting these two equal to each other and multiplying both sides by $6k$ yields $18 - 2k^2 = 3k^2 - 12$, or $k^2 = 6$. Therefore, $k = \pm\sqrt{6} = \pm 2.4$. [5.9].

27. g **D** Enter $7^{\wedge}(x - 1) - 6^{\wedge}x$ into Y_1, and generate an Auto table with TblStart $= -14$ and Δtbl $= 1$. Scan through values of x and observe a change in the sign of Y_1 between $x = 12$ and $x = 13$. Thus, D is the correct answer choice.

An alternative solution is to take $\log_7$ of both sides to get

$x - 1 = \log_7 6^x = x\log_7 6 = x\dfrac{\log 6}{\log 7}$. Therefore,

$x\left(1 - \dfrac{\log 6}{\log 7}\right) = 1$, so $x = \left(1 - \dfrac{\log 6}{\log 7}\right)^{-1} \approx 12.6$.

[4.2].

28. i **B** Probability that an item from the red box is defective and an item from the blue box is good $=$ $\dfrac{3}{8} \cdot \dfrac{3}{5} = \dfrac{9}{40}$. Probability that an item from the red box is good and that an item from the blue box is defective $= \dfrac{5}{8} \cdot \dfrac{2}{5} = \dfrac{10}{40}$. Since these are mutually exclusive events, the answer is $\dfrac{9}{40} + \dfrac{10}{40} = \dfrac{19}{40}$.

[5.3].

29. a   C By the change-of-base theorem, $\log_3 x = \dfrac{\log_5 x}{\log_5 3}$. Therefore $\log_5 x = 3$ and $x = 5^3 = 125$. [4.2].

An alternative solution is to use the change of base theorem to get

$$\left(\frac{\log x}{\log 3}\right)\left(\frac{\log 3}{\log 5}\right) = \frac{\log x}{\log 5} = 3.$$

$\log x = 3 \log 5 \approx 2.0969$. Therefore, $x \approx 10^{2.0969} \approx 125$.

30. g   A Enter f into Y_1; g into Y_2; and h into Y_3. Then return to the Home Screen and evaluate $Y_1(Y_2(Y_3(2)))$ to get the correct answer choice.

An alternative solution is to evaluate each function, starting with h in turn:

$$\sqrt{\left(\sqrt[3]{\sqrt[4]{2+2}+1}\right)} + \sqrt{\left((4)^{\wedge}(1/4)+1\right)^{\wedge}(1/3)} \approx 1.2.$$

[1.2].

31. a   D Use the law of sines: $\dfrac{\sin 45°}{a} = \dfrac{\sin 30°}{8}$.

$\dfrac{1}{2}a = 8\dfrac{\sqrt{2}}{2} \cdot a = 8\sqrt{2} \approx 11$. [3.7].

32. i   C From this form of the equation of the hyperbola, $\dfrac{x^2}{9} - \dfrac{y^2}{4} = 1$, the equations of the asymptotes can be found from $\dfrac{x^2}{9} - \dfrac{y^2}{4} = 0$. Thus, $y = \pm\dfrac{2}{3}x$. [4.1].

33. i   B Since $x = (x - 1 + 1)$, $g(x) = g((x - 1) + 1) = (x + 1)^2 + 2 = x^2 + 2x + 3$. [1.2, 2.3].

34. i   D $f(i) = 3i^3 - 2i^2 + i - 2 = -3i + 2 + i - 2 = -2i$. [2.4, 4.7].

35. a   D In 1 hour, the hour hand moves $\dfrac{1}{12}$ of the way around the clock, or $\dfrac{2\pi}{12} = \dfrac{\pi}{6}$ radians.

$\dfrac{48}{60} \cdot \dfrac{\pi}{6} = \dfrac{2\pi}{15} \approx \dfrac{6.28}{15} \approx 0.4$. [3.2].

36. a   D Diagonals of a rhombus are perpendicular and bisect each other and the angles of the rhombus.

Tan $40° = \dfrac{x}{5}$. $x = 5 \tan 40°$. $x \approx 4.195$.

$A = \dfrac{1}{2}d_1 d_2 = \dfrac{1}{2}(10)(2)(4.195) = 41.95 \approx 42$. [3.1, 6.5].

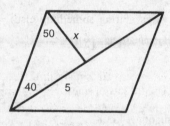

37. g   C Plot the graphs of f and g in an $x\varepsilon[-10,15]$ and $y\varepsilon[-10, 50]$ window. One point of intersection is the origin. Find the other by using CALC/ intersection at $x \approx 9.4$ to get the correct answer choice.

An alternative solution is to set $\sqrt{x^3 - 4x} = 3x$, square both sides, and factor out x to get $x(x^2 - 9x - 4) = 0$. Use the Quadratic Formula with the second factor to find the solutions $x \approx -0.4, 9.4$, and observe that the first does not satisfy the original equation. Thus, the two solutions are 0 and 9.4, resulting in the correct answer choice. [1.2, 2.3].

38. i   B Each element is either in a subset or not, so there are 2 choices for each of n elements. This yields $2 \times 2 \times \cdots \times 2$ (n factors) $= 2^n$ subsets. [5.9].

39. i   B Counterexamples of Choices A, C, and D can be found. Choice E does not follow if the sets of tourists and vacationers are equal. [5.7].

40. a   B If the zeros of $P(x)$ are in the ratio $2:3$, they must take the form $2k$ and $3k$ for some value k, and $(x - 2k)(x - 3k) = x^2 - 5k + 6k^2 = 0$. Dividing $P(x)$ by 5 and equating coefficients yields

$\dfrac{n}{5} = -5k$ and $\dfrac{12}{5} = 6k^2$. Therefore, $k = \pm\sqrt{\dfrac{2}{5}}$.

Since the problem asks for a positive value of n,

we use $k = -\sqrt{\dfrac{2}{5}}$, so $n = -25k \approx 15.8$.

41. g   B With your calculator in radian mode, plot the graphs of $y = \sin^{-1}x$ and $y = 2\cos^{-1}x$ in an $x\varepsilon[-2,2]$ and $y\varepsilon[-2,2]$ window. Use CALC/ intersection to find the correct answer choice.

An alternative solution is to let $A = \text{Arcsin}\, x$ and $B = \text{Arc cos}\, x$. Then $\sin A = \cos B = x$, so A and B are complementary, and $A = 2B$. Therefore, $B = 30°$ and $A = 60°$, and it follows that $x = \dfrac{\sqrt{3}}{2} \approx 0.9$. [3.6].

42. i **A** This is an arithmetic series with $t_1 = 3$, $d = 1$, and $S = 150$. $150 = \dfrac{n}{2}[6 + (n-1) \cdot 1]$. $n = 15$. [5.4].

43. g **B** Complete the square in the equation of the circle: $(x-1)^2 + (y-3)^2 = r^2$. The center of the circle is at $(1,3)$. If the circle is tangent to the line, then the distance between $(1,3)$ and the line $5x + 12y - 60 = 0$ is the radius of the circle, r. Use the program in your calculator to find that distance between the point $(1,3)$ and the line $5x + 12y - 60 = 0$, the correct answer choice.

An alternative solution is to complete the square to find the center of the circle, then evaluate the distance between $(1,3)$ and $5x + 12y - 60 = 0$ by using the formula $D = \dfrac{|Ax + By + x|}{\sqrt{A^2 + B^2}}$. [4.1, 2.3].

44. g **B** Enter 0.4 into your calculator, followed by $2|\text{Ans}| - 1$ five times to get $a_5 = 0.2$.

An alternative solution is to evaluate each a_i in turn:

$a_1 = 2|0.4| - 1 = -0.2$

$a_2 = 2|a_1| - 1 = 2|-0.2| - 1 = -0.6$

$a_3 = 2|a_2| - 1 = 2|-0.6| - 1 = 0.2$

$a_4 = 2|a_3| - 1 = 2|0.2| - 1 = -0.6$

$a_5 = 2|a_4| - 1 = 2|-0.6| - 1 = 0.2$ [5.9].

45. i **D** Find the direction numbers by taking the difference of the corresponding coordinates of the two points: 1, –3, 6. Choice D is the only possibility. [5.5].

46. g **E** The infinite product can be approximated by using your calculator and a "large" value of n. The exponents $\dfrac{1}{3}, \dfrac{1}{6}, \dfrac{1}{12}, \ldots$ form a geometric sequence with a first term of $\dfrac{1}{3}$ and constant ratio of

$\dfrac{1}{2}$. Enter $prod\left(seq\left(3^\wedge\left(\left(\dfrac{1}{3}\right)\left(\dfrac{1}{2}\right)^\wedge x\right), x, 0, 50\right)\right)$ to approximate the product of the first 50 terms as 2.08. Evaluating the product of 75 terms yields the same approximation to 9 decimals, so choose E.

An alternative solution is to recognize that the desired product is equal to $3^{1/3+1/6+1/12+\ldots}$, so the exponent is the sum of an infinite geometric series with $t_1 = \dfrac{1}{3}$ and $r = \dfrac{1}{2}$. Using the formula for the sum of such a series yields $\dfrac{1/3}{1-1/2} = \dfrac{2}{3}$, so the desired product is $3^{2/3} \approx 2.1$. [5.4].

47. a **E** The sum of the zeros $= 4 + 7 + t = -\dfrac{-p}{4}$. $p = 4t + 44$. The product of the zeros $= (4)(7)(t) = -\dfrac{-2p}{4}$. $p = 56t$. Therefore, $44 + 4t = 56t$ and $t = \dfrac{44}{52} \approx 0.85$. [2.4].

48. a **A** Complete the square on the ellipse formula, and put the equation in standard form: $x^2 - 4x + 4 + 4(y^2 + 2y - 1) = 28 + 4 + 4$. $\dfrac{(x-2)^2}{36} + \dfrac{(y+1)^2}{9} = 1$. This leads to the length of the major axis: $2\sqrt{36} = 12$. Therefore, the radius of the circle is 6, and the area $= 36\pi \approx 36 \cdot 3.14 \approx 113$. [4.1].

49. a **B** $\dfrac{F}{Av^2} = K$. First case: $\dfrac{45}{(50)(15)^2} = \dfrac{1}{250} = K$. Second case: $\dfrac{F}{(50)(45)^2} = \dfrac{1}{250}$. $F = \dfrac{(50)(45)^2}{250} = \dfrac{(45)^2}{5} = 9(45) = 405$. $F = 405$. [5.6].

50. g **B** Total horizontal distance traveled $= (4)(8) = 32$. Total vertical distance traveled $= (5)(6) = 30$. If a coordinate system is superimposed on the diagram with A at $(0,0)$, then B is at $(32,30)$. Use the program on your calculator to find the distance between two points to compute the correct answer choice. [2.2].

SELF-EVALUATION CHART FOR MODEL TEST 8

SUBJECT AREA	QUESTIONS	NUMBER OF RIGHT WRONG OMITTED

Mark correct answers with C, wrong answers with X, and omitted answers with O.

Algebra
(8 questions)
Review section

2	3	7	11	23	27	29	49
4.3	2.2	2.4	4.2	2.3	4.2	4.2	5.6

___ ___ ___

Solid geometry
(4 questions)
Review section

5	10	18	45
5.5	5.5	5.5	5.5

___ ___ ___

Coordinate geometry
(6 questions)
Review section

1	24	32	43	48	50
4.7	4.1	4.1	1.5	4.1	2.2

___ ___ ___

Trigonometry
(10 questions)
Review section

6	14	15	17	20	22	31	35	36	41
3.5	3.1	3.4	3.1	3.7	3.4	3.7	3.2	3.1	3.6

___ ___ ___

Functions
(12 questions)
Review section

4	8	12	13	16	25	30	33	34	37	40	47
1.2	2.4	1.3	1.4	4.5	2.3	4.2	1.2	2.4	1.2	2.3	2.4

___ ___ ___

Miscellaneous
(10 questions)
Review section

9	19	21	26	28	38	39	42	44	46
5.4	5.3	5.9	5.9	5.3	5.9	5.7	5.4	5.9	4.2

TOTALS   ___ ___ ___

Raw score = (number right) $-\frac{1}{4}$ (number wrong) = ____

Round your raw score to the nearest whole number = ____

Evaluate Your Performance
Model Test 8

Rating	Number Right
Excellent	41–50
Very good	33–40
Above average	25–32
Average	15–24
Below average	Below 15

SUMMARY OF FORMULAS

APPENDIX

CHAPTER 2:
POLYNOMIAL FUNCTIONS

Linear Functions:

General form of the equation: $Ax + By + C = 0$

Slope-intercept form: $y = mx + b$, where m represents the slope and b the y-intercept

Point-slope form: $y - y_1 = m(x - x_1)$, where m represents the slope and (x_1, y_1) are the coordinates of some point on the line

Slope: $m = \dfrac{y_1 - y_2}{x_1 - x_2}$, where (x_1, y_1) and (x_2, y_2) are the coordinates of two points

Parallel lines have equal slopes.

Perpendicular lines have slopes that are negative reciprocals.

If m_1 and m_2 are the slopes of two perpendicular lines, $m_1 \cdot m_2 = -1$.

Distance between two points with coordinates (x_1, y_1) and $(x_2, y_2) = \sqrt{(x_1 - x_2)^2 + (y_1 - y_2)^2}$

Coordinates of the midpoint between two points = $\left(\dfrac{x_1 + x_2}{2}, \dfrac{y_1 + y_2}{2} \right)$

Distance between a point with coordinates (x_1, y_1) and a line $Ax + By + C = 0 = $
$$\dfrac{\left| Ax_1 + By_1 + C \right|}{\sqrt{A^2 + B^2}}$$

If θ is the angle between two lines, $\tan \theta = \dfrac{m_1 - m_2}{1 + m_1 m_2}$, where m_1 and m_2 are the slopes of the two lines.

Quadratic Functions:

General quadratic equation: $ax^2 + bx + c = 0$

General quadratic formula:
$$x = \dfrac{-b \pm \sqrt{b^2 - 4ac}}{2a}$$

General quadratic function: $y = ax^2 + bx + c$

Coordinates of vertex: $\left(-\dfrac{b}{2a}, c - \dfrac{b^2}{4a} \right)$

Axis of symmetry equation: $x = -\dfrac{b}{2a}$

Sum of zeros (roots) $= -\dfrac{b}{a}$

Product of zeros (roots) $= \dfrac{c}{a}$

Nature of zeros (roots):
If $b^2 - 4ac < 0$, two complex numbers
If $b^2 - 4ac = 0$, two equal real numbers
If $b^2 - 4ac > 0$, two unequal real numbers

CHAPTER 3:
TRIGONOMETRIC FUNCTIONS

$\sin \theta = \dfrac{\text{opposite}}{\text{hypotenuse}}$ $\cos \theta = \dfrac{\text{adjacent}}{\text{hypotenuse}}$

$\tan \theta = \dfrac{\text{opposite}}{\text{adjacent}}$ $\cot \theta = \dfrac{\text{adjacent}}{\text{opposite}}$

$\sec \theta = \dfrac{\text{hypotenuse}}{\text{adjacent}}$ $\csc \theta = \dfrac{\text{hypotenuse}}{\text{opposite}}$

$\pi^R = 180°$

Length of arc in circle of radius r and central angle θ is given by $r\theta^R$.

Area of sector of circle of radius r and central angle θ is given by $\frac{1}{2}r^2\theta^R$.

Trigonometric Reduction Formulas:

1. $\sin^2 x + \cos^2 x = 1$
2. $\tan^2 x + 1 = \sec^2 x$ } Pythagorean identities
3. $\cot^2 x + 1 = \csc^2 x$

4. $\sin(A+B) =$
   $\sin A \cdot \cos B + \cos A \cdot \sin B$
5. $\sin(A-B) =$
   $\sin A \cdot \cos B - \cos A \cdot \sin B$
6. $\cos(A+B) =$
   $\cos A \cdot \cos B - \sin A \cdot \sin B$
7. $\cos(A-B) =$
   $\cos A \cdot \cos B + \sin A \cdot \sin B$ } sum and difference formulas
8. $\tan(A+B) =$
   $\dfrac{\tan A + \tan B}{1 - \tan A \cdot \tan B}$
9. $\tan(A-B) =$
   $\dfrac{\tan A - \tan B}{1 + \tan A \cdot \tan B}$

10. $\sin 2A = 2\sin A \cdot \cos A$
11. $\cos 2A = \cos^2 A - \sin^2 A$
12. $\quad = 2\cos^2 A - 1$ } double-angle formulas
13. $\quad = 1 - 2\sin^2 A$
14. $\tan 2A = \dfrac{2\tan A}{1 - \tan^2 A}$

15. $\sin \frac{1}{2}A = \pm\sqrt{\dfrac{1 - \cos A}{2}}$
16. $\cos \frac{1}{2}A = \pm\sqrt{\dfrac{1 + \cos A}{2}}$
17. $\tan \frac{1}{2}A = \pm\sqrt{\dfrac{1 - \cos A}{1 + \cos A}}$ } half-angle formulas
18. $\quad = \dfrac{1 - \cos A}{\sin A}$
19. $\quad = \dfrac{\sin A}{1 + \cos A}$

In any $\triangle ABC$:

Law of sines : $\dfrac{\sin A}{a} = \dfrac{\sin B}{b} = \dfrac{\sin C}{c}$

Law of cosines : $a^2 = b^2 + c^2 - 2bc \cdot \cos A$

Area $= \dfrac{1}{2}bc \cdot \sin A$

CHAPTER 4: MISCELLANEOUS RELATIONS AND FUNCTIONS

General Quadratic Equation in Two Variables:

$$Ax^2 + Bxy + Cy^2 + Dx + Ey + F = 0$$

If $B^2 - 4AC < 0$ and $A = C$, graph is a circle.
If $B^2 - 4AC < 0$ and $A \neq C$, graph is an ellipse.
If $B^2 - 4AC = 0$, graph is a parabola.
If $B^2 - 4AC > 0$, graph is a hyperbola.

Circle:

$(x - h)^2 + (y - k)^2 = r^2$
   with center at (h,k) and radius $= r$

Ellipse:

$\dfrac{(x-h)^2}{a^2} + \dfrac{(y-k)^2}{b^2} = 1$, major axis horizontal

$\dfrac{(x-h)^2}{b^2} + \dfrac{(y-k)^2}{a^2} = 1$, major axis vertical,

where $a^2 = b^2 + c^2$. Coordinates of center: (h,k).
Vertices: $\pm a$ units along major axis from center
Foci: $\pm c$ units along major axis from center
Minor axis: perpendicular to major axis at center

Length $= 2b$

Eccentricity $= \dfrac{c}{a}$

Length of latus rectum $= \dfrac{2b^2}{a}$

Hyperbola:

$\dfrac{(x-h)^2}{a^2} - \dfrac{(y-k)^2}{b^2} = 1$, transverse axis horizontal

$\dfrac{(y-k)^2}{a^2} - \dfrac{(x-h)^2}{b^2} = 1$, transverse axis vertical, where

$c^2 = a^2 + b^2$. Coordinates of center: (h,k).

Vertices: $\pm a$ units along the transverse axis from center
Foci: $\pm c$ units along the transverse from center
Conjugate axis: perpendicular to transverse axis at center

Eccentricity $= \dfrac{c}{a}$

Length of latus rectum $= \dfrac{2b^2}{a}$

Asymptotes: Slopes $=$

$\pm\dfrac{b}{a}$ if transverse axis is horizontal

$\pm\dfrac{a}{b}$ if transverse axis is vertical

Parabola:

$(x-h)^2 = 4p(y-k)$ opens up or down—axis of symmetry is vertical

$(y-k)^2 = 4p(x-h)$, opens to the side—axis of symmetry is horizontal

Coordinates of vertex: (h,k)

Equation of axis of symmetry:

   $x = h$ if vertical

   $y = k$ if horizontal

Focus: p units along the axis of symmetry from vertex

Equation of directrix:

   $y = -p$ if axis of symmetry is vertical

   $x = -p$ if axis of symmetry is horizontal

Eccentricity $= 1$

Length of latus rectum $= 4p$

Exponents:

$$x^a \cdot x^b = x^{a+b} \qquad \frac{x^a}{x^b} = x^a - b$$

$$(x^a)^b = x^{ab}$$

$$x^0 = 1 \qquad x^{-a} = \frac{1}{x^a}$$

Logarithms:

$$\log_b(p \cdot q) = \log_b p + \log_b q \qquad \log_b\left(\frac{p}{q}\right) = \log_b p - \log_b q$$

$$\log_b p^x = x \cdot \log_b p \qquad \log_b 1 = 0$$

$$\log_b p = \frac{\log_a p}{\log_a b} \qquad \log_b b = 1$$

$$b^{\log_b p} = p$$

$\text{Log}_b N = x$ if and only if $b^x = N$

Absolute Value:

If $x \geq 0$, then $|x| = x$.

If $x < 0$, then $|x| = -x$.

Greatest Integer Function:

$[x] = i$, where i is an integer and $i \leq x < i + 1$

Polar Coordinates:

$x = r \cdot \cos\theta \qquad y = r \cdot \sin\theta$

$x^2 + y^2 = r^2$

De Moivre's Theorem:

If

$$z_1 = x_1 + y_1 i = r_1(\cos\theta_1 + i \cdot \sin\theta_1) = r_1 \text{ cis } \theta_1$$

and

$$z_2 = x_2 + y_2 i = r_2(\cos\theta_2 + i \cdot \sin\theta_2) = r_2 \text{ cis } \theta_2:$$

1. $z_1 \cdot z_2 = r_1 \cdot r_2 [\cos(\theta_1 + \theta_2) + i \cdot \sin(\theta_1 + \theta_2)]$

    $= r_1 \cdot r_2 \cdot \text{cis}(\theta_1 + \theta_2)$

2. $\dfrac{z_1}{z_2} = \dfrac{r_1}{r_2}[\cos(\theta_1 - \theta_2) + i \cdot \sin(\theta_1 - \theta_2)]$

    $= \dfrac{r_1}{r_2} \text{cis}(\theta_1 - \theta_2)$

3. $z^n = r^n(\cos n\theta + i \cdot \sin n\theta) = r^n \text{cis } n\theta$

4. $z^{1/n} = r^{1/n}\left(\cos\dfrac{\theta + 2\pi k}{n} + i \cdot \sin\dfrac{\theta + 2\pi k}{n}\right)$

    $= r^{1/n}\text{cis}\dfrac{\theta + 2\pi k}{n}$, where k is an integer taking on values from 0 to $n-1$.

CHAPTER 5: MISCELLANEOUS TOPICS

Permutations:

$_nP_r = \dfrac{n!}{(n-r)!}$, where $n! = n(n-1)(n-2)\cdots 3 \cdot 2 \cdot 1$

Circular permutation (e.g., around a table) of n elements $= (n-1)!$

Circular permutation (e.g., beads on a bracelet) of n elements $= \dfrac{(n-1)!}{2}$

Permutations of n elements with a repetitions and with b repetitions $= \dfrac{n!}{a! b!}$

Combinations:

$$_nC_r = \binom{n}{r} = \frac{n!}{(n-r)! r!} = \frac{_nP_r}{n!}$$

Binomial Theorem:

There are $n + 1$ terms in $(a + b)^n$.

The sum of the exponents in each term is n.

The exponent on b is 1 less than the number of the term.

Coefficient of each term $= \begin{pmatrix} n \\ \text{either exponent} \end{pmatrix}$

Probability:

$$P(\text{event}) = \frac{\text{number of ways to get a successful result}}{\text{total number of ways of getting any result}}$$

Independent events: $P(A \cap B) = P(A) \cdot P(B)$

Mutually exclusive events: $P(A \cap B) = 0$

   and $P(A \cup B) = P(A) + P(B)$

Sequences and Series:

Arithmetic Sequence (or Progression)

nth term $= t_n = t_1 + (n-1)d$

Sum of n terms $= S_n = \dfrac{n}{2}(t_1 + t_n)$

$\qquad\qquad\qquad = \dfrac{n}{2}[2t_1 + (n-1)d]$

Geometric Sequence (or Progression)

nth term $= t_n = t_1 r^{n-1}$

Sum of n terms $= S_n = \dfrac{t_1(1-r^n)}{1-r}$

If $|r| < 1$, $S_\infty = \lim\limits_{n\to\infty} S_n = \dfrac{t_1}{1-r}$

Vectors:

If $\vec{V} = (v_1, v_2)$ and $\vec{U} = (u_1, u_2)$,

$\vec{V} + \vec{U} = (v_1 + u_1, v_2 + u_2)$

$\vec{V} \cdot \vec{U} = v_1 u_1 + v_2 u_2$

Two vectors are perpendicular if and only if $\vec{V} \cdot \vec{U} = 0$.

Determinants:

$\begin{vmatrix} a & b \\ c & d \end{vmatrix} = ad - bc$

Geometry:

Distance between two points with coordinates
(x_1, y_1, z_1) and $(x_2, y_2, z_2) =$

$\sqrt{(x_1 - x_2)^2 + (y_1 - y_2)^2 + (z_1 - z_2)^2}$.

Distance between a point with coordinates (x_1, y_1, z_1) and a plane with equation $Ax + By + Cz + D = 0 =$

$$\dfrac{Ax_1 + By_1 + Cz_1 + D}{\sqrt{A^2 + B^2 + C^2}}$$

Triangle

$A = \dfrac{1}{2}bh$; $b =$ base, $h =$ height

$A = \dfrac{1}{2}ab \sin C$; a, $b =$ any two sides, $C =$ angle included between sides a and b

Heron's formula:

$A = \sqrt{s(s-a)(s-b)(s-c)}$; a, b, c are the three sides of the triangle,

$s = \dfrac{1}{2}(a + b + c)$

Rhombus

Area $= bh = \dfrac{1}{2}d_1 d_2$; $b =$ base, $h =$ height, $d =$ diagonal

Cylinder

Volume $= \pi r^2 h$

Lateral surface area $= 2\pi rh$

Total surface area $= 2\pi rh + 2\pi r^2$

In all formulas, $r =$ radius of base, $h =$ height

Cone

Volume $= \dfrac{1}{3}\pi r^2 h$

Lateral surface area $= \pi r \sqrt{r^2 + h^2}$

Total surface area $= \pi r \sqrt{r^2 + h^2} + \pi r^2$

In all formulas, $r =$ radius of base, $h =$ height

Sphere

Volume $= \dfrac{4}{3}\pi r^3$

Surface area $= 4\pi r^2$

In all formulas, $r =$ radius

INDEX

Absolute value, 75–77, 299
  graphing calculator to solve problems involving, 128–129
Ambiguous case, 62
Amplitude, 51–52
ANGLE, 133–134
Angle
  definition of, 45–48
  quadrantal, 49
  special types of, 49–50
Arc, 47–48
Arccos, 59
Arcsin, 59
Arithmetic means, 94
Arithmetic sequence, 93, 299–300
Asymptotes, 69
Average, 94, 105–106
Axis of symmetry, 297

Binary operation, 108
Binomial theorem, 88–90, 299

Calculator
  absolute value problems, 128–129
  algorithms, 122–123
  ANGLE program, 133–134
  applications of, 122–131
  basic operations using, 119–121
  combinations, 130
  commands, 120–121
  computation, 120
  description of, 117–118
  D2P, 132, 134
  DPL, 133–134
  editing using, 120
  equations solved using, 127
  factorial command, 130
  function evaluations, 124–125
  good window, 123–124
  greatest integer function problems, 128–129
  iteration of function using, 130
  keys, 119
  menu keys, 121
  MIDPT, 133–134
  mode, 121
  nonfunction equations, 126
  parametric equations, 131
  permutation, 130
  programs, 132–135
  QUADFORM program, 133, 135
  real mode, 121

  real zeros, 125–126
  screens, 119
  sequence generated using, 129
  support functions, 121
  symmetry, 127
  syntax, 122
Circle, 68, 298
Circular permutation, 87
Coefficients, 38
Cofunctions, 46
Combinations, 87–88, 130, 299
Combined variation, 103
Common ratio, 94
Complementary angles, 46
Composition of function, 26
Cone, 300
Conic sections, 68–72
Conjugate axis, 69
Conjunction, 103
Constant of proportionality, 103
Constant of variation, 103
Coordinates of vertex, 297
Cos^{-1}, 59
Cosecant, 49, 52
Cosine, 49, 52
Cotangent, 49, 52
Cube roots, 82
Cylinder, 300

Degenerate case, 71
De Moivre's theorem, 81, 299
Dependent events, 91
Depressed equation, 40
Descartes' rule of signs, 38, 40
Descriptive statistics, 105
Determinants, 300
Diagnostic test
  answer sheet, 3
  directions for, 5–15
  self-evaluation chart, 21
Direction cosines, 98–99
Direction numbers, 98
Directrix, 69
Direct variation, 102
Discriminant, 35
Disjunction, 103–104
Distributive property, 56
Domain, 25
Dot product, 98
Double-angle formula, 54
Double zero, 40

D2P, 132, 134
DPL, 133–134

Eccentricity
    of ellipse, 69
    of hyperbola, 69
Editing, 120
Ellipse, 68–69, 298
Equations
    depressed, 40
    description of, 53–58
    graphs for solving, 127
    parametric, 79–80, 131
Even function, 28–30
Exponents, 72–75, 299

Factorial command, 130
Factor theorem, 37
Finite sequence, 93
Foci, 69
Focus, 69
Formulas
    general quadratic equation in two variables, 298–299
    polynomial functions, 297
    quadratic functions, 297
    trigonometric functions, 297–298
    trigonometric reduction functions, 298
Frequency, 51–52
Function
    composition of, 26
    definition of, 25
    even, 28–30
    graphing calculator evaluation of, 124–125
    greatest integer, 77
    inverse, 27–28, 59–61
    iteration of, 130
    logarithmic, 72–75
    multivariable, 30–31
    notation, 26
    odd, 28–30
    period of, 50
    polynomial. see Polynomials
    rational, 77–79, 128
    relative maximum of, 126
    relative minimum of, 126
    trigonometric. see Trigonometric functions

General quadratic formula, 35, 298–299
Geometric means, 95
Geometric sequence, 94, 300
Geometry, 96–102, 300
Good window, 123–124
Graphing calculator
    absolute value problems, 128–129
    algorithms, 122–123
    ANGLE program, 133–134
    applications of, 122–131
    basic operations using, 119–121
    combinations, 130
    commands, 120–121

computation, 120
description of, 117–118
D2P, 132, 134
DPL, 133–134
editing using, 120
equations solved using, 127
factorial command, 130
function evaluations, 124–125
good window, 123–124
greatest integer function problems, 128–129
iteration of function using, 130
keys, 119
menu keys, 121
MIDPT, 133–134
mode, 121
nonfunction equations, 126
parametric equations, 131
permutation, 130
programs, 132–135
QUADFORM program, 133, 135
real mode, 121
real zeros, 125–126
screens, 119
sequence generated using, 129
support functions, 121
symmetry, 127
syntax, 122
Graphs
    description of, 50–53
    equations solved using, 127
    exponents, 73
    logarithmic functions, 73–74
Greatest integer function, 77, 299
    graphing calculator to solve problems involving, 128–129

Half-angle formula, 54
Harmonic sequence, 108
Heron's formula, 300
Hyperbola, 69, 298

Identities, 53–58
Implication, 104
Independent events, 91
Inequalities, 41–42, 53–58, 127–128
Infinite sequence, 93
Inner product, 98
Inverse function, 27–28, 59–61
Inverse variation, 102
Iteration of function, 130

Latus rectum, 69
Law of cosines, 61–62, 298
Law of sines, 61–62, 298
Linear functions, 32–34, 297
Logarithmic functions, 72–75, 299
Logic, 103–105

Major axis, 68
Mean, 129–130

Mean proportional, 95
Median, 105, 129–130
MIDPT, 133–134
Minor axis, 68
Mode, 105
Model tests, 138–296
Multivariable functions, 30–31
Mutually exclusive, 91

Negation, 103–104
Norm, 97

Odd function, 28–30, 37
Odds, 91
Odds and ends, 107–109

Parabola, 69, 299
Parallel lines, 32
Parameter, 79
Parametric equations
    description of, 79–80
    graphing of, 131
Periodic functions, 50
Period of function, 50
Permutation, 86–88, 130, 299
Perpendicular lines, 32
Phase shift, 51–52
Point of discontinuity, 77
Polar coordinates, 80–82, 299
Polynomials
    definition of, 32
    higher-degree, 37–41
    inequality, 127–128
    linear functions, 32–34
    quadratic functions, 34–37
    zeros, 127
Probability, 90–93, 299
Product, 129–130
Product of zeros, 297
Pythagorean identities, 54

QUADFORM, 133, 135
Quadrantal angle, 49, 52
Quadratic functions, 34–37, 297

Radians, 48
Radius, 47–48
Range, 25, 105
Ratio, 94
Rational functions, 77–79, 128
Rational polynomial, 32
Rational zero theorem, 37
Real polynomial, 32
Real zeros, 125–126
Rectangular hyperbola, 69
Recursion formula, 93
Reference angles, 46
Relation, 25
Relative maximum, 126

Relative minimum, 126
Remainder theorem, 37
Resultant, 97
Rhombus, 300

Sample space, 90
Secant, 49, 52
Sequence
    arithmetic, 93, 299–300
    definition of, 93
    geometric, 94, 300
    graphing calculator to create, 129
Series, 93
Sigma, 93
Sin^{-1}, 59
Sine, 49, 52
Slope-intercept, 32
Solid figure, 99
Sphere, 300
Standard position, 45
Statistics, 105–106
Sum and difference formulas, 54, 129–130
Sum of zeros, 297
Symmetry, 127
Synthetic division, 38

Tangent, 49, 52
Tautology, 104
Three-dimensional coordinate geometry, 98
Trace, 98
Transformation, 96
Translation, 96
Transverse axis, 69
Triangles, 61–64, 300
Trigonometric form, 81
Trigonometric functions
    angles, 49–50
    arcs, 47–48
    definitions, 45–47
    equations, 53–58
    graphs, 50–53
    identities, 53–58
    inequalities, 53–58
    inverse functions, 59–61
    triangles, 61–64

Variation, 102–103
Vector, 97, 300
Venn diagram, 107

y-intercept, 32

Zero of multiplicity, 40
Zeros
    coefficients and, 38
    definition of, 35, 37
    lower bounds on, 39
    real, 125–126
    upper bounds on, 39

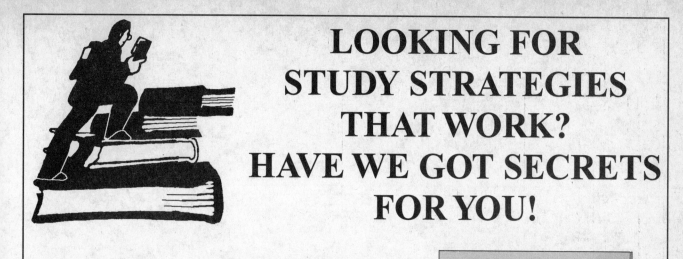
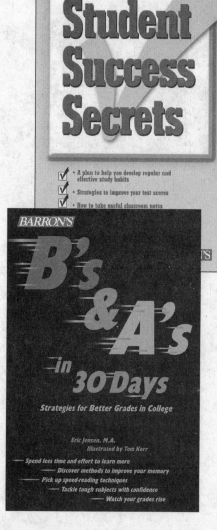